Bioethics

Bioethics

An Introduction for the Biosciences

Ben Mepham
University of Nottingham

OXFORD
UNIVERSITY PRESS

Great Clarendon Street, Oxford OX2 6DP

Oxford University Press is a department of the University of Oxford.
It furthers the University's objective of excellence in research, scholarship,
and education by publishing worldwide in

Oxford New York

Auckland Cape Town Dar es Salaam Hong Kong Karachi
Kuala Lumpur Madrid Melbourne Mexico City Nairobi
New Delhi Shanghai Taipei Toronto

With offices in

Argentina Austria Brazil Chile Czech Republic France Greece
Guatemala Hungary Italy Japan Poland Portugal Singapore
South Korea Switzerland Thailand Turkey Ukraine Vietnam

Oxford is a registered trade mark of Oxford University Press
in the UK and in certain other countries

Published in the United States
by Oxford University Press Inc., New York

First published 2005

British Library Cataloguing in Publication Data
Data available

Library of Congress Cataloging in Publication Data
Data available

Typeset by Newgen Imaging Systems (P) Ltd., Chennai, India
Printed in Great Britain
on acid-free paper by
Ashford Colour Press Ltd, Gosport, Hampshire

ISBN 0-19-926715-4 978-0-19-926715-6

1 3 5 7 9 10 8 6 4 2

THE AUTHOR

Professor Ben Mepham is Director of the Centre for Applied Bioethics at the University of Nottingham. He lectured in physiology to bioscience undergraduates for more than thirty years, and also has more than twenty years' experience of teaching bioethics. He was a founder member of the Government's Biotechnology Commission (AEBC), the European Society for Agricultural and Food Ethics (EURSAFE), and the Food Ethics Council (of which he was the first Executive Director).

ACKNOWLEDGEMENTS

I am most grateful to those who have kindly read chapters of the draft manuscript, and offered constructive advice on how to improve its accuracy and clarity. In particular, my thanks go to Ariane Willemsen, Bob Combes, Charlotte Augst, David Mepham, Geoff Tansey, George Mann, Hilde Wisløff Nagell, Jim Mepham, Pauline Rolfe, Ruth Chadwick, Susan Carr, and Tom MacMillan. Over the years, I have benefited greatly from conversations with Kate Miller, Sandy Tomkins, Colin Moore, and Veronique Delpire at the Centre for Applied Bioethics, and with my colleagues on the Biotechnology Commission (AEBC), the Food Ethics Council, the European Society for Agricultural and Food Ethics, and the EU EBTA project (especially Matthias Kaiser and Ellen-Marie Forsberg). And, not least, thanks are due to generations of undergraduates in the School of Biosciences at Nottingham University, who have challenged me to express unfamiliar, and sometimes complex, ideas with greater clarity - and provided the main motive for writing this book.

I am also indebted to three anonymous reviewers who made constructive comments on the draft manuscript: any errors and omissions are, of course, entirely my responsibility. Finally, my sincere thanks go to Jonathan Crowe, Sally Lane, Nicola Bateman and their colleagues at OUP, who have been a source of sound advice and encouragement; and above all to Bar Masters, without whose patience and unstinting support this project could not have been completed.

Ben Mepham*
Southwell, Nottinghamshire

* The first syllable of the author's surname is pronounced as in 'left'

CONTENTS

PART FIVE **Bioethics in practice**

PREFACE

How to use this book

Most people have a fairly good idea of what they expect from a textbook. It should contain a clear account of current knowledge in the field, and provide a reliable resource for supplementing lectures, essay writing, and project work. If there is a single word that most accurately defines its desired content it would be *information*. This book certainly aims to provide relevant information; but as a textbook for the biosciences it has an unusual emphasis – because rather than provide factual descriptions, its main purpose is to stimulate thought and suggest new ways of thinking.

So readers should not expect to find here 'everything you wanted to know about bioethics'. In practical terms, it would simply be impossible to cover all the ethical issues raised by the biosciences in so short a book. But, more to the point, it would be counter-productive to try to address every possible ethical concern, especially as they tend to crop up with such remarkable rapidity. Dealing with new ethical challenges is all about acquiring the skills of ethical reasoning; unlike most biological knowledge, which tends to undergo frequent revision, these skills should equip you for a lifetime.

Some people think ethics is an 'optional extra'. But it is almost impossible to avoid having an ethical position on many issues in biology, just as it is impossible to opt out of a decision on what to eat or wear. Ethics is so fundamental (whether or not people realize it) that everyone is constantly making ethical decisions. Bioscientists often have additional ethical responsibilities because their subject matter is *life*, which means they are faced with questions of how to weigh the significance of factors affecting different living beings, some human and some non-human.

One reaction is to throw up your hands in despair, because such issues seem too complicated, and not really your business: why not leave such questions to the philosophers and politicians? And in fact, there is evidence of a growing *amorality* in society. That is to say, many people opt out of ethical reasoning, because they think almost everyone else does the same; and in any case there are no absolute ethical standards. Unfortunately, the consequence is that the agendas of the powerful (governments, corporations, wealthy individuals) almost inevitably go by default. These days, a so-called *liberal education* really does entail examining the ethical implications of knowledge and its application.

However, it is perfectly understandable for students to want to avoid spending too much time on ethical thinking if it is not assessed in academic courses. Life is short and priorities have to be set. But, fortunately, the crucial importance of bioethics is increasingly being recognized, and awareness of bioethical issues and competence in ethical reasoning are now specified benchmarks (see Box on page xii) for bioscience degree courses in UK universities.[1] This book has been written specifically for students taking these new bioethics courses as part of bioscience degree programmes; however, it may also be a suitable introduction for other students (in both the sciences and humanities), their lecturers, those involved professionally in the biosciences, and indeed for members of the public who are interested in ethical dimensions of biology. Because biologists are the target readership, a general level of understanding of the biosciences has been presumed; although because they cover a wide range of disciplines some further explanations are given when it seems appropriate. Even so, it is assumed that readers will also have easy access to basic information on the biosciences and biotechnologies from other sources.

The method

The best way to get to grips with bioethics is to engage in discussion and debate. It is *remotely* possible that you could become proficient in ethical reasoning simply by reading widely and deeply. But supplementing your reading by engaging in ethical argument is a far surer way of doing that – and much more enjoyable. A word about argument is necessary. Some people think arguments are quarrels – which is of course one meaning of the word. But here its other meaning is intended, namely 'putting forward a coherent series of reasons to support a viewpoint'. If this deteriorates into a quarrel, something has gone wrong.

Unfortunately, many otherwise well-educated people have not been trained to argue philosophically – to propose and criticize new ideas in an objective, open-minded way. But some readable, inexpensive guides to philosophical reasoning for non-philosophers are now available,[2,3] and it is worth acquiring this 'transferable skill' for all sorts of reasons apart from studying bioethics. In discussions, it is a good idea to always keep in mind the 'principle of charity', to *'try to find the best – the most reasonable or plausible – (rather than the worst) possible interpretation of what we read and hear.'*[4] This is especially good advice in discussions, because on the spur of the moment it is not always easy to express ideas about questions that may not have been seriously considered before.

In fact, debate is an excellent way of clarifying ideas. It possibly had its origins in the *dialectical* process employed by the ancient Greek philosophers Socrates and Plato; and Plato's *Dialogues* remain, more than two millennia later, models of philosophical argument.[5] However, a potential problem with emphasizing debate is that sometimes everybody in a particular group holds more or less the same view, because the answer to an ethical problem seems so *obvious*. It is in these circumstances that you sometimes need to 'think the unthinkable'.

A device often adopted in this book to encourage that approach is practised by an officer of the Sacred Congregation of Rites in the Roman Catholic Church. This is the *Advocatus Diaboli* (devil's advocate), whose duty it is to prepare all possible arguments against the beatification or canonization of someone proposed for sainthood. Of course, readers can be assured that the arguments presented here have no such ecclesiastical objectives, but the *method* is generally applicable. It consists in putting forward as strongly as is reasonable all the arguments *against* a view, to test the strength of the case in its favour. For example, some biotechnologies that bioscientists may find themselves drawn to may be subject to ethical criticism; so it is important for those developing or employing them to be aware of these objections to enable them to prepare counter-arguments if they wish to justify their use.

Readers may sometimes feel that arguments challenging the orthodox views of bioscientists and biotechnologists receive more prominence than those supporting their views. No apology is offered for this: if readers do not sometimes feel a little *uncomfortable* the book may be failing in its aim of encouraging readers to think. Moreover, views tempered by withstanding valid criticism are surely the stronger for that; listening respectfully to the other person's point of view, even if it is then rejected, is the mark of a civilized debate.

It is important to appreciate that there is often an asymmetry in the arguments advanced *for* and *against* a proposed course of action. For example, the argument in favour of a particular crop biotechnology might be stated very simply – to increase the yield of a desired product; but arguments *questioning* it will typically refer to food safety, ecology, and issues of social justice, as well as the acceptability of fundamentally changing the nature of the crop species. Sometimes, the process of ethical reasoning will lead to people changing their minds: and indeed to begin it with a mind already made up would undermine the rationality which characterizes the scientific attitude.

It is commonly believed that science and ethics are worlds apart, the one dealing with facts, the other with opinions. But in reality there are many similarities between them. Ethics, like science, is based on theory and relates to the real world. Both are rational activities, which can be explained and discussed openly. And both employ some technical terms which, when they are understood, help the reasoning process. Just as scientific theory is applied practically in technologies, so ethical theory is applied practically in personal behaviour, and in laws and codes of practice. An ethical theory that prescribed norms of behaviour that people could not attain would be a bad theory: we are not ethically required to do what we *cannot* do. But, like science, ethics develops in the light of new understanding: and the impacts cut both ways. For example, as we understand more about brain function, some forms of criminality are reinterpreted as mental disease. And as we deepen our ethical sensitivities, some formerly approved ways of treating animals are now considered unacceptable.

The structure of the book

The book opens with three chapters that provide an introductory philosophical background to bioethics, and explain some of the technical terminology. The next nine chapters discuss bioethical issues concerning people (4–6), animals (7–9) and crops and the environment (10–12). Obviously, *all* the chapters concern people, but focusing on the separate aspects in different chapters provides a helpful structure. Like the first three chapters, the last three (13–15) raise some generic matters of relevance to all the other subjects discussed.

The best way to use this book would be to read and discuss it from beginning to end, because it has been designed to progressively develop ethical understanding. But the next best way would be to read the first three chapters, and then those that are particularly relevant to your specialist interests. The last three chapters are also relevant to most specialist fields. Wherever you start, there is generous cross-referencing (indicated by listing relevant section numbers in brackets), so that in reading about your chosen subject you may well find yourself taking up the option of moving from chapter to chapter to pursue a particular line of argument. Ethical reasoning is not easily compartmentalized, so to avoid undue repetition some critical issues may appear in chapters that at first sight seemed outside your field of interest. A series of exercises is given at the end of each chapter, which can form the basis of essays, discussions or project work.

Keeping up with events

The pace of developments in both the biosciences and the political responses to them means that bioethics is in a constant state of flux. This state of affairs is almost certain to continue unabated, so that, without knowing where to look for recent developments, it would be easy to get out of date. Fortunately, there are now a number of websites that list relevant scientific and bioethical publications, reports, policy statements and public debates. Some of the more useful (often providing links to sources) are:

- Bioethics Today http://www.shef.ac.uk/bioethics_today/
- CCELS News http://www.ccels.ac.uk/whatsnew.html
- Bioethics.com http://bioethics.com

In conclusion, it is worth recognizing that many students of the biosciences discover that, despite their scientific interests, they are not inclined on graduation to seek employment as professional (laboratory) scientists. But if they subsequently wish to use their scientific training at all, bioethics provides an invaluable link to the wider social and political factors that will shape our future world.

To the lecturer

The importance of the many ethical issues raised in this book suggests a need for bioethical theory and practice to form part of the training of all students of the biosciences. Most students who opt for bioethics courses enjoy them so much that they believe that such courses should be a standard component of all bioscience curricula. The Quality Assurance Agency for Higher Education (QAAHE) benchmarks (see Box) should encourage appropriate developments, but a significant constraint is the availability of staff sufficiently motivated and qualified to run the new courses.

QUALITY ASSURANCE AGENCY FOR HIGHER EDUCATION

Academic standards: biosciences

'We have reached a point in the earth's history where a knowledge of biology is essential for a viable human future. It is therefore important for leaders of society whether in government, industry, business or education to appreciate this and for an informed electorate to understand the scope and limitations of biological knowledge and techniques. Only then can we face the challenging social, ethical and legal problems posed by new developments such as stem cell cloning, gene patenting and gene therapy while working to maintain biodiversity and a stable and sustainable environment.'

The ***good standard*** for all honours graduates in the biosciences entails that they:

'be able to construct reasoned arguments to support their position on the ethical and social impact of advances in the biosciences'

(The ***threshold standard*** requires 'some understanding' of these issues.)

The following points may be useful advice in using this book as a course text:

- In order to meet the modular nature of many undergraduate course, all chapters are roughly the same length. The content of each chapter is probably sufficient to form the basis of two to three sessions (lectures or seminars) if required.
- Updates to the chapters of the book, and suggested new exercises, are posted at the beginning of each semester on the book website at http://www.oup.co.uk/isbn/0-19-926715-4
- It is usually profitable to alternate lectures with seminars or tutorials. Buzz groups (of four to six students, seated so as to facilitate face-to-face conversation) are a good way of building up individual students' confidence before opening up an issue for whole-class discussion. Random eavesdropping and intervention by the lecturer in buzz group discussions can maintain focus and draw out the shy.
- Essays and project work enable students to get to grips with an issue in depth, and suggestions for such exercises are given at the end of each chapter. Group work on a poster is an effective way of encouraging students to work together, gather information (e.g. from websites) and agree on how differing perspectives should be presented in a balanced way. The combined posters of a whole class can provide a useful resource, and a talking point for other students, staff and visitors.

An important venture in the UK is the Bioethics Special Interest Group of the Centre for Bioscience of the Higher Education Academy.[6] It holds regular meetings, maintains a website, and publishes briefing papers to promote sound bioethics teaching at university level. Several publications also provide useful insights on bioethics teaching.[7–10]

NOTES

1. Quality Assurance Agency for Higher Education: Biosciences. http://www.qaa.ac.uk
2. Warburton N (1996) Thinking from A to Z. London, Routledge
3. Blackburn S (1999) Think: a compelling introduction to philosophy. Oxford, Oxford University Press
4. Hursthouse R (2000) Ethics, humans and other animals. London, Routledge, p. 4
5. Plato (1942) [approx. 400 BC] Five dialogues. (Introduction: Lindsay A D). London, Dent
6. Centre for Bioscience of the Higher Education Academy. http://www.bioscience.heacademy.ac.uk
7. Levinson R and Reiss M J (2003) (Eds) Key Issues in Bioethics: a guide for teachers. London, Routledge Falmer
8. Bryant J, Baggott la Velle L, and Searle J (2002) (Eds) Bioethics for Scientists. Chichester, Wiley
9. Elster J and von Troill H (2003) (Eds) How Best to Teach Bioethics. Copenhagen, Nordic Council of Ministers
10. Myser C (1998) How bioethics is being taught. In: A Companion to Bioethics. (Eds) Kuhse H and Singer P. Oxford, Blackwell, pp. 485–500

ACRONYMS AND ABBREVIATIONS

Several UK Government advisory committees not included here are listed in Table 14.2 and a number of international organizations are listed in Table 14.3.

AAT	alpha-1 antitrypsin
ACTH	adrenocorticotrophic hormone
ADI	acceptable daily intake
AEBC	UK Agriculture and Environment Biotechnology Commission
ACNFP	UK Advisory Committee on Novel Foods and Processes
ACRE	UK Advisory Committee on Releases to the Environment
AI	artificial insemination
AIDS	acquired immune deficiency disease
AMRIC	UK Animals in Medical Research Information Centre
APC	UK Animal Procedures Committee
APHIS	US Animal and Plant Health Inspection Service
ART	assisted reproductive technology
ASC	adult stem cells
ASPA	UK Animal (Scientific Procedures) Act 1986
ATP	adenosine triphosphate
BBSRC	UK Biotechnology and Biological Sciences Research Council
BMA	British Medical Association
BMI	body mass index
BR	birth rate
BrR	Brundtland Report (World Commission on Environment and Development)
BSE	bovine spongiform encephalopathy
bST	bovine somatotrophin
Bt	*Bacillus thuringiensis*
BUAV	British Union for the Abolition of Vivisection
CAM	causal analogue model
CAP	EU Common Agricultural Policy
CAT	computer-assisted tomography
CFC	chlorofluorocarbon
CHD	coronary heart disease
CIWF	UK Compassion in World Farming
vCJD	new variant Creutzfeldt–Jakob disease
CoPUS	UK Committee on the Public Understanding of Science
CST	UK Parliament Council for Science and Technology
DC	developed country
DDT	dichlorodiphenyltrichloroethane
DEFRA	UK Department for the Environment, Food and Rural Affairs
DfES	UK Department for Education and Skills
DFID	Department for International Development
DI	donor insemination
DIPS	deliberative and inclusionary processes
DNA	deoxyribonucleic acid
rDNA	recombinant DNA
DR	death rate
DTI	UK Department of Trade and Industry
ECNH	Swiss Ethics Committee on Non-Human Gene Technology
ECVAM	EU Centre for the Validation of Alternative Methods
ELSI(A)	ethical, legal and social implications (aspects)
EPA	UK Environmental Protection Agency
EPC	European Patent Convention
EPO	European Patent Office
ESC	embryonic stem cells
EU	European Union
FAO	UN Food and Agriculture Organization
FAWC	UK Farm Animal Welfare Council
FDA	US Food and Drug Administration
FF	functional food
FISH	fluorescent *in situ* hybridization
FMD	foot and mouth disease
FOSHU	'foods for specified health uses' (in Japan)
FRAME	UK Fund for the Replacement of Animals in Medical Experiments
FSA	UK Food Standards Agency
FSE	UK farm-scale evaluations of genetically modified crops
GCT	germinal choice technology
GDP	gross domestic product
GH	growth hormone
GIFT	gamete intrafallopian transfer
GM	genetically modified
GMA	genetically modified animal
GMO	genetically modified organism
GNP	gross national product
GR	Golden Rice
HACCP	hazard analysis and critical control point system
HAM	hypothetical analogue model
HDI	human development index
HEFC	UK Higher Education Funding Council

HFE UK Human Fertilisation and Embryology (Act)
HFEA UK Human Fertilisation and Embryology Authority
HGA 'human generations ago'
HGC UK Human Genetics Commission
HGP Human Genome Project
HIV human immunodeficiency disease
HT (R) herbicide tolerant (resistant)
ICH International Conference on Harmonisation
ICSI intracytoplasmic sperm injection
IFM integrated farm management
IFOAM International Federation of Organic Agricultural Movements
IGF-I insulin-like growth factor-I
IPM integrated pest management
IPPC Intergovernmental Panel on Climate Change
IPR intellectual property rights
IVF *in vitro* fertilization
JHCI UK Joint Health Claims Initiative
kcal kilocalorie
LEAF UK Linking Environment and Farming
LDC less developed country
MAFF UK Ministry of Agriculture, Fisheries and Food
MJ megajoule
MNC multinational corporation
MND motor neurone disease
MOET multiple ovulation/embryo transfer
MP member of the UK House of Commons (Parliament)
MRC UK Medical Research Council
NAVS UK National Anti-Vivisection Society
NBD 'narrow but deep' element of the UK GM Nation? debate
NERC UK Natural Environment Research Council
NGO non-governmental organization
NHS UK National Health Service
NOEL no-effect level
Nuffield UK Nuffield Council on Bioethics
OECD Organization of Economic Cooperation and Development
OST UK Office of Science and Technology
p.a. per annum
PCB polychlorinated biphenyls
PCR polymerase chain reaction
PGD pre-implantation genetic diagnosis
PERV porcine endogenous retrovirus
PKU phenylketonuria
PMI pronuclear microinjection
PND prenatal genetic diagnosis
POST UK Parliamentary Office of Science and Technology
PP Precautionary Principle
PUS public understanding of science
PZD partial zonal drilling
QAAHE Quality Assurance Agency for Higher Education
RSPCA UK Royal Society for the Prevention of Cruelty to Animals
R&D research and development
SARS severe acute respiratory syndrome
SCNT somatic cell nuclear transfer
sic 'so' (Latin): confirming the accuracy of the quotation
SRP UK GM Science Review Panel
SUZI subzonal insemination
Three Rs reduction, replacement and refinement
TFR total fertility rate
TS Tay–Sachs disease
UK United Kingdom
UKXIRA UK Xenotransplantation Interim Regulatory Authority
UN United Nations
UNDP United Nations Development Program
UNEP UN Environment Program
UNICEF UN Children's Fund
UNFPA UN Population Fund
US(A) United States (of America)
USSR Union of Soviet Socialist Republics
VAD vitamin A deficiency
WHO UN World Health Organization
WTO World Trade Organization
ZIFT zygote intrafallopian transfer

PART ONE
The theoretical background to bioethics

Philosophy is not the underlabourer of the sciences, but rather their tribunal; it adjudicates not the truth of scientific theorising, but the sense of scientific propositions. Its aim is neither to engage nor abjure science, but restrain it within the bounds of sense.

P M S Hacker (1996)
Wittgenstein and Analytical Philosophy

1

The nature of bioethics

OBJECTIVES

When you have read and discussed this chapter you should:

- be aware of the remit of bioethics and its various definitions
- appreciate the significance of historical and philosophical perspectives to a sound understanding of how scientific knowledge is acquired and used
- understand how biosciences and biotechnologies can be ethically abused
- be aware, in general terms, of the ethical responsibilities of bioscientists
- be aware of different theories concerning the biological basis of altruism

1.1 Introduction

If the word *bioethics* crops up in general conversation, most people's initial reaction is to look blank. But mention of certain buzz words usually triggers a seemingly knowing response; and bioethics is usually instantly recognizable if it is related to questions such as:

- should we allow the use of cloning by somatic cell nuclear transfer (the technique used to produce Dolly) to help an infertile couple have a child?
- should genetically modified (GM) crops be grown in the UK, where it will be very difficult for organic farmers to maintain their required GM-free status?
- should GM pigs be used to provide organs (xenografts) for transplantation into human patients, with a much reduced risk of immunological rejection?

These are just three examples of high-profile biotechnologies which raise profound ethical concerns. But, in fact, they are rather deceptive examples if they seem to imply that ethics is only involved when some startling, headline-grabbing proposal is announced. For, as we shall see, bioethics is involved at all stages where facts and values interact, and there are few occasions when they do not. So bioethics also concerns less prominent issues such as:

- whether meat-eating is ethically acceptable
- whether people should be able to choose the sex of their children
- how we need to modify our lifestyles to ensure that future generations inherit a world worth living in.

The problem is that ethics is often interpreted in a rather narrow way. For example, in discussions of the impacts of certain technological innovations it is not unusual to see these listed as 'economic, safety, environmental, and *ethical*'. But the logic of that approach implies that it might be acceptable to countenance *unethical* economics, *unethical* safety, and *unethical* environmental protection measures. Isolating ethics in such a way risks limiting its significance to sentiment, gut-feelings, or religious scruples. Undoubtedly, these concerns matter, but they do not define ethics.

In introducing a new subject it is often necessary to question assumptions that have become second-nature. Within the purely scientific domain, taking certain assumptions for granted usually presents few difficulties. In fact, there are good practical reasons for thinking that questioning the orthodox scientific views presented in formal education is at best time-wasting, and at worst risky. With all due modesty, you might conclude that your lecturers know far more than you about their subjects, so that it would be presumptuous to engage in fruitless questioning. A better degree result is likely to be obtained by learning the facts as they are taught and the orthodox theories that support them. Questioning, it might be considered, can come later – perhaps after graduation (in that illusory period when the pressures are off!) or when pursuing postgraduate studies.

However, one important distinction between education in bioscience and in bioethics is that for the latter questioning assumptions is both *critical* and *indispensable.* While it is dependent on an adequate understanding of the relevant biology, bioethics cannot be taught dogmatically. Students must be encouraged to use their minds, not to have their minds made up for them. Consequently, readers of this book will find that they are presented with a range of arguments, many of which may seem to challenge views commonly accepted in the scientific community. But no attempt is made to provide *the answers* to ethical dilemmas: that is a task for readers themselves.

So, in practice, studying bioethics effectively almost invariably involves discussion and debate, for which books like this can be useful catalysts and guides. The distinguished geneticist Conrad Waddington described books as *tools for thinking with*, which, allowing for the ungainly phraseology, is what this book aims to be.

1.2 Defining bioethics

What do we mean by the term *bioethics*? Perhaps surprisingly, the answer is not straightforward. According to one definition it is *'the study of the moral and social implications of techniques resulting from advances in the biological sciences'*,[1] although, as we shall see in two further definitions below, different writers emphasize different aspects. But we might reasonably regard it as a sub-field of the branch of philosophy called *ethics*, sometimes also called **moral philosophy**. In this book the words **morality** and **ethics** will generally be used interchangeably (as they often are in common speech), although some differences will be explored in 2.1.1. In any event, ethics should not be thought of as an abstract 'pie in the sky' activity. Moral philosopher Bernard Williams defined a theory of ethics as *'a philosophical structure, which together with some degree of empirical fact, will yield a decision procedure for moral reasoning'*. Or to put it more simply, ethics seeks to answer the question *'What should I do, all things considered?'*[2] So ethics has important practical consequences, in that it aims to help us decide what to do in morally puzzling circumstances.

In fact, assigning this broad remit to ethics can be seen as the attempt to answer one of two big questions which we face as human beings. When we become aware of the world intellectually (for most people in their teens) we start asking questions which, in essence, are of two types: 'What is all this, and how does it work? - the answers to which are sought by *science*; and 'What should we do?' - the answers to which are sought in ethics. Of course, the answers to the two questions are often closely related, not least because what we should do may be largely dependent on *the way things are*, for example in terms of our human (and biological) natures. Science and ethics are also much more closely interrelated in the process of scientific discovery than many people realize, and exploring this relationship makes a good entry point into the study of bioethics, as we shall see in sections 1.3–1.4.

The word *bioethics* seems to have first been used in the 1970s by Van Rensselaer Potter, an American medical scientist, who defined it, rather more expansively than the definition given above, as: *'a new discipline which combines biological knowledge with a knowledge of human value systems, which would build a bridge between the sciences and the humanities, help humanity to survive, and sustain and improve the civilised world.'*[3] People who are professionally committed to developing this approach are called **bioethicists**. The underlying assumption is that in considering bioethical questions, biological understanding is an important part of the reasoning process. Deciding what to do, 'all things considered', cannot be left entirely to pure philosophers because they might miss scientific insights that are essential to a full understanding of the issues.

Because it is such a new field of enquiry, few people will initially have had training in all the relevant disciplines. Some bioethicists started out as philosophers and acquired knowledge of the biosciences later; some, like the author of this book, started out as biologists and subsequently developed an understanding of ethical theory. In fact, the pluralism evident in the ranks of bioethicists has led philosopher Onora O'Neill to suggest that *'Bioethics is not a discipline...It has become a meeting ground for a number of disciplines, discourses, and organisations concerned with ethical, legal, and social questions raised by advances in medicine, science, and biotechnology.'*[4] From some perspectives this might be seen as a weakness, because many (if not most) 'bioethicists' lack at least some of the skills necessary to address rigorously the issues raised by bioethics. Even so, collectively, the varied backgrounds of bioethicists can often result in ethical judgements which are far more than the 'sum of the parts'. This is why so much bioethical deliberation takes place within committees such as government commissions and ethics councils, whose members are chosen to complement each others' skills.

Despite these observations, and whatever the exact definition of the word 'discipline', there would seem to be much merit in seeking to integrate the insights of the different subjects that underpin bioethics if it is to advance beyond the 'talking shop' stage. Consequently, it is an important aim of this book to lay the foundations of a new understanding of the ways bioscience and values interact, which, in Potter's words, means building *'a bridge between the sciences and the humanities'*.

All this raises an interesting question now being addressed by sociologists: *'How did bioethics come to supplant literature, law, and religion as a source of moral instruction and arbitration'* in these matters? For centuries, people looked to novels, plays, and poetry for advice on how to behave ethically - and more than 100 years ago many writers (such as Mary Shelley in *Frankenstein*[5]) were doing what would now be called bioethics - raising questions about the proper use of science. *'Why do we now turn to bioethicists (rather than doctors, the clergy, or lawyers) to help us decide what is right?'*[6] This is a question with no straightforward answer, but being aware of it will remind the reader of the many ways in which bioethical questions permeate our lives.

It should be appreciated that, despite the all-inclusive account of bioethics assumed here, its original focus was on medical issues. After the 1939–45 World War, Nazi doctors and scientists who had carried out research on Jews, gypsies, prisoners and disabled people were tried for their crimes against humanity (1.6.1). As a result, a code of practice for all future research on human subjects was drawn up - the Nuremberg Code (Box 15.2). But matters have moved on over the last 60 years, and it is now widely appreciated that non-medical aspects of biology also raise many ethical concerns. Because medical ethics has become a somewhat specialized branch of bioethics, with its own extensive literature, this book will concentrate on non-medical issues, and relate to the biosciences as distinct from the medical sciences. Even so, the precise remit of medicine is debatable, and certain issues that some people would class as medical ethics are discussed in chapters 5 and 6.

1.3 History and philosophy

A good way to approach this subject is to consider the history of biology. Superficially, history might seem to have little to contribute to scientific understanding. It is, after all, by definition, 'out of date'. But, in reality, a study of the history of scientific ideas reveals much about our current ideas and, equally importantly, about those assumptions that seem so obvious that we don't even question them. By allowing us to stand back and view events dispassionately, history provides a route into the philosophy of science, which is crucial for an appreciation of bioethics. **Philosophy** is a word some science students find a little daunting, even though it is an ambition of many to become PhDs - doctors of philosophy. But philosopher Bertrand Russell's description of philosophy as just *'an unusually obstinate attempt to think clearly'*[7] should dispel any sense of mystique.

Few scientists would deny that the history of their subject can be interesting, but most scientists' knowledge of history tends to be limited to the outline of 'great discoveries' which often features in the introductory chapters of science textbooks. This orthodox scientific history usually represents earlier scientists as operating in a fog of ignorance, largely influenced by superstition and old wives' tales, which has gradually given way to the discovery of the more and more accurate facts that underpin the modern, truly scientific account of the subject. The theory of heredity is a useful case study for exploring these ideas, because it has implications for almost all the rest of biology.

1.3.1 A brief 'orthodox' history of heredity

Some of the earliest ideas about heredity were that embryos grew from miniature organisms (called 'homunculi' in the case of humans) that were preformed in the father's spermatozoa or mother's ova. Development thus consisted simply of the growth of the preformed individual. The logical consequence of this theory was that the genitalia of the first man - Adam - contained in miniature all the future generations of mankind. This bizarre notion (as we now consider it) was replaced in the nineteenth century by the theory of **blending inheritance**, in which the characteristics of offspring were a mixture of those of their parents, that could, moreover, be modified by environmental factors affecting the parents during their lifetime. Such views were shared by Charles Darwin, who developed the modern theory of evolution in the nineteenth century;[8] and he incorporated them into his theory of **pangenesis**. This theory proposed that all the organs of the body produced particles (**gemmules**) that collected in the genitalia and were transferred to the offspring following copulation.

Only later in the nineteenth century were chromosomes identified as the structures involved in conveying information between generations. This led to the re-discovery of the work of the Augustinian monk Gregor Mendel, first reported in 1865, who had stated the laws of inheritance now bearing his name. According to this account, around about 1900 three European scientists independently realized that Mendel's mathematical principles, derived from a study of garden peas, provided a sound basis for explaining heredity as a whole. Later, work with the fruit fly *Drosophila* showed that in reproduction chromosomes behave exactly as predicted by Mendel's laws.

Genetics could thus now build on the concept of genes as discrete material units, which code for parental characteristics and pass them to their offspring. So some of the original difficulties of Darwin's theory were resolved, and a new genetic theory of **natural selection** became a fundamental biological principle. In 1953, after the discovery that genes were made of deoxyribonucleic acid (DNA), and not protein, James Watson and Francis Crick showed that DNA's double-helical structure provided a means of explaining gene function and the physical basis of heredity. The discoveries of the last half-century have only served to confirm their observations and have allowed the complete sequencing of the human genome.

This orthodox history of heredity emphasizes *discovery*, on the assumption that the scientific method ensures that objective truth is progressively uncovered as the tools and methods of investigation get better and better. And if further proof were required that this is what happens, the technologies devised on the basis of this scientific knowledge seem to provide adequate demonstration: the proof of the pudding is in the eating.

1.4 A new history of biology

However, closer historical analysis suggests that the above account is simplistic, and can be challenged at three levels, as described by historian of biology Peter Bowler.[9]

1.4.1 The conceptual level

The old idea that scientific knowledge grows by a simple accumulation of facts is now considered naïve. 'Facts' only appear as such to people with a particular frame of mind. For example, Mendel's laws only seemed plausible when it was accepted that characters are transmitted from one generation to the next as separate units of 'information' – a view that even Darwin did not accept.

In fact, and more remarkably, historians point out that in these terms Mendel himself was not what we would now call a 'Mendelian'! He was chiefly concerned with the hybridization of species as an alternative to evolution, so that his discovery of regularities in the inheritance of characters was simply a by-product of a line of research that would not have made sense to the scientists who 'rediscovered' Mendel's laws.[10] It follows that acceptance of Mendelism did not depend on the discovery of facts but on the creation of a new conceptual scheme in which such laws could 'make sense'.

It is also important to note that, as pointed out by statistician and geneticist Ronald Fisher in 1936, *'in statistical terms Mendel's results are a little too good to be true'*, because he would have been very lucky indeed to have hit on the ratios he did by chance. Some people attribute the result to an overenthusiastic assistant, or to an unconscious bias in counting. Either way, it seems Mendel got the results he *wanted.*

1.4.2 The professional level

It is important to appreciate that science is a social activity. This means that a new idea will only attain 'factual' status when the group of scientists who make up the specialist field *agree* to accept it. In turn, this means scientists have to be members of a 'club' through which they secure research grants, get their work accepted for publication in the recognized journals, and achieve status in the academic community, e.g. as lecturers or professors.

It is now believed that two important reasons why Mendel's work remained unappreciated for so long were that he was not a recognized academic scientist (working as he did in a monastery) and that he published his results in an obscure journal. But even within the more formal academic channels, acceptance of a new theory is not a straightforward affair: it often depends on the authority or debating skills of the key decision-makers. Research is an expensive activity, and if the grant-awarding authorities are not convinced of its value, a research proposal will not get funded.

As pointed out by Bowler, although science is assumed to be a completely international activity it does have certain national 'flavours' – and in some cases this results in quite large differences in what counts as an important scientific contribution. Such differences may well explain the fact that genetics achieved much prominence in the USA and Britain, but was less important in Germany and France.

1.4.3 The ideological level

Because scientific progress involves some major rethinking by members of a professional community, it is not surprising that it is also affected by the 'mood of the time' and by politics. For example, it has been argued that Darwin's theory of evolution fitted in with the social values of Victorian England, when the dominant capitalist ideology saw life as a competitive struggle in which the industrious and virtuous achieved justified success, whereas the 'idle masses' deserved the poverty they had to endure. In fact, Darwin's ideas were exploited, as so-called **social Darwinism**, by sociologists like Herbert Spencer in order to justify such beliefs.[11]

The effect of this attitude in facilitating acceptance of Mendelism is graphically illustrated by the dominant views of the respective roles of **nature** and **nurture** in determining human character. *Nature* refers to a person's genetic inheritance and *nurture* to the physical and cultural environment in which they grow up. If character is largely determined by a person's genes, it follows that education and upbringing will have little influence. In the late nineteenth century there was growing concern that the birth rate of people who were considered both mentally and physically inferior exceeded that of those thought to be fitter and more talented people, thus diminishing the overall quality of society.

According to the advocates of **eugenics** (a term introduced by Darwin's cousin, Francis Galton), strict controls were necessary to prevent progressive enfeeblement of the nation. For Galton, who based many of his views on a study of distinguished families, eugenics became a kind of moral crusade. This led him to advocate, on the one hand, **negative eugenics**, including confinement in lunatic asylums or actual sterilization of the 'weak-minded' and rigorous immigration controls to prevent dilution of the nation's genetic legacy, and on the other hand, **positive eugenics**, by which the professional classes should be encouraged by tax incentives to have more children. Galton believed that knowledge of the workings of heredity imposed a *'moral duty... to further evolution, especially that of the human race.'*[12]

Perhaps the starkest instance of the influence of ideology on biology was the belief in the inheritance of acquired characteristics that was official government policy in the Soviet Union (USSR) from the 1930s to the 1960s. In 1940, the Soviet leader Josef Stalin, who rejected Mendelism because it did not conform to the Marxist belief that permanent change could be achieved by environmental influences (i.e. as a result of political changes), sacked the Mendelian geneticist Nikolai Vavilov as president of the Lenin Academy of Agricultural Sciences, and replaced him with Trofim Lysenko. But Lysenko was *'a fanatical charlatan (who) was allowed absolute dictatorship and control over both research in biology and practical agriculture.'* It is a remarkable fact that a country capable of developing a nuclear potential rivalling that of the USA, and being at the forefront of space research, could have been so subject to political control in the field of genetics.[13]

BOX 1.1 THE SOCIAL CONSTRUCTION OF BIOLOGY

Every age constructs a model of the living world, built up from theories, and the social and political imagery of the day, that highlights or emphasizes particular aspects of our understanding.

In the eighteenth century, an age of classification in botany and zoology, the emphasis was on harmony and systemic order. Nature was a catalogue of organic forms, each fashioned by an ingenious creator, each with a place on a 'Great Chain of Being' that stretched from inanimate matter to God. The scientist's task, confronted by this majestic scheme, was to classify its elements, to contemplate the subtlety of the connections that held it together, and to reveal the harmonious functioning of particular parts.

In the nineteenth century, the picture changed with the idea of dynamic, evolutionary change, based on competition and struggle. 'Nature red in tooth and claw' was the image for a new age of rapid industrialization, aggressive business practices, and intensifying struggles between capital and labour. Organisms were approached in a different light as the products, not of design, but of millennia of competition with other species, in which the better adapted eventually outbreed their competitors.

The dominant image of the second half of the twentieth century, deepened by insights of genetics, is less reverential than that of the eighteenth, and places less emphasis on competition and struggle than that of the nineteenth. Nature is a system of systems. Organisms function, reproduce, and evolve as systems ordered by their genes, managed by the programme in their DNA. Life is the processing of information.

(From Yoxen, 1983)

Philosopher Jennifer Trusted has pointed out that, by comparison with physics and chemistry, biology has been particularly prone to the influences of *'ethical, religious, social, cultural, and philosophical beliefs as to the nature of life and our human place in the natural world'.*[14] This is graphically illustrated by Ed Yoxen's concise overview (written in 1983) of two centuries of biology, shown in Box 1.1.[15]

Acknowledging that these social and cultural factors play a much more important role in the development of biology than is often realized, it is also true that advances in methodology are often a critical factor. A good example is provided by the history of cytology, which was greatly influenced by advances in microscopy, as described in the engaging account provided by the distinguished pathologist Henry Harris.[16]

1.4.4 Epistemology

In summary, 'great discoveries', such as those of Mendel, were only possible when people had adopted attitudes about the nature of life that allowed the underlying theories to 'make sense'. These ideas could only gain widespread support when the scientific community was won over, and that was influenced by factors such as loyalty to colleagues, acceptance of the established hierarchy, and the need to be successful in the competition for research funding.

This means that scientific progress depends on the scientific community adopting (some might say, *inventing*) those new models that appear best able to explain the observations made. Often, in devising such models, there is appeal to metaphor, i.e. figures of speech that imply likeness or analogy. Prominent examples are the references to **selfish genes**[17] (in which a chemical is ascribed human characteristics) and to adenosine triphosphate (ATP) as the **energy currency** of the cell[18] (with allusions to current and deposit bank accounts). But it is important to appreciate that such models are embedded in the dominant mind set of the particular time and place in which they are formulated, i.e. they are *theory-laden*. There is no guarantee that the models proposed are the *right*, or the *only*, ones that could be taken to explain the observations.

The important questions raised by this discussion fall into the branch of philosophy called **epistemology**. Defined as the 'theory of knowledge', epistemology is concerned with what it means to say we know something, how we know it, and what the limits to knowledge are. Knowledge is of two broad types - **explicit** knowledge, when the person is aware of the knowledge, and can express it in words, pictures, mathematical formulae, and so on; and **tacit** knowledge, which may still be considered genuine but is not capable of being described.[19] A simple way of explaining tacit knowledge is to say that although dogs presumably have such knowledge of many things (their surroundings, their owners, the smells associated with food, etc.) they cannot express this knowledge. In this sense, by analogy, tacit knowledge resembles many of our psychological states. If, as is claimed by many philosophers, tacit knowledge is a critical element of our understanding of the world, it is certain to have important implications for ethics.

1.5 The scientific method

One reason science is considered to be so successful in explaining the natural world is that it is believed to have a rigorous, objective method - the **scientific method**. For many years the way new scientific laws were established was thought to depend on a process of **induction**, which, as the name implies, is the opposite of **deduction**. That is to say, by amassing accurate data and noting the similarities and differences between related observations it is possible to induce the underlying scientific principles by a process rather like informed guesswork. For example, if every carefully made observation suggests that, at constant temperature, the pressure exerted by a gas is inversely related to its volume it might be induced that this inverse relationship will *always* apply - giving rise in this case to Boyle's Law. So pursuing this approach, it was considered that scientists should seek to confirm scientific laws by accumulating more and more supporting data.

But the belief that induction is *the* scientific method was first questioned by scientists in the nineteenth century, and is now widely discredited. For example, the fact that scientific ideas change so frequently over time, as the above account of ideas about heredity shows (1.3 and 1.4), casts doubt on the claim.

BOX 1.2 KEY PHILOSOPHERS OF SCIENCE

- **Karl Popper** (1902–1994), an Austrian by birth, emigrated to New Zealand in 1937 and subsequently to England, where he became professor of logic and scientific method at London University. In *The Logic of Scientific Discovery* (1954), he claimed that a scientific theory cannot be proved simply by adding confirmatory evidence. Rather one should attempt to disprove (falsify) hypotheses. *'The method of science is the method of bold conjectures and strenuous and severe attempts to disprove them.'*
- **Thomas Kuhn** (1922–1996), an American historian of science, argued in *The Structure of Scientific Revolutions* that Popper's prescriptive approach to scientific method is rarely followed in practice. Instead, most scientists work (rather uncritically) within a set of accepted norms and suppositions, extending and articulating the *paradigm* into new areas of application (*normal science*). When inconsistencies build up, certain (exceptional) scientists challenge the paradigm and, if successful, bring about a **scientific revolution** – which then becomes the basis of the new paradigm.
- Despite their different approaches, each theory can contribute to an understanding of the aims and methods of science.

1.5.1 Popper's rational approach to scientific method

More realistically, science provides only *provisional* knowledge. According to the distinguished philosopher of science Karl Popper (Box 1.2), the distinctive feature of science is its method of critical testing; testing that should consist of attempts to prove a theory *wrong*. In his words, *'the method of science is the method of bold conjectures and strenuous and severe attempts to refute them.'*[20] No theory can *ever* be proved true, not least because the evidence is always incomplete. But theories that do not accord with the facts *can* be **falsified** or **refuted**.

The time-honoured illustration of Popper's view is the 'bold conjecture' that *'All swans are white.'* This can never be *proved*, because even in the unlikely event that we had examined every living swan and found it white, tomorrow a black one might be hatched. Seeking confirmatory evidence for the hypothesis is thus not a sound way of testing it; a better strategy would be to look for non-white swans. In essence, no number of observations of white swans can ever prove the hypothesis is true but, in principle, discovering one black one can *disprove* it. Thus, an important feature of Popper's methodology of science is its dependence on the **hypothetico-deductive method**, that is, the formulation of *rich hypotheses* from which the predicted effects that can be deduced are then subjected to rigorous testing.

Interestingly, a biological metaphor is useful in emphasizing the point. Just as in the struggle for existence only the fittest species survive and propagate their kind, so in the world of ideas only the best theories survive (i.e. those that, because they are more objectively true, best resist the attempts to refute them). Popper regarded this as the logical, rational approach to science.

The problem with induction (1.5) is that it confuses the **context of discovery** with the **context of justification**. Basing your hypothesis on reasoning by analogy is fine for suggesting (and perhaps discovering) new lines of enquiry but, as Popper stressed, justifying a hypothesis demands a sterner test – that of surviving the attempt to knock it down.

1.5.2 Kuhn and normal science

But a strong challenge to Popper's vision of scientific objectivity was made by the historian and sociologist of science Thomas Kuhn (Box 1.2). While not denying that the major advances in science, such as Watson and Crick's elucidation of the structure of DNA, conform to Popper's prescription, Kuhn argued that science is a social phenomenon in that the vast majority of scientists follow the same sorts of codes of behaviour that govern other human groups.[21] For Kuhn, most scientists, most of the time, are not trying to refute theories (which for Popper is the only true method of science) but are doing **normal science**. This is an activity where scientists try to extend the **paradigm** (pronounced 'para-dime'), e.g. by seeing whether a theory established, say for bacteria, also applies in tomatoes or rats. (The paradigm is the basic set of assumptions accepted by a scientific community that seems to 'work' over a period of time.) The paradigm thus exerts a very powerful influence because it determines not only what are considered the sensible questions to ask, but also what count as acceptable answers to those questions. Anyone who doesn't accept these standards tends to be regarded as a crank.

More generally, the paradigm defines the subject matter of a science, how it should be investigated, and how academic standards should be assessed. Kuhn argued that our observations are invariably **theory-dependent**, that is to say, what we see depends in large measure on what we expect to see. For example, it is usually (always?) the case that when people first use a microscope they have difficulty in making out the essential features of the material examined. It is only after they have been instructed on 'what to look for' that they can make sense of the image. Figure 1.1 is a now 'classical' illustration

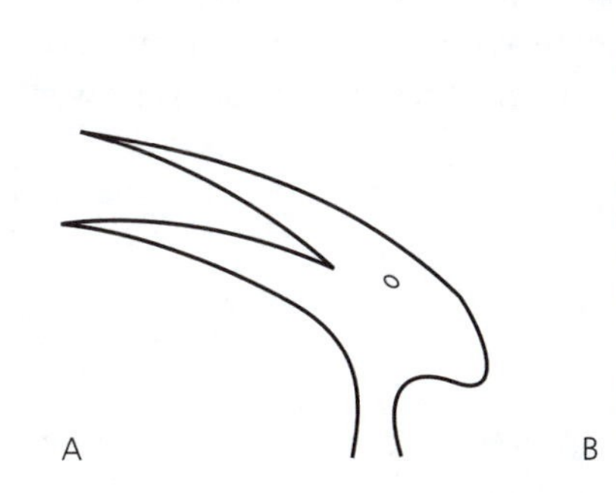

Figure 1.1 The antelope–bird image, indicating the phenomenon of theory dependence. For details, see 1.5.2. [From: Hanson N R (1958) Patterns of Discovery. Cambridge, Cambridge University Press]

of this point. If you had been told figure A was an antelope you would accept, unquestioningly, that it was no different from the other heads in figure B. But when you are told that these other heads are meant to represent birds, closer inspection will reveal that they *are* different.[22] The sense data are the same but the 'explanations' given are different. To see how this affects biological explanations, refer again to the different interpretations of Mendel's results given by Mendel himself and by later scientists (1.4.1).

If you took a cynical view, scientific education might thus be regarded as a process of indoctrination, in which students are presented with a whole new vocabulary (including such esoteric concepts as transcription, homeostasis, and organogenesis) and the approved solutions to otherwise apparently perplexing problems. If students demonstrate they have mastered these adequately, they are awarded degree certificates, ceremoniously allowed to wear appropriate academic dress, and admitted to the inner circle of qualified scientists. In fact, Kuhn, without irony, claims a parallel between the training of scientists and that of novices in a monastery.

Such ideas can come as something of a shock, because they seem to imply that scientists are just as much subject to custom as are people in any other job. Theories, Kuhn suggested, are held largely as matters of fashion or convention rather than due to the demands of logic; and social factors are likely to have a much bigger influence on science than its claimed objectivity would allow. Some of these factors were explored in 1.4.1–1.4.4.

Most science students learning of the competing theories of Popper and Kuhn appreciate that each contains elements of truth. Perhaps *the true* situation lies somewhere between the two extremes described. After all, the observation of one black swan need not necessarily disprove Popper's illustrative hypothesis: someone might have dyed it black, or have mistaken another bird for a swan, or had too much to drink! But it is most unlikely that the social factors Kuhn identified have absolutely no influence on what scientists consider scientific truth. As Bowler points out: *'the establishment of a successful new theory involves the complex and often unpredictable interaction of a host of ideological and professional pressures'.*[23] And it is precisely because these 'external' factors are so important that recognizing their potential influence is essential for anyone who wishes to understand what their subject is really about. But such insights are also crucial for those who want to know how they should act, both as scientists and as members of a society which exploits biological knowledge. It follows that these are critical issues when considering bioethics.

1.6 Abuses of science

If the objectivity of science is one cherished idea that needs to be re-examined, another is the oft-claimed sharp distinction between science (viewed as knowledge) on the one hand, and technology (the practical application of scientific knowledge) on the other. Maintaining that there is a distinction is, of course, often claimed by scientists who wish to pursue their researches untroubled by ethical concerns that might be raised about the way their discoveries are subsequently used. But, to cite a dramatic case, J R Oppenheimer (the 'father' of the Atom Bomb, which was dropped on Japan in the 1939–45 World War), when recalling his work on the Manhattan project to develop the bomb, admitted: *'In some sort of crude sense which no vulgarity, no humor, no overstatement can quite extinguish, the physicists have known sin, and this is a knowledge which they cannot lose.'*[24]

Explaining how scientists had tried to square their consciences with working on the project, he wrote: *'When you see something that is technically sweet you go ahead and do it and you argue what to do about it only after you have had your technical success.'*[25] As a matter of historical fact, many of the scientists who worked on the Manhattan project subsequently switched their research to molecular biology, a new field which they saw as untainted by military involvement.

1.6.1 **Abuses of biology**

But biologists have no reason to feel complacent, because it was members of the biomedical community in Germany during the 1939–45 World War who perpetrated what were surely the most ethically indefensible acts in the name of biology. Historian of science Paul Weindling has described how, justifying their actions as necessary to protect the nation's 'genetic treasury', people classed as 'undesirable' (Jews, gypsies, black German half-castes, and carriers of genetic diseases, such as schizophrenia and muscular dystrophy) were subjected to coercive euthanasia and sterilization. This, then, was a particularly aggressive form of the eugenic programme first described by Galton (1.4.3). Biologists even found a perverse justification for the Holocaust (the mass extermination of Jews) by exploiting the opportunity to conduct experiments on people before their death, e.g. in one case, inducing seizures in children in low-pressure chambers to test a new drug for epilepsy; in another, *'killing gypsies to obtain heterochromic pairs of eyes'*.[26]

The lurid enthusiasm for such research shown by some scientists is illustrated by the statement of the then director of the Institute of Brain Research at Berlin. When relating to American officers at the end of the war his conversations with officials at an extermination camp, he had said:

I went up to them: 'Look here now, boys, if you are going to kill all these people at least take their brains out, so the material could be utilised.'...I gave them fixatives, jars, and boxes, and instructions for removing and fixing the brains and they came bringing them like the delivery van from a furniture company. There was wonderful material among those brains, beautiful mental defectives, malformations, and early infantile diseases.[27]

Such abhorrent acts may now seem a long time ago, but sixty years is well within living memory for many. Indeed, in more recent years, the horrors of the programme of ethnic cleansing in the former Yugoslavia showed strong parallels with the Nazi atrocities. Whether or not we consider that these are current dangers in our own society, such events serve as a poignant reminder of how science *can* be recruited for the most unethical purposes. In the present context, we will need to examine whether there is a distinction between bioscience and biotechnology. And a critical question concerns whether there is a danger that scientists pursuing programmes that are 'technically sweet' might unwittingly promote technologies that turn out to be unethical.

1.7 Bioscience and biotechnology

One way that ethical control over the application of science as biotechnology might, in theory, be encouraged would be to maintain a rigid distinction between academic research, pursued to increase scientific understanding, and its commercial or political exploitation. But, in practice, the distinction would prove almost impossible to enforce; and, most probably, no government now seeks to do so.

For example, in the UK, government-sponsored research in non-medical biology is conducted by the Biotechnology and Biological Sciences Research Council (BBSRC), a title which conflates scientific and technological objectives. One of BBSRC's stated missions is to contribute to *'the economic competitiveness of the United Kingdom'*,[28] and although this may well be ethically justifiable (see chapter 14), it demonstrates that government support for bioscience and biotechnology is not *just* about increasing knowledge, but is explicitly directed to particular social ends. Indeed, the government often actively collaborates with commercial companies in seeking to achieve these objectives. The need to ensure that bioscience is employed in ethically acceptable ways thus becomes a challenging task not only for governments but also for scientists 'at the laboratory bench' (chapter 15).

The conflation of science and technology is even more problematical when the limitations of science are not appreciated. In a speech to the Royal Society in 2002, Prime Minister Tony Blair expressed the view that *'Science is just knowledge... It allows us to do more but doesn't tell us whether doing more is right or wrong.'*[29] What this common misunderstanding reveals is a lack of appreciation of the conceptual, professional, ideological and epistemological influences (1.4.1–1.4.4) which shape scientific knowledge. It is highly questionable that science is ever neutral in the way claimed; rather, it is embedded in society's current value system.

1.8 The importance of bioethics for the biosciences

At this point it will be useful to take stock. What does all this mean for the biosciences? The above considerations would seem to suggest that bioethics is important for students of the biosciences (in the widest sense – since lecturers are also 'students' in the pursuit of their professional interests) in a number of ways.

Some of these ways are specific to their being educated specialists, while others relate to their being members of societies in which science and technology play such a large role culturally, economically and politically. And some affect decisions in their personal or family lives. These ethical dimensions are listed in Box 1.3.

BOX 1.3 HOW BIOETHICS AFFECTS BIOSCIENTISTS

- At the practical/experimental level:
  - the procedures entailed in investigating biological phenomena often involve experimenting on, confining, or killing animals, and altering the natures and environments of both plants and animals in ways that have ethical impacts
  - the Quality Assurance Agency for Higher Education (QAAHE) benchmarks include among *intellectual skills:* 'recognising the moral and ethical issues of investigations and appreciating the need for ethical standards and professional codes of conduct'
- At the epistemological level (1.4.4):
  - the scientific study of life is constantly revealing new ways of understanding living organisms, which often lead to a reappraisal of our ethical assumptions and the acceptable norms of human behaviour
- For professional biologists or biotechnologists ethical concerns raised by the following will influence their decisions, either positively or negatively, to work on particular projects:
  - the implications of biotechnology for the treated organisms (animals, plants, or microbes) and for people and animals affected by use of the technologies, both now and in future
  - as many living organisms colonize the wider global environment, technologies which might induce permanent change (e.g. by genetic modification) raise important ethical issues concerning sustainability and biodiversity
  - the economic implications of biotechnology as a consequence of its domination by a small number of large multinational commercial organizations
- Like all responsible citizens, bioscientists need to address the extent to which developments in the biosciences and biotechnology affect personal lifestyle choices, concerning:
  - their own and their family's health
  - consumption patterns, notably of food
  - reproductive choices
  - financial investments

1.9 Biological dimensions of ethics

An important question for many biologists is 'Where did the concept of ethics come from?', and this chapter concludes with a brief discussion of whether the apparently innate, unselfish character of ethical behaviour is consistent with the ways in which biologists currently understand human life.

An old idea, which seemed to gain support from those aspects of evolutionary theory that emphasized the 'struggle for existence', was that ethics applies specifically to humans living in society and is based on **egoistic prudence**. According to this view, in the earliest stages of human society the population density was so low that there was no real competition for resources. But as numbers increased, frequent conflict was inevitable, and individuals became engaged in a bitter struggle for survival (competing for food, shelter and sexual mates), in which only the fittest survived. This idea was thought to be consistent both with Darwin's theory of natural selection, and with the views of the seventeenth-century philosopher Thomas Hobbes, who claimed that the 'state of nature' was *'a warre, as is of every man, against every man'* for whom life is *'solitary, poore, nasty, brutish, and short'*.[30]

Consequently, survival and social order were only possible when a reluctant bargain was struck, in which people promised not to harm or steal from each other. That is, in their own selfish interests it was prudent to abide by a set of ethical rules or **norms**. (This notional **social contract** is, of course, symbolic, but it seemed quite plausible.) The contract has since been institutionalized in laws, so that responsibility for maintaining it is largely a role of the State, and, for many people, it is the threat of punishment that ensures that they behave in accordance with the notional contract. That people do not always do so willingly is demonstrated by the looting and rape which often follow the breakdown of law and order in the periods immediately following the collapse of totalitarian regimes.

However, more recently this pessimistic view of human nature has appeared to be inconsistent with the facts of biology; nor does it seem to be a view held by Darwin himself. In 1859, he wrote: *'any animal whatever, endowed with well-marked social instincts... would inevitably acquire a moral sense or conscience, as soon as its intellectual powers had become as well, or nearly as well, developed as in man'*, and he backed up his claim with numerous examples of apparently **altruistic** (unselfish) behaviour performed by baboons, dogs, cattle, birds, and even insects.[31]

1.9.1 Altruism

The biological basis of altruism has been the subject of much speculation; but we have to start with a problem of definition. An altruistic act is usually defined as one that 'benefits another organism at a cost to the actor, where cost and benefit are defined in terms of reproductive success'. So, by definition, altruism always entails helping others; but, going beyond reproductive success, does it always have to be detrimental to the altruist? In common parlance, altruism includes those acts that provide some sort of reward to the altruist, even if it is only the satisfaction of having behaved 'honourably'. Indeed, it would be strange to count as 'altruistic' only acts that were grudgingly performed merely out of a sense of duty. In the following discussion, this, broader, definition of altruism will be used.

In discussing the biological basis of altruism, it is important to appreciate that animals are targets of selection in three different contexts, and these correspond to three different types of 'altruistic' behaviour.

Kin selection entails types of behaviour that enhance the fitness of the genotype that is shared by the altruist and those benefiting from the acts. Parental care is probably the most prominent example of this type of 'altruism' – and the image of the nursing mother and her child has become an icon of human love and devotion at its most profound.

Reciprocal altruism amounts to the mutual exchange of favours, as in 'you scratch my back and I'll scratch yours', which might bear some relationship to Hobbes' reluctant bargain (1.9), but need not amount to anything so calculating. Even the longer-term benefits of building a reputation for 'friendliness' might be deemed sufficient justification for investing time and effort in such acts.

Group altruism. The first two types of altruism have evolved through selection pressure on the individual, and because the individual benefits directly from these types of behaviour, it might be questioned whether they count as genuine altruism. But group altruism is represented by social norms and ethical behaviours that have emerged as a result of selection pressure on human cultural groups, and in this case the individual might actually suffer for the benefit of the group. So this is a more authentic form of altruism.

Most animal associations cannot serve as a target for group selection, but the cooperation that characterizes social animals, like humans, provides the appropriate conditions for its emergence. Historian of biology Ernst Mayr argues that the co-evolution of two factors, a larger brain and a larger social group, made possible the emergence of two aspects of this form of altruistic behaviour, viz.

- natural selection: which, working through group selection, rewards those unselfish traits that benefit the group even though they might be detrimental to the individual
- humans' increased reasoning capacity, which allows us to actively chose behaviour benefiting the group rather than relying on instinctive selfishness.[32]

(We might also note that a related phenomenon, group *loyalty*, can have more mundane or more sinister aspects, e.g. when tens of thousands support their football team, or when millions are aroused to aggressive nationalistic feelings by a demagogue's rhetoric.)

1.9.2 Where do ethical norms come from?

Some biologists have attempted to explain ethics and to derive ethical norms from what they consider to be biological facts. Notable is the American sociobiologist Edward O Wilson, who claims that *'the time has come for ethics to be removed temporarily from the hands of philosophers and biologicised'*[33] because *'ethical precepts are reached by consensus under the guidance of the innate rules of mental development'*.[34] But bioethicist Peter Singer argues that this approach is mistaken, because although understanding the biology of altruism is often necessary, it is not sufficient to formulating ethical norms.[35]

Indeed, to some philosophers the attempt to reduce all actions to consequences of genetic programming seems simplistic. According to Mary Midgley, people often act from *'a sense of justice, from friendship, loyalty, compassion, gratitude, generosity, sympathy, family affection and the like'*[36] and it does not seem possible to reduce all these to unconscious attempts to propagate one's genes. To cite just two examples – very large sums of money are often raised in response to mass-media charity appeals (e.g. following the Asian tsunami disaster), while acts of personal kindness, even to strangers (e.g. when people act as blood donors), are quite common.

For ethologist Robert Hinde, moral codes are a product of culture.[37] According to this view, moral codes are constructed, maintained, transmitted and amended by human beings interacting with each other. They depend both on human nature and on experience in the physical, psychological and cultural environments of development. In the same vein, philosopher of science Alex Rosenberg suggests that Darwinian theory is progressively contributing to an understanding of morality, as philosophers show *'how nature may have selected both for cooperative norms and for the emotions that express our commitment to these norms'.*[38]

1.9.3 **Dispositions and ethics**

From the preceding discussion, we might reasonably deduce that as a consequence of evolution humans have acquired certain natural social dispositions, and that these have acted as the raw material which has then given rise to ethics. (The alternative to this 'naturalistic' explanation of ethics – that morality has been handed down from a supernatural authority – is discussed in 2.2.) But the important distinction between dispositions and ethical behaviour is that the latter entails conscious reflection. Although we cannot always avoid having gut reactions to circumstances, these are not a sound basis for action, not least because they often suggest that we should take actions that are contradictory. So the reason for, and challenge of, ethics would seem to be to arbitrate between the different dispositions, and to derive rational, coherent and consistent codes of behaviour. And, to a large degree, consistency implies impartiality – acting altruistically towards others, irrespective of age, gender, race – even, some would say, species.

The distinguished palaeontologist Gaylord Simpson suggested that three conditions must be met before we can meaningfully talk about our actions being 'ethical':

- there are alternative courses of action
- we are capable of judging the actions in ethical terms
- we are free to choose what is considered to be ethically 'right'.[39]

It follows that this sequence of steps depends on a fourth (which is probably unique, certainly in degree, to humans) – the capacity to *predict* the results of our actions.

As in other aspects of human activity, formerly tacit knowledge (1.4.4) has been to a large degree replaced by explicit knowledge. We no longer act on hunch, but for *reasons*, and those reasons can be explained, discussed and criticized. Often the reasoning leads to the formulation of *principles*, generalizations which help us make ethical decisions in novel circumstances, but which are nevertheless of a kind previously encountered. But these modes of reasoning appear to be specifically human; for, as Darwin remarked: *'of all the differences between man and the lower animals, the moral sense or conscience is by far the most important... a sense that is summed up by that short but imperious word* ought.'[40]

THE MAIN POINTS

- Bioethics is a relatively recent field of academic enquiry that deals with the ethical, legal, social and cultural implications of the biosciences and their application in biotechnology.
- Historical and sociological analysis demonstrates that progress in the biosciences is not simply a matter of accumulating objective data. The conceptual models employed to explain scientific observations are shaped by the cultural and political environment, and hence influenced by human values. This is even more the case for biotechnology.
- The specialized knowledge possessed by bioscientists places them under particular ethical obligations in their professional roles.
- The current interpenetration of the biosciences in the academic and commercial worlds has important ethical implications for the way knowledge is produced and used.
- There are competing theories about the biological basis of ethics, but there is general agreement that ethics developed from altruistic dispositions that are exhibited by non-human species.

EXERCISES

These can form the basis of essays or group discussions:

1. Investigate the history (1.4), over the past 150 years, of a specific area of biology (anything from the role of the cell nucleus, to the biochemistry of hormones, to reproduction in plants, to the action of artificial fertilizers in agriculture), identifying the extent to which social and cultural factors influenced scientific theories.
2. Assess the relative influence of Popperian and Kuhnian accounts of scientific theory (1.5) on your specialist area of the biosciences.
3. Is Oppenheimer's warning about the dangers of scientists being lured by the attractions of the 'technologically sweet' (1.6) relevant to modern biosciences; and if so identify where you consider the main dangers lie?
4. The 'mission statements' of all UK research councils include the aim of improving the 'the economic competitiveness of the UK' (1.7). What are the pros and cons of this objective?
5. Does evolutionary theory (1.9) help us to understand the origins and meaning of ethics or are cultural factors much more important?

FURTHER READING

The following provide readable and informative introductions to the history, philosophy and sociology of biology, respectively:

- The *Mendelian Revolution: the emergence of hereditarian concepts in modern science and society* by Peter J Bowler (1989). Baltimore, John Hopkins University Press. A engrossing account which shows the relevance of history to our current understanding of biology.

- *Thinking about Biology* by Stephen Webster (2003). Cambridge, Cambridge University Press. A readable 'practical manual for the thinking student'.
- *Real Science: what it is and what it means* by John Ziman (2000). Cambridge, Cambridge University Press. A very useful analysis written by a theoretical physicist turned sociologist of science.

USEFUL WEBSITES

Websites provide some accessible material, but it is important to be wary of propaganda dressed up as philosophy. Useful sites are:

- **http://www.dartmouth.edu/~bio1/** *History of Biology* (Michael R Dietrich): several articles on Darwinism and evolutionary theory.
- **http://plato.stanford.edu/emtries/popper/** *Stanford Encyclopedia of Philosophy*: contains a useful collection of articles on Sir Karl Popper's philosophy of science.
- **http://www.royalinstitutephilosophy.org/think/articles.php** The journal *Think* aims to provide to a non-specialist readership some highly engaging and accessible writing by leading philosophers, and to counter the popular impression that philosophy is pointless and detached from everyday life. Several articles from the journal appear on the Royal Institute of Philosophy website.

NOTES

1. Oxford Companion to Philosophy (1995) (Ed.) Honderich T. Oxford, Oxford University Press, p. 93
2. Williams B (1993) Ethics and the Limits of Philosophy. London, Fontana, p. 6
3. Gillon R (1998) Bioethics, overview. In: Encyclopedia of Applied Ethics. (Ed.) Chadwick R. San Diego, Academic Press. Vol. 1, pp. 305–317
4. O'Neill O (2002) Autonomy and Trust in Bioethics. Cambridge, Cambridge University Press, p. 1
5. Shelley M (1999) [1818] Frankenstein. Ware, Wordsworth Editions Ltd
6. DeVries R and Subedi J (1998) Bioethics and Society: constructing the ethical enterprise. http://www.stolaf.edu/people/devries/socdocs/preface.html
7. Russell B (1976) Unpopular Essays. London, Unwin, p. 56
8. Darwin C (1883) The Descent of Man. New York, Appleton and Company
9. Bowler P J (1989) The Mendelian Revolution. Baltimore, John Hopkins University Press, pp. 6–12
10. Olby RC (1966) Origins of Mendelism. London, Constable
11. Munz P (1998) Darwinism. In: Encyclopedia of Applied Ethics. (Ed.) R Chadwick. San Diego, Academic Press. Vol. 1, pp. 701–716
12. Galton F (1883) Enquiries into Human Faculty. London, J M Dent
13. Medvedev ZA (1969) The Rise and Fall of TD Lysenko. New York, Columbia University Press
14. Trusted J (2003) Beliefs and Biology (2nd edition). Basingstoke, Palgrave Macmillan
15. Yoxen E (1983) The Gene Business. London and Sydney, Pan Books, p. 30

16. Harris H (2000) The Birth of the Cell. New Haven and London, Yale University Press
17. Dawkins R (1976) The Selfish Gene. New York, Oxford University Press
18. Rose S and Rose H (1971) In: The Social Impact of Modern Biology. London, Routledge and Kegan Paul, pp. 215–224
19. Polanyi M (1969) Knowing and Being. London, Routledge and Kegan Paul
20. Popper K R (1979) Objective Knowledge (revised edition). Oxford, Oxford University Press, p. 81
21. Kuhn T S (1970) The Structure of Scientific Revolutions (2nd edition, enlarged). Chicago and London, University of Chicago Press
22. Hanson N R (1958) Patterns of Discovery: an enquiry into the conceptual foundations of science. Cambridge, Cambridge University Press, p. 13
23. Bowler P J (1989) The Mendelian Revolution. Baltimore, John Hopkins University Press
24. Cited by Mackay A L (1977) The Harvest of a Quiet Eye. Bristol and London, Institute of Physics, p. 113
25. Ibid., p. 114
26. Weindling P (2002) The ethical legacy of Nazi medical war crimes. In: A Companion to Genethics. (Eds) Burley L and Harris J. Oxford, Blackwell, pp. 53–69
27. Galton D (2001) In Our Own Image: eugenics and the genetic modification of people. London, Little, Brown and Co., p. 99
28. Biotechnology and Biological Sciences Research Council (2004) http://www.bbsrc.ac.uk/about/
29. Blair A (2002) Science matters (Speech; 10 April). London, Royal Society
30. Hobbes T (1914) [1651] Leviathan. London, J M Dent, pp. 64–65
31. Darwin C (1883) The Descent of Man. New York, Appleton and Company. Chapter 4
32. Mayr E (1997) This is Biology. Cambridge, Mass., Harvard University Press, p. 254
33. Wilson E O (1975) Sociobiology: the new synthesis. Harvard, Belknap Press
34. Wison E O (1998) Consilience. London, Little, Brown and Company, p. 275
35. Singer P (1983) The Expanding Circle. Oxford, Oxford University Press
36. Midgley M (1991) The Origin of Ethics. In: A Companion to Ethics. (Ed.) Singer P. Oxford, Blackwell, pp. 3–13
37. Hinde R A (2002) Why Good is Good: the sources of morality. London, Routledge, p. 13
38. Rosenberg A (2003) Darwinism in moral philosophy and social theory. In: The Cambridge Companion to Darwin. (Eds) Hodge J and Radick G. Cambridge, Cambridge University Press, p. 331
39. Simpson G C (1969) Biology and Man. New York, Harcourt, Brace and World, p. 146
40. Darwin C (1883) The Descent of Man. New York, Appleton and Company. Chapter 4

2

Theories of ethics

OBJECTIVES

When you have read and discussed this chapter you should:

- be aware of important discussions on ethical theory, including its relationship with religion and the concepts of free will and natural law
- understand the main features of three major ethical theories: utilitarianism, deontology and virtue theory
- understand how ethical theory can be applied to a specific issue in the biosciences – the use of animals in research
- appreciate the strengths and weaknesses of the different ethical theories and the challenges presented to the biosciences in applying them to present-day concerns

2.1 Introduction

Chapter 1 discussed the general remit of bioethics and indicated why, because it is concerned with human values, it permeates the scientific and technological aspects of the biosciences. But there was no attempt to explore the meaning of the word ethics in any detail - beyond the fact that, first, ethics concerns the question of what we *should do*, and, second, that bioethics relates specifically to what we should do with regard to the biological knowledge and to the skills and techniques developed in biotechnology. We now need to take a closer look at ethical theory.

2.1.1 Ethics and morality

While **morality** refers to general attitudes and standards of behaviour, **ethics** is usually taken to mean the disciplined and systematic enquiry into the nature of morality. Apart from a very small number of people whom the rest of us would regard as psychopaths, this sense of morality seems to be an innate and defining characteristic of human beings (1.9.3). We believe, that is to say, that certain types of behaviour are **right** and others wrong, and/or that we should pursue the **good** and avoid the bad.

It doesn't follow that we know instantly what we should do in all circumstances, or that we always do what we believe we should: complete awareness and moral perfection are certainly beyond all of us. But it is generally expected that every sane and humane person should be aware of the moral dimensions of human life, and take measures to act accordingly. Putting it bluntly, if witnessing someone else mugging an elderly person in the street, or someone cheating to improve their assessment grades, didn't bother you, most people would consider you were almost as morally defective as the perpetrator of the actual acts.

It would seem that everyone who aspires to lead a life that does not consist simply of mindless reactions to events is compelled to reflect on ethics. Our *sense* of morality may be to a large degree innate and greatly influenced by our upbringing, but our human ability to reason requires us to submit this moral sense to the discipline of rational thought. The process involved is **ethical deliberation**, where 'deliberation' may be defined as 'the careful discussion and consideration of an issue'.[1]

Of course, having deliberated, some people may decide to act in ways that the majority of others consider unethical: there is no guarantee that everyone will agree on *the* ethical course of action in any particular circumstances. Nor is there any guarantee that the majority view is the correct one: ethics is not determined by opinion polls. But an important point is that for behaviour to be considered ethical requires that it be justified – to ourselves, and/or to others – and/or, for some people, to God. That is to say, we need to (be able to) give *reasons* for our actions. This is a necessary condition, although it may not of itself be a sufficient reason to class any particular form of behaviour as ethical.

2.1.2 Ethics and free will

The fact that ethical decisions need to be justified implies that we have some choice in the matter. For out-and-out **determinists**, who deny that people have any **free will**, all behaviour is predetermined. And because, for them, all human actions are simply the inevitable consequences of earlier events, there is no room for ethics. Of course, if we don't have free will, those who do not believe in it cannot do otherwise – so there is no reason to take any notice of what they say. Believers in free will are at least more logically consistent, even if their belief is actually false. This is a fascinating but ultimately insoluble problem! But the reality is that everyone *behaves* in ways that assume that they do have some measure of free will, and it would be very difficult to live at all (certainly as a member of society) if we thought we were incapable of making any real choices. You might think that the fact of free will (e.g. as simply demonstrated by your ability to raise your finger or not, as you desire) is one of the most certain things you *do know* in a world where so much else is doubtful. (Those with an appetite for such questions might find philosopher Daniel Dennett's ideas stimulating, if not exactly 'light reading'.[2])

The assumption made when discussing ethics is that people do have a significant degree of freedom of action. There are clearly some things that we must do to survive (such as breathe and, over a longer time-scale, eat), but the exercise of choice and creativity would seem essential to living a life which aspires to be ethical to any degree. Indeed, if there were no free will it is not only ethics that would be in trouble: all our objective knowledge of the world, which it is the aim of science to discover, would be undermined – because we should be predetermined to believe what we do regardless of its objective truth.

In fact, the almost universally held belief that we can exercise a measure of choice can be said to be the most crucial factor underlying ethics. Because if we do have choice we are faced with the question of what precisely we *should do* with it – which is what ethics is all about. It also raises another critical philosophical question, which we can hardly avoid coming to a conclusion about, even if we only end up with a working hypothesis. The question is: 'Who or what is the *I* that makes the choices?' Because fully satisfactory answers to that question in scientific terms cannot be given, philosophers explore possible answers in the study of **metaphysics**, which may be defined as 'the philosophical investigation of the nature, constitution and structure of reality'.[3] Non-philosophers will hardly bring the same rigour to their reasoning, but almost inevitably virtually everyone arrives at some sort of answer to the question, which is bound to influence their views on life, including their ethical opinions.

2.1.3 Ethics and others

Another important feature of ethics is that it is largely about our relations with others, where 'others' refers to people, animals, some would say plants and the environment generally, and some would say God. So ethics can be said to be principally about how much one person's interests should take precedence (if at all) over others' interests – or more generally, the order in which different individuals' (or groups') interests should be prioritized.

For someone living on a desert island, the occasions when ethical choices have to be made must be far fewer than for people living in society or in a family. And for some people – such as a doctor, who has to balance the needs of her many patients with those of her family life, who may have to make decisions about telling a patient he has a fatal condition, or notifying the police of information obtained in the consulting room which has public health or criminal implications – the exercise of ethical judgement is a constant and critical demand.

2.2 Ethics and religion

It was suggested in 1.9.3 that ethics may have had its origins in the dispositions inherited from non-human species. But many people assume that ethical behaviour has a different type of origin, being intimately bound up with religion, or that it is part of the legacy of an earlier age when religion was a dominant force. Consequently, accounts are often presented of ethical theories that are claimed to gain their authority from particular scriptural sources, giving rise to Christian Ethics, Buddhist Ethics, Islamic Ethics, etc. Thus, it is widely believed that the required standards of moral behaviour for people following a particular religion are determined by the necessity to obey God's will as prescribed in the sacred texts, such as the Christian Bible or the Islamic Qur'an.

However, as was demonstrated by the philosopher Plato in ancient Greece, this merely shifts the problem, because it raises the question as to why we should obey God's will in the first place. If, say, a supernatural being had created us for his own *bad* purposes, we should presumably not feel it right to obey his will. So it would seem to follow that things cannot be good or bad simply because God commands or prohibits them: there must be some other reason, and in that case the appeal to God's will would be unnecessary.

Moreover, there are a number of other problems with assuming that religion is the source of ethics. First, adherents of the same religion frequently interpret the scriptures differently. Some Christians support 'just wars', meat-eating, and gay rights, while others derive the opposite conclusions from their reading of the scriptures. This presumably has much to do with the fact that the scriptures were written at times when the dominant outlooks of their authors were totally different from those prevailing today in the variety of twenty-first-century cultures. (And we might reasonably ask why it should be assumed that the insights of ancient writers necessarily have greater worth than our own, because, unless we are abandoning responsibility for our actions, ultimately it is *we* who have to make the judgement on whether to accept them.) Moreover, because different religions differ in their ethical stances, and because large numbers of people are agnostics or atheists, if ethics were the preserve of the religious, the possibility of meaningful dialogue between people not sharing particular religious beliefs would be much diminished.

However, it needs to be stressed that the above arguments do not seek to challenge the fact that many sincerely religious people find the motivation and justification for their ethical views in their religious beliefs. The insights derived from the mental or spiritual experiences of people engaged in prayer, meditation, or contemplation can have the profoundest effect on their beliefs about the way they should live. Some exceptional individuals act in ways most people, including non-believers, would agree were 'saintly'. (On the other hand, mere ritualistic observance of rules, such as permitting the eating of fish only on Fridays, would seem to owe little to spiritual experience.)

2.2.1 Newer ideas on religion and ethics

But the influence of religious ways of thinking on ethics cannot be categorized easily. Aldous Huxley wrote of what the philosopher Leibniz called the **Perennial Philosophy**, as the *'highest common factor in all preceding and subsequent theologies . . . the ethic* [of which] *places man's final end in the knowledge of the immanent and transcendent Ground of all being.'*[4] Such metaphysical beliefs might well inform the ethical positions of people who subscribe to different religions, or even to none. Indeed, if we consider the psychological origins of religion, belief in God might be seen as an expression of a Popperian 'bold hypothesis' (1.5.1), and rather, as Voltaire put it: *'If God did not exist it would be necessary to invent him'.*[5] Reversing cause and effect in the same way, it might be said that *'Religions traditionally reflect and reinforce a culture's deepest ideals.'*[6]

Moreover, the influence of religious ways of thinking on ethics is now subject to radical reappraisal. For example, the idea that traditional religion is best seen as a phase in human spiritual development has been explored by theologian Don Cupitt, who argues that belief in God as a *'very big and friendly unseen Superperson'* is *'neither clearly statable nor rationally defensible'*. Accordingly, he commends a postmodern non-realist conception of God, in which *'what we should pick out as valuable and try to salvage will be certain forms of religious existence; that is, certain forms and practices of selfhood, certain modes of consciousness and ways of expressing one's life.'* This, he claims, will bring about *'the end of morality and the return of ethics'*.[7]

That said, since it is widely acknowledged that religious insights cannot be *demonstrated* objectively to others, religious experience might be best viewed as a form of tacit knowledge (1.4.4). And if so, it would seem to follow that discussion of ethics is more generally meaningful when conducted in terms that do not appeal to scriptural authority, but instead refer to fundamental aspects of human nature acknowledged by believers and non-believers alike.

2.3 Natural law and the naturalistic fallacy

The Cambridge philosopher G E Moore identified an influential concept in ethical theory more than 100 years ago (although it had been raised by David Hume in the eighteenth century). Moore argued that in the past many people discussing factual matters had confused **facts** with **values** by slipping into ethical language without acknowledging the transition. For, he said, you *'can't get an ought from an is'*, and people who attempted to do so were guilty of committing the **naturalistic fallacy**.[8] Putting it another way, the way things *are* is not necessarily a guide to how they *should* be: for example, the fact of male dominance in most human societies does not mean that we ought to accept this situation as 'right'.

But the naturalistic fallacy is also important in a 'technical' sense in ethical reasoning. For example, the fact that many people in some African countries are severely malnourished does not lead inevitably to the conclusion that they ought to be sent food aid. There has to be another step in the reasoning process, which will most likely take the form of a value judgement that those who are well fed and in a position to do so ought to help other people who are starving. So, the sequence of steps from *is* to *ought* takes the form: (*a*) the facts of a situation need to be established; (*b*) the relevant ethical principles need to be identified; (*c*) the appropriate actions ought to be applied to the situation.

The validity of the naturalistic fallacy has achieved widespread recognition, but it is important to appreciate its limitations. For example, philosopher John Searle noted that if you have made a promise, you *ought* to keep it, because saying you ought not to keep promises would be self-contradictory. So in this case you can get an 'ought' from an 'is' – when the 'is' consists of the fact of having made a promise.[9] It follows that we need to distinguish between brute facts, like 'sparrows have wings' and social facts such as 'Jack promised to marry Jill'.

Even allowing for such exceptions, Moore's claimed separation of fact and value is not universally acknowledged. For example, to certain religious believers God created the universe for a particular purpose, so that the way things *naturally* exist represents a God-ordained order which it is ethically wrong to seek to alter. A prominent example is the position of the Roman Catholic Church on the role of sexuality. If it is accepted that the sexual attraction between men and women is intended solely to create children, then all practices that frustrate this aim (everything from homosexual acts, to contraception, to abortion) will be seen as offending **natural law**.

But belief in natural law is by no means confined to those who subscribe to religious beliefs, and for many people the idea that 'Nature knows best' challenges the concept of the naturalistic fallacy. For example, it might well be claimed that the desire to procreate, and the caring behaviour shown by parents towards their children, are based on natural moral imperatives; consequently, artificial means ought to be made available to help sub-fertile people have children (5.3.1). Indeed, natural law thinking is acknowledged worldwide. For example, it is enshrined in the United Nations (UN) Declaration of Human Rights, as when Article 16 states that all men and women 'of full age' have *'a right to marry and found a family'*, asserting that *'the family is the natural and fundamental group unit of society'*. It is also evident in the widespread rejection of the prospect of human cloning by nuclear transfer (5.5.3).

A leading scientific critic of the naturalistic fallacy is sociobiologist Edward O Wilson (1.9.2). According to Wilson, *'Ought is the product of material processes*... [a conclusion] *which points the way to an objective grasp of the origin of ethics.'*[10] A danger of this approach is that it can be seen as politically reactionary, e.g. if it seems to accept the current social circumstances of different countries (such as the disparities between the 'haves' and the 'have-nots' - see chapter 4) as the natural outcome of evolutionary processes. More generally, a rigorous observance of natural law might mean a fatalistic acceptance of everything that happens, on the assumption that all disease, suffering, and natural disasters are simply 'what Nature intended'.

2.4 Moral acts and ethical theory

Very few acts have no ethical implications. Even sitting in a chair, when you could be doing something else such as raising money for charity, is the result of a decision that has ethical implications. Taken to an extreme, you could imagine sinking into a condition of guilt-ridden helplessness if you were to agonize over every act or state of inactivity. But it would surely be irresponsible if, just because we can't do everything we ought, we decided not to bother about acting ethically at all. Many of our actions are, of course, almost intuitive. But acting deliberately and ethically would seem to require that we draw up some general **rules** to help us make decisions about individual cases, as for example when facing the dilemma of whether to give money to a famine relief charity or spend it on ourselves.

On reflection, these rules usually turn out to be specific instances of **principles** with wider applicability. In turn, these may often be justified by appeal to a **theory** of ethical behaviour which we have probably adopted, perhaps largely unconsciously, as a combined result of our upbringing, a process of reasoning and some mental reflection. We can thus envisage that the individual act (e.g. giving or not giving money to charity) is the tip of an **ethical iceberg** (Figure 2.1), the bulk of which is hidden from view - or, in terms of our action, not usually consciously thought about or discussed. But while this

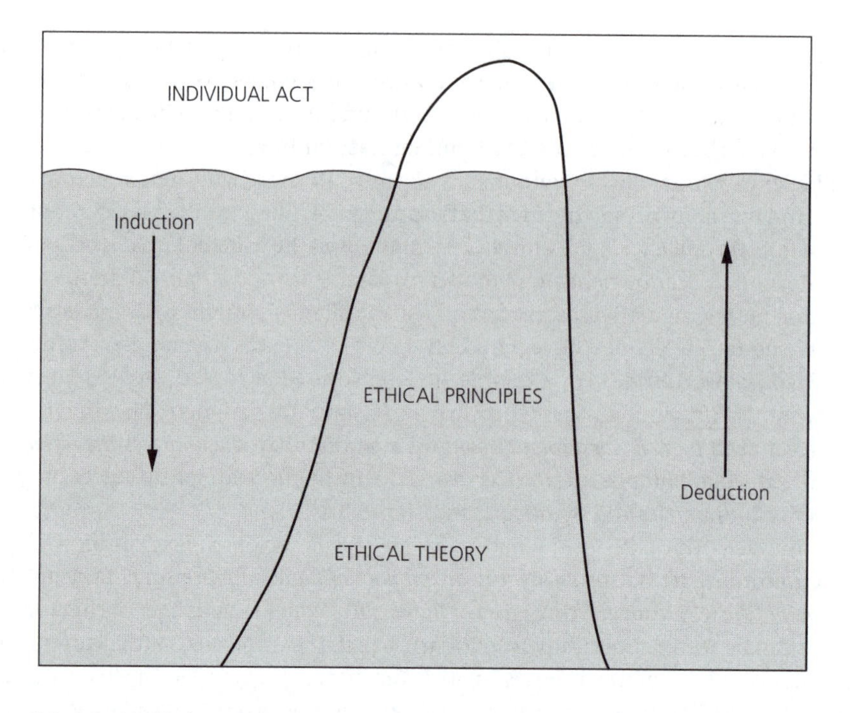

Figure 2.1 An *ethical iceberg*, indicating the relationship between individual acts and ethical theories. In *reflective equilibrium* ethical principles are the outcome of a combination of inductive and deductive processes.

may suffice for ethical concerns at the personal level, such as deciding what to do in the case of donating to a charity, there are many instances when we need to be explicit about the ethical justification for our actions - and this is typically the case with bioethical issues, which often affect society and/or the environment at large.

There are different opinions about the relationship between ethical theory, principles, rules, and actions. It might be as a result of responding to individual circumstances that we decide on rules and principles, and then develop a theory that accommodates them coherently. This might be called the **inductivist** view (see 1.5). Or our theoretical approach might come first, the principles and rules being deduced from the theory. An intermediate position is that people adopt both inductivist and **deductivist** strategies at different times in order to try to achieve a coherent and consistent ethical outlook. That is to say, experience of life, together with an intuitive sense of what seems right and the exercise of reasoning, all contribute to the outlook in a process called **reflective equilibrium** (Figure 2.1). Whatever the exact relationship between the different steps, the scheme outlined does seem to go some way to explaining the interactions between ethical theory and our actions.

The sheer complexity of the different ethical questions that life throws up has led philosophers, going back at least 2500 years, to try to simplify our ethical decision-making by devising explanatory theories. The aim of the next sections is to summarize the main theories; and it may be useful to refer back to these sections when considering specific bioethical issues later in the book. The emphasis here will be on three main theories: **utilitarianism, deontology**, and **virtue theory**. It needs to be appreciated that the fact that they were formulated hundreds, or even thousands, of years ago, suggests not only that they have proved their worth over the years, but also that some of the assumptions made at the time they were proposed may appear dated. A challenge for bioethics is to interpret these theories in the light of our views of the world revealed by modern bioscience.

2.5 Utilitarianism

The ethical approach called **consequentialism** is the view that our actions should be ethically determined by the consequences likely to result from them. The desired consequences clearly need to be specified to make this a useful theory, and in the most prominent form of consequentialism, called *utilitarianism*, the aim is to *'produce the greatest good for the greatest number.'*[11] The theory was introduced by Jeremy Bentham in the eighteenth century and developed by John Stuart Mill (Box 2.1). In modern terminology, utilitarianism employs the methodology of **cost–benefit analysis.** We need to weigh up the costs

BOX 2.1 FOUNDERS OF UTILITARIANISM

Jeremy Bentham (1748–1832) was a person of outstanding intellectual ability. Beginning to study Latin at 4 years of age and French at 5, he entered Oxford University when only 12 and took his degree at 15. He subsequently spent his life writing and advocating changes to the legal system along utilitarian lines.

He exerted a strong influence on changes to the British law of evidence, on the abolition of laws permitting imprisonment for indebtedness, and in the reform of parliamentary representation. His major work was *An Introduction to the Principles of Morals and Legislation* (1789).

In accordance with his directions, when he died his body was dissected in the presence of friends, and the skeleton preserved in an 'auto-icon', dressed in the clothing of the times. To this day, this effigy can be seen in its glass cabinet at University College London, which was established along Benthamite lines in 1828.

John Stuart Mill (1806–1873) was the son of the Scottish philosopher James Mill, himself a leading advocate of reforms based on Bentham's utilitarianism. Like Bentham, J S Mill was intellectually precocious, beginning the study of Greek at 3 years of age.

Throughout his life Mill propagated principles that he perceived essential for human happiness, which extended from moral principles, to principles of political economy, to the principles of logic and metaphysics.

He wrote several very influential philosophical books, e.g. *System of Logic* (1843), *On Liberty* (1860) and, as an early feminist, *The Subjection of Women* (1869). He was a Member of Parliament from 1865 to 1868.

of doing something and assess the resulting benefits; if the surplus of the latter over the former is thought likely to be maximized then we ought to proceed with it.

There are, however, some problems in defining 'benefit'. According to Bentham, *'Nature has placed mankind under the governance of two sovereign masters, pleasure and pain'*, so that 'good' is what tends to maximize pleasure and at the same time minimize pain. Giving it a mathematical spin, we can refer to this as the **hedonic calculus** (from the Greek meaning 'pleasure'). Bentham considered that pleasure (or pain) could be assessed in terms of factors such as its intensity, duration, degree of certainty, and whether it had a chance of being succeeded by sensations of the opposite kind. But the hedonic calculus runs the risk of reducing ethics to a question of satisfying what might be quite trivial pleasures: e.g. does someone's pleasure at getting very drunk really outweigh the distress caused to those who have to take him home and perhaps to his family (not to mention the effects on his liver)? Moreover, if a sadist derived great pleasure from torturing people the theory might be considered to perversely justify such acts.

Mill tried to avoid such problems by adding a qualitative criterion to Bentham's purely quantitative measures.[12] For example, he maintained that pleasures of the mind are higher pleasures than those of the body (so the joy of playing Beethoven might count more than the exhilaration of playing badminton). In essence, he suggested that rather than pleasure, the good should be defined as the satisfaction of **preferences**, and such preferences might not usually be thought of as pleasurable. Rather than seeking enjoyment, someone might prefer to spend their time visiting people in a hostel for the homeless.

Another refinement was the proposal of a role for **rules** within the utilitarian calculus.[13] Thus, although in the short term it might seem beneficial to break rules (such as that requiring us to tell the truth), ultimately this will lead to a breakdown of trust which will undermine the aim of maximizing happiness. This **rule utilitarianism** contrasts with the **act utilitarianism** that requires each case to be treated separately.

2.5.1 Illustrations of utilitarian reasoning in the biosciences

An example from the biosciences will illustrate the use of the theory. Consider a utilitarian justification for the use of mice in experiments performed to test a new drug for people suffering from obesity. Such people might suffer much physical discomfort, often have to endure other people's unsympathetic or insulting comments, and be at greater risk of suffering from other conditions such as heart disease. On one interpretation of utilitarian theory, the benefits to obese people of the research in question would clearly greatly outweigh the harm to a few mice. The mice, after all, do not have intelligence approaching that of people; they might easily otherwise be caught by cats; those particular mice only exist at all because they were bred for experimental use; and they are, in any case, protected by strict animal welfare laws. This is the form of the standard defence of animal use in medical research, and it is an argument endorsed by successive governments and subject to legal regulations (8.3).

But the utilitarian argument could produce a quite different result if the assumptions made above were not valid. For example: if the mice suffered appreciable pain in the experiments (so that the question of whether they were intelligent was irrelevant); if the conditions in which they were kept, e.g. crowded together in cages in a sterile, featureless environment, seriously reduced their ability to express their normal patterns of behaviour; if a very large number of them were used in the experiments; and if the laws were so loaded that any amount of animal suffering was deemed acceptable provided that some obese people received, perhaps only minor, relief – then the original judgement might need to be revised.

Moreover, if it were possible to obtain the information sought in the experiments by means which did not involve animals at all, e.g. using cells in tissue culture, then the cost–benefit analysis might prove to be deceptive. Indeed, there might be evidence that drug treatment was not the only way of treating the distressing condition of obesity, and that equally or more effective treatments could be achieved by dietary or lifestyle changes. It might thus be possible, solely by appealing to utilitarian reasoning, to decide that these particular animal experiments should *not proceed.* (It is interesting that even in the eighteenth century, Jeremy Bentham included the interests of animals in the utilitarian calculus when he wrote: *'The question is not, Can they reason? nor Can they talk? But, Can they suffer?'*)

The point of the above example is not to argue the case either way (a question to be explored more extensively in chapter 8), but to demonstrate how the theory might be used in justifying an ethical position. It is clear that deciding which is the more acceptable conclusion will depend to a large degree on *evidence,* i.e. about how many people might be helped and how much; how many animals might suffer and how badly; and whether alternative forms of treatment *do* exist. But even if such evidence were available (which it often is not), no definite judgement on the ethical acceptability of the animal experiments could be produced because there is no universal formula for deciding how to assess the relative happiness of mice and men.

2.5.2 Scientism

Utilitarianism is widely used to justify aspects of science and technology where risks are involved – and very few types of behaviour do not carry risks of some kind (chapter 13). But those who employ it as a form of ethical reasoning need to be aware of its theoretical limitations.[14] In a sense, much of the discussion in this book will entail a critique of utilitarian reasoning. This should certainly not be seen as a 'science-bashing' exercise, because undoubtedly science and technology can often be used in ethically acceptable ways. Rather, it should be regarded as a corrective to what might be called a naïve positivist approach (also termed **scientism**). The scientistic approach assumes that relying on a utilitarian justification for science and technology is the *only* way of assessing matters ethically – either because it is assumed that science only deals with *facts* (which are considered to have no ethical content) or because it is assumed technology is necessarily beneficial (or no one would bother to develop it). The discussion in 1.5–1.6 seriously questions both those assumptions.

2.5.3 Some limitations of utilitarian reasoning

One of the serious problems with naïve forms of utilitarianism is that because they measure ethical acceptability on the basis of *net* costs and benefits they can be held to justify actions which offend normally accepted behaviour. To illustrate the point dramatically, on such a view of utilitarianism circumstances might possibly arise a situation in which (to maximize the surplus of good over harm) it would be acceptable for 49% of the population to live in abject misery as long as 51% were very happy. Indeed, since it is only *consequences* which count as a measure of ethical acceptability, *anything* would strictly speaking be permitted if the *net* amount of happiness were to be maximized. A case could be made on this basis for lying, theft, even murder, if more people ended up happy than were made miserable by the actions (but Mill's refinements could rule out at least some of these actions). A classical 'thought experiment' suggests that killing one person to provide vital organs for transplantation into two patients, saving two lives for the loss of one, would be endorsed by utilitarian theory[15] – particularly if the 'donor' was someone with no friends or family.

A second problem is that, because it is *future* outcomes that decide what should be done, all ethical decisions must be to a degree speculative. Of course, we do have a fairly certain idea of the consequences of many actions (e.g. medicines usually do work for most people), but the more unusual the proposed action, the less likely is it that we will be able to predict how things will turn out. Often, it is more appropriate to talk about a risk-benefit analysis than a cost-benefit analysis (13.1) – although risks can, of course, result both from action and inaction.

A third difficulty concerns the **scope** of the cost or risk analysis. Who or what should count? Do we limit our analysis to people currently alive in our own country or region? Are unborn generations to count? Or the very early human embryo? Are animals and plants to count, or the biosphere as a whole? Many challenges to decisions based on utilitarian reasoning are concerned with such issues. But attempting to include all the relevant interests in the analysis is very difficult; and no one should be in any doubt as to the complexity of the task and the potential loopholes in any analysis that can, realistically, be performed. Indeed, taking all such difficulties into account, philosopher John Mackie described utilitarianism as **'fantasy ethics'**.[16] In fact, in common usage, utilitarianism is often interpreted quite loosely (and inaccurately) as simple cost-benefit analysis, where a practice is considered justified if some (limited) benefits are held to simply *exceed* some (limited) costs.

2.6 Deontology

Deontological theory, which had its origins in the ideas of philosopher Immanuel Kant in the eighteenth century (Box 2.2), refers to the rights and duties we have as individuals with respect to other individuals (**deontology** is derived from the Greek for *duty*). In essence, the theory is based on the observation that, however wisely we try to act, the results are subject to circumstances beyond our control; so the morality of actions ought not to be judged by consequences but by their *motivations*. Accordingly, it is only the **intention** of an act that is good, not the outcome: people should act out of a sense of duty – a principle that is right, in and of itself. To the extent that pursuit of the *right* and the *good* do not necessarily amount to the same thing, deontology prioritizes the right, particularly if the good were to be defined simply as pleasure.

BOX 2.2 IMMANUEL KANT

Immanuel Kant (1724–1804) is commonly regarded as the most outstanding figure in Western philosophy since the ancient Greeks. From the early 1780s he produced a number of highly influential books which left their mark on philosophy for many years, including *Critique of Pure Reason* (1781), *Fundamental Principles of the Metaphysics of Ethics* (1785), *Critique of Practical Reason* (1788), *Critique of Judgement* (1790) and *The Metaphysics of Morals* (1797).

He was born, and spent almost the whole of his life, in Königsberg, East Prussia, where he led a life governed by familiar routine – such that it was said that the people of the town could set their clocks by the punctuality of his daily walk and habits. However, despite his conventional lifestyle, he was an amusing conversationalist and, although he never married, he enjoyed company and never dined alone. He was renowned for his brilliance as a lecturer.

Kant argued (controversially) that although scientific explanations of events appear to suggest that everything is caused by a prior event, as well as this *phenomenal* world there also exists another, *noumenal*, world (a sort of parallel universe), where we make decisions that affect our actions, and where it is permissible to ascribe praise or blame to actions. But only individuals who understand the reasons for so doing can act morally, so that morality is only possible for rational beings.

Kant then proceeded to argue for a system of ethics based on reason, drawing the parallel with science. This led him to claim that the fundamental rule of morality, which, like scientific laws, must be universal, was the categorical imperative: *'Act only according to maxims you can will also to be universal laws.'*

2.6.1 **Features of Kant's theory**

Kant's main aim was to construct ethical principles which were based on rational procedures. Rather than assuming what 'the good' is, he tried to establish principles that would apply regardless of other peoples' desires or social relations; and this meant that nothing could be a moral principle for one person that could not at the same time be a principle for everybody else. The distinctive features of this theory might be summarized as:

- Each person has a duty to respect the inherent dignity (or **autonomy**) of other people and treat them as *ends in themselves* and *not instrumentally*, i.e. merely as a means to one's own ends.
- Morality consists of performing the right actions, which can be described as **categorical imperatives** (e.g. 'do not tell lies'; 'do help the needy').
- If an ethical right applies to us as an individual, it also applies to everyone else, i.e. it is a **universal right**, which thus places us all under a *duty* to respect it in other people.[17]

So in formulating principles which were based on **reason**, Kant concluded that the only way to live an ethical life was to be guided by what amounted to a **moral law**. In essence, the approach corresponds to the so-called Golden Rule: *'Do as you would be done by'* – a rule sometimes claimed by Christianity as its own, but which is found in most ethical traditions, including that of the Chinese philosopher Confucius (fifth century BC).[18]

An important aspect of Kant's thinking is that no account is taken of the particular consequences of actions. For example, telling lies is categorically wrong because it undermines the trust that has been developed by other people telling the truth: the liar behaves as a parasite on society. Moreover, performing beneficial acts for the wrong reasons - like participating in a sponsored charity event out of self-interest (because you enjoy it) or to avoid likely criticism if you did not participate - does not count as ethical in Kant's view, although to a utilitarian that would be irrelevant if the event raised money for charity.

2.6.2 Limitations of Kant's deontology

As with utilitarianism, there are some serious problems with deontological theory. One concerns the difficulties that arise when there is a conflict between the duties to act in accordance with different categorical imperatives, whose consequences might be inconsistent. If, for example, you are approached by an apparently demented person, wielding an axe and asking you the whereabouts of your friend, you are faced with the dilemma of wishing both to tell the truth and to protect your friend from likely harm. Your desire to protect the friend by telling a white lie will entail ignoring the categorical imperative to 'always tell the truth', but there are no rules to guide or justify this decision. (Kant appeared to suggest that lying would be wrong even in such circumstances, but others argue that certain exceptions to the rule against lying would not be inconsistent with his beliefs.[19])

There is also a problem in the formulation of the categorical imperatives. Although these are intended to apply to everyone, they might be framed in such a way that they in fact affect only a small section of society. For example, if stealing by hungry people were deemed a categorical imperative, a small minority of hungry people could feel it right to steal: but there are no precise rules for deciding who is 'hungry'.

2.6.3 Kantianism

Over the last 200 years Kant's approach has been adopted and adapted by others to produce forms of ethical reasoning which are rather different from those Kant himself proposed. Nowadays, philosophers often talk of **Kantian ethics** when they want to stress deontological theory as opposed to utilitarian theory, or more generally lay emphasis on actions rather than results.[20]

Two characteristics of a modern Kantian approach are:

- a strong opposition to deception, as being incompatible with respect for persons
- the limitation of acts of coercion to cases that are compatible with respect for persons, such as emergency aid to rescue victims of crime.[21]

But within the broad span of modern Kantian ethics differences of emphasis often become apparent, such as the high priority given to the autonomy of the individual in much bioethical literature in the USA. For some philosophers, especially in Europe, this places too much emphasis on rights and too little on duties.[22] Perhaps the most prominent modern form of Kantian theory is that developed by John Rawls, which is discussed in 3.3.

2.6.4 Kantian ethics and animals

We have seen that Bentham regarded the interests of animals as ethically relevant because, although they could not reason or talk, undoubtedly, they could suffer. In contrast, for Kant, the ability to reason is a decisive factor as to whether a being has **ethical standing** (3.4). Rejecting the instrumental use of other people, he wrote: *'Unlike objects or animals, humans are never to be used as a means to another's ends.'* However, *'Animals must be regarded as man's instruments...as a means to an end.'* So Kant argued that we have no direct obligations to animals but only indirect ones. People should not be cruel to animals, not because we have any duties to them but because it might offend the animals' owners and/or it might encourage people to be cruel in dealing with other people.

In recent times Kant's views on animals have been challenged by those who, encouraged by developments in evolutionary biology, see the sharp distinction drawn between animals and humans as arbitrary. Philosopher Tom Regan has adapted Kant's view that people are ends in themselves by claiming that animals (at least those with a sufficiently developed nervous system) also have ethical standing because they are **subjects of a life**, a fact which assigns them **inherent value**. According to Regan, all the animals we eat, hunt, trap and exploit in sport and science have a life of their own quite apart from their utility to us. *'They have a biography and not just a biology. They are somebody not something.'*[23] This Kantian line of reasoning suggests that animals should be assigned rights, analogous if not identical to those we ascribe to humans. If we refer back to the case of animal experimentation discussed above (2.5.1), Regan's belief in the rights of animals persuades him that this, along with all other forms of animal exploitation, is ethically impermissible.

2.7 Virtue theory

Both utilitarianism and deontology seek to define ethics by referring to a single dominant principle. But a prominent recent ethical theory adopts a different starting point, one based on **virtue** and **character**. In fact, although the formulations of so-called *virtue theory* are recent, they have their origins in ancient writings, notably those of Aristotle[24] (Box 2.3).

Virtue theory puts emphasis on the person who performs the actions and makes the choices, rather than on the situations in which choices have to be made. Aristotle believed that the goal of life is to live virtuously to attain a state of happiness. This was to be achieved as a result of the exercise of reason and entailed choosing a **golden mean** between two extremes. For example, exercising the virtue of courage meant acting in a way that was intermediate between rashness on the one hand and extreme

BOX 2.3 ARISTOTLE

Son of a court physician, **Aristotle** (384–322 BC) went to Plato's Academy (effectively the first university – which gave rise to our word 'academic') in Athens when 18 years old and stayed there for almost 20 years. For the next 12 years, after Plato's death, he was tutor to the young Alexander the Great, later returning to Athens, where he set up his own 'university', the Lyceum. With Plato, and Plato's teacher Socrates, Aristotle was one of three philosophers who can be said to have laid the foundations of all subsequent philosophical enquiry.

But his achievements were not limited to what we now class as philosophy. He studied and wrote extensively on physics, cosmology, and mineralogy, and for his anatomical, embryological, and ethological insights he has, in recent times, been called 'the first biologist'.

His ethical ideas are contained in two books, of which the *Nicomachean Ethics* (based on his lectures, which were compiled and edited by his son, Nicomachus) is regarded as the most important.

For Aristotle, the key human attribute is *reason*, which is not only necessary for understanding, but also for right action. Reason is the route to achieving the goal (*telos*) of human life (i.e. happiness, which he called *eudaimonia*), while the skill of considering proposed actions and matching them with *telos* is called prudence (*phronesis*).

timidity on the other. Other natural virtues were wisdom, justice, and temperance; and overall, morality was considered to be a matter of **practical wisdom**. For Aristotle, then, the wise or virtuous person has an 'eye' for what is appropriate in any given situation, and deciding what to do is arrived at by a process of training in virtuous behaviour. Subsequently, Christian thought added three theological virtues to the natural virtues - faith, hope, and charity (love), claiming that the latter was 'the greatest' of these.

In recent years, philosopher Alasdair MacIntyre has suggested that ethics should be less concerned with individual moral decisions (the preoccupation of utilitarianism and deontology) and more with the health and welfare of the community; and he argues that a return to Aristotle's ethical philosophy might be the best way to achieve this.[25] (It is perhaps worth stressing that it is Aristotle's *style* of reasoning that is here referred to: some of his actual beliefs, such as those approving of the roles of slaves and the subjugation of women, are certainly not now generally considered ethically acceptable.) MacIntyre claims that the language and practice of morality are currently in a state of disorder because *'ill-assorted fragments of various parts of our past are deployed together in private and public debates'*. If people could be encouraged to behave virtuously, almost as it were by instinct, a society might be evolved in which citizenship became a priority for all.

2.7.1 Some limitations of virtue theory

'Obviously, the world would be a better place if more lawyers were taught the virtue of honesty, politicians integrity, physicians beneficence, and professors humility.'[26] But despite such attractions, some significant limitations of the theory have been identified. For

example, although Aristotle assumed that happiness is the reward of a virtuous life (virtuous people 'flourish', in his terms), this is not necessarily the case. Some fairly unpleasant, selfish people seem to have a whale of a time, and yet do so without suffering from a guilty conscience (although they may, of course, just be very clever at concealing their real feelings).

Moreover, critics of this approach question what the virtues of this new Aristotelianism might be. The 'good life' is viewed differently in different cultures, and this is surely likely to have important implications for what are regarded as the virtues. For example, in the Muslim world a virtuous man may have up to four wives, but Judaism and Christianity permit only one. Or some might regard euthanasia as an act of compassion, but others as an act of murder. Can a modern virtue theory, with its appeal to ancient wisdom, cope with such different attitudes?

Another aspect of that problem is the application of identified virtues in practice. If, for example, it is agreed that justice and courage are virtues, we are still left with the question of how and when to practise these virtues in specific circumstances, and how to balance competing claims.

2.7.2 **Virtue theory and animals**

For philosopher Rosalind Hursthouse, virtue theory provides a basis not only for our dealings with other people, but also with animals. She suggests that the theory requires the virtuous person to pursue virtues such as kindness, compassion and responsible behaviour, and condemn *vices* such as callousness, cruelty and irresponsible behaviour; and this demands serious respect for the lives of sentient animals that others might use in experiments, for food or in sport. For Hursthouse this has entailed becoming a vegetarian.

But, as she admits: *'although virtue ethics does tell us what we ought to consider, it does not thereby make our moral decisions easy or straightforward.'*[27] The case is made tellingly by her analysis of the position of another philosopher, Roger Scruton, who is also an adherent of virtue theory but, unlike her, does not consider that this conflicts with his support for fox-hunting or angling.[28]

2.8 **Other ethical theories**

2.8.1 **Normative theories**

Normative theories address the fundamental question 'How ought we to live?' That is to say, such theories prescribe, imply or explain certain standards (norms) of conduct that are considered justified or required. In this chapter we have briefly discussed three important normative ethical theories (2.5–2.7), which can be summarized by saying that they place emphasis respectively on *well-being* (utilitarianism), *autonomy* (deontology), and *character* (virtue theory).

But some ethicists stress the value of other normative approaches. These include, for example:

- **communitarianism** - which emphasizes the importance of the interests of whole communities rather than focusing on the concerns of individuals[29]
- **feminist ethics**[30] - which suggests that traditional masculine-oriented ethics has not only laid undue emphasis on *justice* to the detriment of an ethic of *care*, but has also endorsed imbalances of power, such as those between men and women, rich and poor, healthy and disabled, and white people and those of other skin colours
- **ethical relativism**[31] - which claims that the rightness or wrongness of actions is entirely determined by the cultural and social environment in which you live or were brought up - so that *no* opinions can be said, in themselves, to be ethically right or wrong.

In the latter case, it is clear that social norms (like those derived from traditional religious beliefs) have an important impact on notions of right and wrong, and result in sincere disagreements between well-meaning people, e.g. over issues such as abortion and euthanasia. But there are two important counter-arguments, which are implicit in the earlier discussion in this chapter. First, anthropologists conclude that there is much fundamental agreement about ethical codes between different cultures,[32] and second, formal acceptance of ethical relativism would prevent *any* criticism of others' behaviour, even of brutal and intolerant practices such as slavery, torture, female genital mutilation and religious persecution.

2.8.2 Non-normative theories

In addition to normative ethical theories there are two broad categories of non-normative ethics, *descriptive ethics* and *metaethics*.

Descriptive ethics consists of the factual description of moral behaviour and beliefs. It includes sociological, psychological, legal and political accounts, and is represented by public policies, professional codes and common attitudes and beliefs. A number of examples are discussed in this book.

Metaethics addresses more abstract questions such as: 'What is morality?' and 'In what sense can moral judgements be said to be true or false?' Philosophers who address such questions consider theories such as *realism*, *intuitionism*, *naturalism*, *subjectivism*, and *emotivism*.[33] Although a rigorous philosophical enquiry almost inevitably leads into metaethical questions, this book will largely be confined to analysis at the level of normative ethics.

However, by way of illustration we might profitably consider just one metaethical theory - **emotivism**. This theory challenges the whole enterprise of attempting a rational analysis of ethics. This is because emotivists do not believe that ethical statements express anything other than emotional reactions, as if you were to say 'hurray' to things you like and 'boo' to things you don't. Attributed historically to the eighteenth-century Scottish philosopher David Hume, emotivism became prominent in the last century through the writings of the English philosopher A J Ayer.[34] More recently, it has been subjected to much criticism. Undoubtedly, there is a connection between feelings and ethical decisions (and we might question the humanity of someone who, for example, did *not* get angry or disturbed by witnessing or learning of acts of moral depravity), but it is difficult to believe that matters like compassion and justice *only* amount to such feelings. This is because *'emotions can be responses to already discriminated moral properties; and crucially they can (and ought) themselves to be judged morally appropriate or perverse.'*[35] That is to say, we *can* overrule our emotions.

2.9 Can't we make moral judgements?

One of the commonest initial reactions encountered by the author, over many years of teaching bioethics to bioscience students, is that ethics is simply a matter of opinion – often couched in terms such as 'it all depends on what *you* think'. This appears to support the findings of the American psychologist Lawrence Kohlberg, who conducted studies on the moral development of children and adults, and found that people whose ethical ideas are grounded in the norms of their society often pass through a phase of relativism before arriving at a considered ethical position.[36]

A variation on this attitude is the statement by a student at another establishment: *'But surely it's always wrong to make moral judgements?'* – which stimulated philosopher Mary Midgley to respond by writing a book with the title of this section.[37] If you reflect on the student's rhetorical question you soon realize that it is, itself, a moral judgement. Although some justification for it might lie in the view that we should not impose our moral standards on others, in reality, as Midgley notes:

> *getting outside morality would be rather like getting outside the atmosphere. It would mean losing the basic social network within which we live and communicate with others. . . . a state where, although intelligence can still function, there is no sense of community with others, no shared wishes, principles, aspirations or ideals, no mutual trust or fellowship with those outside, no preferred set of concepts, nothing agreed on as important.*

As we saw in 1.9.3, the predisposition to altruistic behaviour and its rationalization in ethical norms seems to be a characteristic of human societies, which has resulted in attempts to devise ethical schemes and apply them by means of principles and rules. But it is apparent from the accounts given in this chapter that the theories proposed have only been partially successful in arriving at an explanation of, and justification for, an ethical life. Each theory appears to satisfy some important aspects of commonly perceived ethical behaviour, but it also has some serious drawbacks in terms of practical application.

In chapter 3 we consider how the various theories contribute to a *common morality*, and how this might form a basis of assessments of the bioethical concerns discussed in subsequent chapters. But each of us, as individuals, might also weave a distinctive ethical garment from those diverse threads of ethical theory that appeal to our own inner convictions. It is a garment that will remind us how we believe we should act, and inform others where we stand.

THE MAIN POINTS

- Ethical deliberation is a rational process, albeit sometimes informed by religious and/or emotional considerations, which principally concerns our relationships to others, both human and non-human.
- In making ethical decisions, a sound approach is to appeal to a set of principles, grounded in ethical theory.
- Utilitarianism is the most prominent consequentialist theory, which in seeking to achieve the 'greatest good for the greatest number', applies cost–benefit analyses.
- Deontological theory emphasizes motives rather than the results of actions. An important principle is to 'Do as you would be done by'.
- Virtue theory stresses the importance of character in ethical decision-making and often attaches priority to the well-being of the community.

EXERCISES

These can form the basis of essays or group discussions:

1. Is the idea of *free will* (2.1.2) compatible with a scientific view of human nature?
2. Does *natural law* thinking (2.3) play any part in your ethical reasoning about whether an infertile person should be allowed to have a child by reproductive cloning using the technique of nuclear transfer? What other arguments, if any, are important in your ethical reasoning?
3. Make a table of pros and cons of the three theories described (utilitarianism, deontology and virtue theory: 2.5–2.7) as they apply to the case of medical research involving experiments on anaesthetized rabbits to study heart disease.
4. How do you respond to the statement 'Ethics is just a matter of opinion'? (2.9) Give detailed reasons for agreeing or disagreeing with the statement.
5. Can any ethical distinctions be made between the motives and activities of people who threaten others with violence in their opposition to: (*a*) abortion (5.2.4), (*b*) proposals to ban fox-hunting (7.8.2), and (*c*) animal experiments (8.9)?

FURTHER READING

Many books on ethical theory are written for philosophers, and will make difficult reading for the non-specialist. But there are several books written for the general reader, which will be useful in amplifying points made in this chapter. Examples are:

- *Ethics: a contemporary introduction* by Harry J Gensler (1998). London, Routledge. An engaging approach to ethical theory for the non-specialist.

- *Being Good: a short introduction to ethics* by Simon Blackburn (2001) Oxford, Blackwell. Oxford, Oxford University Press. Also appears as *'Ethics'* in the 'A Very Short Introduction' series.
- *A Companion to Bioethics* edited by Helga Kuhse and Peter Singer (2001). Oxford, Blackwell. A valuable guide (with 46 articles by leading philosophers) for those who want to pursue the subject more deeply. Its emphasis is on medical bioethics.

USEFUL WEBSITES

For articles on ethical theory, applied ethical issues, and biographies of key philosophers, consult:

- **http://ethics.acusd.edu/** *Ethics Updates* (edited by Lawrence M Hinman): a useful guide to ethical theory.
- **http://www.edu/research/iep** *The Internet Encyclopedia of Philosophy* (edited by James Fieser): a handy guide to philosophers and their works.
- **http://www.earlham.edu/** *A guide to philosophy on the internet*: a large database of links to useful sites (edited by Peter Suber).

NOTES

1. Longman Dictionary of the English Language (1984). London, Longman
2. Dennett D C (2003) Freedom Evolves. London, Penguin
3. The Cambridge Dictionary of Philosophy (2nd edition). (Ed.) Audi R. Cambridge, Cambridge University Press, p. 563
4. Huxley A (1946) The Perennial Philosophy. London, Chatto and Windus, p.1
5. Voltaire [*circa* 1750] The Oxford Dictionary of Quotations (2nd edition) (1953). Oxford, Oxford University Press, p. 557
6. Pluhar E (1998) Animal rights. In: Encyclopedia of Applied Ethics. (Ed.) Chadwick R. San Diego, Academic Press, Vol. 1, pp. 161–172
7. Cupitt D (1997) After God: the future of religion. London, Weidenfeld and Nicolson
8. Moore G E (1903) Principia Ethica. Cambridge, Cambridge University Press
9. Searle J F (1967) How to derive an 'ought' from an 'is'. In: Theories of Ethics. (Ed.) Foot P. Oxford, Oxford University Press, pp. 101–114
10. Wilson E O (1998) Consilience. London, Little, Brown and Co., p. 280
11. Bentham J (1948 [1823]) A Fragment on Government and Principles of Morals and Legislation. Oxford, Blackwell
12. Mill J S (1910) [1863] Utilitarianism, Liberty and Representative Government. London, J M Dent
13. Smart J J C (1990) Act-utilitarianism and rule-utilitarianism. In: Utilitarianism and its Critics. (Ed.) Glover J. London, MacMillan, pp. 199–201
14. Blackburn S (2001) Being Good: a short introduction to ethics. Oxford, Oxford University Press
15. Harris J (1986) The survival lottery. In: Applied Ethics. (Ed.) Singer P. Oxford, Oxford University Press, pp. 87–95

16. Mackie J L (1977) Ethics: inventing right and wrong. London, Penguin, pp. 129–134
17. Kant I (1932 [1785]) Fundamental Principles of the Metaphysic of Ethics. London, Longmans, Green and Co
18. Blackburn S (2001) Being Good: a short introduction to ethics. Oxford, Oxford University Press, p. 117
19. Potter N (1998) Kantianism. In: Encyclopedia of Applied Ethics. (Ed.) Chadwick R. San Diego, Academic Press. Vol. 3, pp. 31–38
20. O'Neill O (1993) Kantian ethics. In: A Companion to Ethics. (Ed.) Singer P. Oxford, Blackwell, pp. 175–185
21. Potter N (1998) Kantianism. In: Encyclopedia of Applied Ethics. (Ed.) Chadwick R. San Diego, Academic Press. Vol. 3, pp. 31–38
22. O'Neill O (2002) Autonomy and Trust in Bioethics. Cambridge, Cambridge University Press, p. 36
23. Regan T (1985) The case for animal rights. In: Defence of Animals. (Ed.) Singer P. Oxford, Blackwell, pp. 13–26
24. Aristotle (1925) [*circa* 350 BC] The Nicomachean Ethics (trans. W D Ross). Oxford, Oxford University Press
25. MacIntyre A (1985) After Virtue (2nd edition). London, Gerald Duckworth and Co
26. Louden R B (1998) Virtue ethics. In: Encyclopedia of Applied Ethics. (Ed.) Chadwick R. San Diego, Academic Press. Vol. 2, p. 494
27. Hursthouse R (2000) Ethics, Humans and Other Animals. London, Routledge, p. 149
28. Ibid., pp. 157–162
29. Frazer E (1995) In: Oxford Companion to Philosophy. (Ed.) Honderich T. Oxford, Oxford University Press, p. 143
30. Ainley A (1995) In: The Oxford Companion to Philosophy. (Ed.) Honderich T. Oxford, Oxford University Press, pp. 270–272
31. Hepburn R W (1995) In: The Oxford Companion to Philosophy. (Ed.) Honderich T. Oxford, Oxford University Press, p. 758
32. Konner M (1982) The Tangled Wing: biological constraints in the human spirit. Harmondsworth, Penguin
33. A Companion to Ethics. (Ed.) Singer P. Oxford, Blackwell, pp. 399–490
34. Ayer A J (1982) Language, Truth and Logic. Harmondsworth, Penguin
35. Hepburn R W (1995) In: The Oxford Companion to Philosophy. (Ed.) Honderich T. Oxford, Oxford University Press, p. 226
36. Kohlberg L and Kramer R (1969) Continuities and discontinuities in childhood and adult moral development. Hum Dev *12*, 93–120
37. Midgley M (1991) Can't We Make Moral Judgements? Bristol, The Bristol Press

3

A framework for ethical analysis

OBJECTIVES

When you have read and discussed this chapter you should:

- understand how modern theories of utilitarian, deontological, and virtue ethics are reflected in the *common morality*
- understand the importance of Rawls' theory of justice to modern liberal democracies
- appreciate the theoretical basis of the framework called the *ethical matrix*
- understand how the ethical matrix can be applied to ethical reasoning in a specific example in the biosciences – the use of a hormone preparation to stimulate the milk yield of dairy cattle
- appreciate the range of ways the ethical matrix can be used and the nature of its limitations

3.1 Introduction

This chapter describes a framework for analysing bioethical issues and for helping in, but *not determining*, ethical decision-making. Such issues cover a very wide spectrum; and the people involved might range from an individual (such as a bioscience student) deciding, for example, whether to become a vegetarian or whether to apply for a job with a particular biotechnology company - to a government committee deciding whether to advise that a specific reproductive technology should be legalized. Typically, such questions take the form of dilemmas (problems that initially, at least, seem insoluble), which can sometimes be perplexing. Bioethical dilemmas are often characterized by one or more of the following features:

- good reasons are proposed both for supporting and for opposing a particular course of action
- the ethical acceptability of a course of action depends to a significant degree on scientific evidence, which may be complex and/or incomplete and/or debatable
- a decision has to be made by, and/or for, society as a whole, in which a significant number of people (sometimes the majority) may oppose the opinion held by most scientific experts.

We saw in chapter 1 that ethical issues are crucially important in the biosciences, in relation to the process of discovery as well as in the application of biological knowledge as biotechnology. Chapter 2 considered the principal ethical theories which philosophers have proposed with the aim of systematizing, and hence facilitating, ethical decision-making. But we also saw that the complexity of the issues that ethics seeks to address seems to defeat all attempts to come up with a single, all-inclusive, but widely acceptable, theory. For example, both Kant's *categorical imperative* to 'tell the truth' under all circumstances, and Bentham's *hedonic calculus*, that appears to allow the violation of individual human rights in pursuit of net happiness for society, seem liable on occasion to result in acts of inhumanity and injustice. Even so, despite their limitations, modern versions of utilitarianism, deontology (Kantianism), and virtue theory do seem to identify important ethical concerns. So it is natural to ask whether a way forward in ethical analysis and decision-making might be found in combining elements of the different theories; or whether the differences between them are so large as to make any synthesis impossible.

3.1.1 The common morality

Three factors suggest that a synthesis is, at the very least, feasible. In the first place, the *outcome* of ethical reasoning based on the different theories is often quite similar, if not identical. For example, contributing to famine relief when you are able to do so is an ethical requirement of all three theoretical positions, while speaking honestly would be advocated by ethical reasoning based on deontology, virtue theory and most forms of rule-utilitarianism.

Second, the fact that the main ethical theories were formulated many years ago has both strong and weak implications. Their continued relevance suggests that they appeal to fundamental, and timeless, aspects of the human condition, in that they are concerned with what it means to be human. But it also suggests that they may need some revision in the context of the modern world, in which, for example: advances in genetics have transformed our scientific understanding of life; enormous advances in health and educational achievement have been made; and many people live in multicultural, pluralist societies. Indeed, chapter 2 described several ways in which the original theories have evolved over the years.

Third, it is possible to identify elements of all three theories in the norms of behaviour and ethical belief generally accepted by society, which together constitute the so-called **common morality**. This may be defined as the ethical code shared by members of society in the form of unreflective common sense and tradition. As the 'lowest common denominator' of society's ethical norms, it might not seem to be a very promising basis for ethical analysis and decision-making - but we should avoid undue cynicism, and not judge the standards of society's morals by their portrayal in tabloid newspaper headlines. Most people do act (or, at least, believe they should) in accordance with ethical norms such as honesty, compassion, and fairness - even though there are many temptations to act deceitfully, selfishly and greedily. So, although rarely described as such, these **ethical principles** are important factors in most people's behaviour. If, then, we were to begin our search for a synthesis of the major ethical theories at the level of the principles they give rise to, there is a chance that the different approaches will begin to converge.

To avoid misunderstanding, it must be stressed that although our reasoning usefully *begins* with the common morality, it doesn't end there. People in casual conversation frequently make ethical judgements – on issues such as the justification for going to war, whether some footballers deserve their high incomes or whether a modern biotechnology should be banned. But such views rarely pass the rigorous tests required of a philosophically valid theory. We will consider below some of the factors which are necessary in an authentic bioethical analysis – conditions such as the need for factual accuracy, coherence and consistency.

3.2 Ethical principles

The use of a principled approach in bioethics (sometimes called **principlism**) was first developed by the medical ethicists Tom Beauchamp (pronounced *Be-cham*) and James Childress (pronounced *Chill-dress*)[1] in the USA, whose aim was to help doctors, nurses and other healthcare professionals facing ethical issues in treating their patients. Many complex ethical issues are raised by modern medicine, including, for example, those involving decisions about euthanasia, experimentation on human subjects, organ transplantation and issues of justice in allocating scarce resources. These concerns are dealt with in the specialized field of **medical ethics**, and generally do not feature in this book. But some medical biotechnologies (such as infertility treatment) have wider implications and form the basis of chapters 5 and 6. However, because the *principles* introduced in medical ethics have more general applicability, they are discussed further in this chapter.

The Oxford philosopher David Ross noted in 1930 that an effective way of dealing with the problem of conflicting principles was to recognize that Kant's categorical imperatives and Bentham's attempt to maximize happiness were too rigid. What were needed were *conditional* principles (he called them ***prima facie*** **principles**, meaning 'at first appearance'), which allow a stronger case to overrule a weaker one in a particular circumstance.[2] This compromise position does not favour either the utilitarian or the deontological case, but accepts that neither duties nor consequences can be ignored.

Beauchamp and Childress built on Ross's approach by proposing that decisions in medical ethics should be based on a process of deliberation that involves considering the impacts of proposed actions in the light of four *prima facie* principles. Thus, in treating patients a doctor is regarded as having ethical duties to, respectively:

- cause no harm (based on the fourth-century BC Greek 'Hippocratic Oath') **[non-maleficence]**
- effect a cure (or at least provide palliative treatment) **[beneficence]**
- respect patients' autonomy (not regarding them merely as 'cases') **[autonomy]**
- treat patients fairly (e.g. without sexual or racial discrimination) **[justice]**

The first two principles, named in square brackets, are mainly utilitarian, and the last two mainly deontological.

It has been claimed that the principles '*are general guides that leave considerable room for judgement in specific cases and that provide substantive guidance for the development of more detailed rules and policies*'. According to medical ethicist Raanan Gillon, the principles provide a set of '*substantive moral premises upon which to base reasoning in health care ethics*' and offer '*a transcultural, transnational, transreligious, transphilosophical framework for ethical analysis*' by allowing differences of emphasis within a scheme of universal applicability.[3] These are perhaps crucial features of any ethical approach that seeks to be relevant to today's multicultural, pluralist societies.

An important feature of this approach is that abstract principles are **specified** in concrete terms, providing **action guides** to be applied in specific clinical circumstances. The specified principles also need to be balanced, which can entail one *prima facie* duty overriding another, or some duties being only partially discharged. Sometimes, the rule of **the double effect** is invoked, which states that '*a single act having two foreseen effects, one good and one harmful (such as death) is not always morally prohibited if the harmful effect is not intended.*'[4] For example, if, as seems likely, suicide rates are higher in countries with higher literacy rates, schoolteachers should not in consequence be considered guilty of driving people to take their own lives!

The combination of different principles appears to be a useful strategy in that it attempts to achieve a synthesis of perspectives. But there are different opinions on the most suitable principles to employ. For example, the Danish bioethicist Peter Kemp has proposed that biomedical issues might best be considered in relation to the principles of *autonomy*, *dignity*, *integrity*, and *vulnerability*.[5]

3.3 The principle of justice

We have noted (3.2) that the fact that the theories underpinning much ethical reasoning are often ancient (even though they have sometimes been updated) might imply that their relevance is limited by modern developments, both scientifically and socially. So it is reasonable to ask whether any important modern theories significantly amend or complement the traditional theories. One recent theory that has received considerable attention, and seems particularly relevant here, is the **theory of justice** advanced by philosopher John Rawls.[6] The theory is a development of the notion of the social contract, an early form of which was proposed by Hobbes (1.9). But much more recently, the idea of the social contract has been important in the establishment of the modern democratic state. In this political system, certain constraints on personal liberty (such as taxes and speed limits), decided by our representatives in government, are implicitly agreed to by the electorate in the interests of maintaining a safe and orderly community where justice prevails.

The term *justice* is open to many interpretations, depending on whether it is seen as a matter of deserts (which might entail punishing criminals) or entitlements (which might involve providing free healthcare). The related term **distributive justice** refers to the fair, appropriate or equitable distribution of resources between members of society. But here again, the words can mean different things: for example, goods might be considered justly allocated according to criteria such as: need, effort, merit, equality, or by market forces or lottery.

3.3.1 Rawls' concept of 'justice as fairness'

Rawls' concept of **justice as fairness** is an egalitarian theory that he claimed would be acceptable to '*free and rational persons concerned to further their own interests*'.[7] His case rested on a reinterpretation of the social contract as a hypothetical one, made under conditions that would guarantee that the chosen principles of justice would be *fair*. So he suggested a set of rules that would guarantee this outcome. Arguing that it would be wrong to decide what was fair if we already knew our positions in society and the talents we possessed, Rawls proposed that our personal characteristics (age, sex, race, intelligence, physical attributes, wealth, etc.) should be screened out by placing us behind an imaginary **veil of ignorance**. With this hypothetical device in place (which he called the **original position**), the parties choosing the principles would do so impartially, uninfluenced by their own interests or prejudices. Putting it another way, when we try to work out what will be a fair way of operating in a liberal democracy (the political system most people reading this book will live in), we shall need to *forget who we are* and imagine the possibility that we might belong to a highly disadvantaged group. Rawls argued that in such circumstances it would be most rational for people to adopt two specific principles of justice (Box 3.1).

This account of Rawlsian justice has so far focused on the individual, but Rawls saw the principle as equally applicable to organizations. In a telling comparison with Popper's methodology of science (1.5.1), Rawls emphasized the point thus: '*Justice is the first virtue of social institutions as truth is of systems of thought. A theory, however elegant and economical, must be rejected or revised if it is untrue; likewise laws and institutions, no matter how efficient or well arranged, must be reformed or abolished if they are unjust.*'[8,9]

Rawls' more recent work, building on his theory of justice, considered the problems faced in modern democratic societies, in which '*a plurality of incompatible and irreconcilable doctrines – religious, philosophical, and moral – coexist within the framework of democratic institutions*'.[10] Since these political and cultural factors impact on bioscience and biotechnology no less than other aspects of modern life, there is a good case for examining their relevance to our bioethical analyses.

BOX 3.1 RAWLS' TWO PRINCIPLES

- **Principle I: equal liberties for all**

This states that each person should have as much liberty as is consistent with other people having the same amount of liberty.

- **Principle II: the difference principle**

This would ensure fair equality of opportunity, while restricting social and economic inequalities to those that would benefit the *least advantaged* members of society. The principle would, for example, accept that it might be fair to pay surgeons disproportionately high salaries (to encourage them to undergo the lengthy and demanding training involved) if people in general, including the least advantaged, benefited from their surgical skills.

3.3.2 The status of justice as an ethical principle

Rawls considered that his development of ideas on justice as fairness was in the Kantian tradition of deontological ethics, and this is how most other philosophers have interpreted his theories. But justice can also be considered to be an important aspect of utilitarian theory. The point was made forcibly by no less a person than John Stuart Mill (Box 2.1). The argument rests on the claim that to experience injustice is to be harmed, so that it legitimately becomes part of the cost–benefit analysis that forms the basis of utilitarian reasoning. According to Mill, justice is: '*The highest abstract standard*' which '*is involved in the very meaning of Utility or the Greatest Happiness Principle*'. Moreover, '*the equal claim of everybody to happiness . . . involves an equal claim to the means of happiness*'.[11]

Accordingly, in the following discussion the principle of justice as fairness will be considered a hybrid principle which is pertinent to both deontological and utilitarian reasoning.

3.4 Ethical standing

Up to this point, the discussion has focused almost exclusively on ethical impacts on people, and in the case of medical ethics, on people as patients. But it is clear that when considering new biotechnologies the people affected often fall into some fairly distinct categories (often called **stakeholder groups**), whose specific interests it would be sensible to consider separately in any ethical analysis. Moreover, many issues in the biosciences also concern non-humans – both animals and plants. This suggests that the approach described by Beauchamp and Childress needs to be modified to take account of the relevant interests of other living beings.

The term **ethical standing** refers to the claim that someone or something is a subject of ethical consideration in their/its *own right*, independently of usefulness as a means to some other end. So an important question arises in the biosciences as to which of all the entities that might be affected by any procedure or technology should be considered to possess ethical standing, and how this will affect the ways in which we act. Kant excluded animals as having ethical standing because they are incapable of rationality (2.6.4): any consideration he might give them would be aimed at the sensibilities of their owners or the effects that our treatment of them might have on other people. Similarly, Rawls excluded animals, because they are incapable of acting as rational agents in the social contract forming the basis of his ethical theory.

But it is clear that if we take the common morality as a guide, not only do many animals possess ethical standing, but so also does the biosphere, both in the global sense and, more specifically, as defined aspects of the natural environment such as the rainforests or the biodiversity of a region. A useful term here is the **biota**, defined as 'the plant and animal life of a region', with the implied concern for any ecological implications of human activities. (Environmental ethicists often refer to the **biotic community**.[12]) Indeed, the claimed rights of animals and the perceived need to preserve the integrity of the environment have both now become important areas of applied ethics. It is worth noting that other phrases are also used in philosophical literature to indicate possession of ethical standing, such as *moral standing*, *ethical status*, and being *ethically considerable*. For our purposes, they all amount to the same thing. Indeed, it facilitates consistency if we refer to all groups possessing ethical standing as **interest groups**, recognizing that this does not always imply that they are capable of conscious thought.

3.5 The ethical matrix

In analysing the ethical impacts of any procedures and technologies that arise in and from the biosciences, it is useful to establish a framework to assist us in our deliberations. There would seem to be two necessary ingredients of this framework – a set of relevant *prima facie* principles, and a list of the agents that have 'interests'. Appeal to principles reminds us of the overarching considerations that need to be taken into account. The list of agents with interests will depend on the nature of the issue to be analysed, but it will usually include: different human interest groups (such as consumers, patients or farmers); animals, where they are used in experimental, agricultural, or biotechnological procedures; and the biota where the procedures have ecological effects. Some people might want to include other interest groups. For example, if a proposed biotechnology involved fundamental effects on the genotypes or phenotypes of plants, e.g. by genetic modification, the crop might be regarded as a legitimate interest group; but others would draw the line at sentient (conscious) beings and probably attribute any such concerns to the consumer group.

A practicable framework for ethical analysis is a compromise between competing requirements. It needs to: a) be based in established ethical theory to give it authenticity; b) be sufficiently comprehensive to capture the main ethical concerns; c) employ user-friendly language as far as possible.

Such considerations have led the author to introduce a framework called the **ethical matrix**, which has been used in research and teaching programmes over a number of years. In the ethical matrix, three principles are employed. These are 'respect for':

- **well-being**
- **autonomy**
- **fairness**

which are applied to the interests of the different groups relevant to the issue being analysed. The three principles were chosen to represent the major traditional ethical theories; i.e. respect for well-being represents the major utilitarian principle; respect for autonomy represents the major deontological principle; and respect for fairness is important to both the utilitarian and deontological traditions (3.3.2), but also, and importantly, it encompasses the fundamental tenet of modern social contract theory.

Inclusion of all three *prima facie* principles in the framework acknowledges the plurality of perspectives that sincere people bring to an ethical analysis, and provides a means of registering the importance of each principle in any particular context. (It should be noted that in this scheme *well-being* combines respect for beneficence and non-maleficence, which are given separate identity by Beauchamp and Childress.[13]) Virtue theory (2.7) is less easily characterized, but the special place of virtues in moral life does not mean they take precedence over obligation-based principles.[14] Those whose primary motive is to live a life of virtue still have to decide what to do in order to act virtuously, and it is here that a principled approach is often valuable.

The term *matrix* derives from the fact that arranging the three principles and the different interest groups in a table produces a regular grid, in the cells of which the principles are specified in language appropriate to the group in question. Thus, respect for the well-being of animals is specified as **animal welfare**, while respect for consumer autonomy is specified as **consumer choice**, for example in making purchase decisions. The word 'matrix' also has another, biological, meaning, which has relevance here. It refers to 'a substance or environment' within which something else 'develops' – an apt metaphor to describe use of the ethical matrix in facilitating the development of a reasoned ethical analysis.

3.6 An example using the ethical matrix: the case of bovine somatotrophin

To illustrate the use of the ethical matrix, the following sections refer to a particular example – a biotechnology used in animal agriculture. This involves a hormone called **bovine somatotrophin** (bST), which when injected subcutaneously into dairy cattle stimulates their milk yield. The hormone, which is produced by recombinant DNA (rDNA) technology in *Escherichia coli*, was the first genetically modified (GM) product to be used (in the USA) in animal agriculture.

We need first to consider some facts. By injecting cows with bST every two weeks, farmers can expect an average increase in yields of 12–15%; and, although slight changes in nutrient content can be produced, their overall concentrations in bulked milk are probably unaffected. However, because higher metabolic demands may lead to increased rates of illness, there is an increased risk that the welfare of injected cattle will be diminished. The treatment also leads to an increase in the milk concentration of insulin-like growth factor-I (IGF-I), which is a potent mitogen. If the increased milk concentration of IGF-I was physiologically significant and if it were to remain biologically active at the level of the gut mucosa (a claim which is contested by some scientists), it might pose a public health threat to people consuming the milk or dairy products.

Figure 3.1 identifies the different steps of the process – from bST manufacture to impacts on the milk consumer, and on the farm environment. It also raises some of the alleged benefits and concerns over bST use. This is a useful case study because it deals with an ethically contentious issue, which has resulted in quite different political

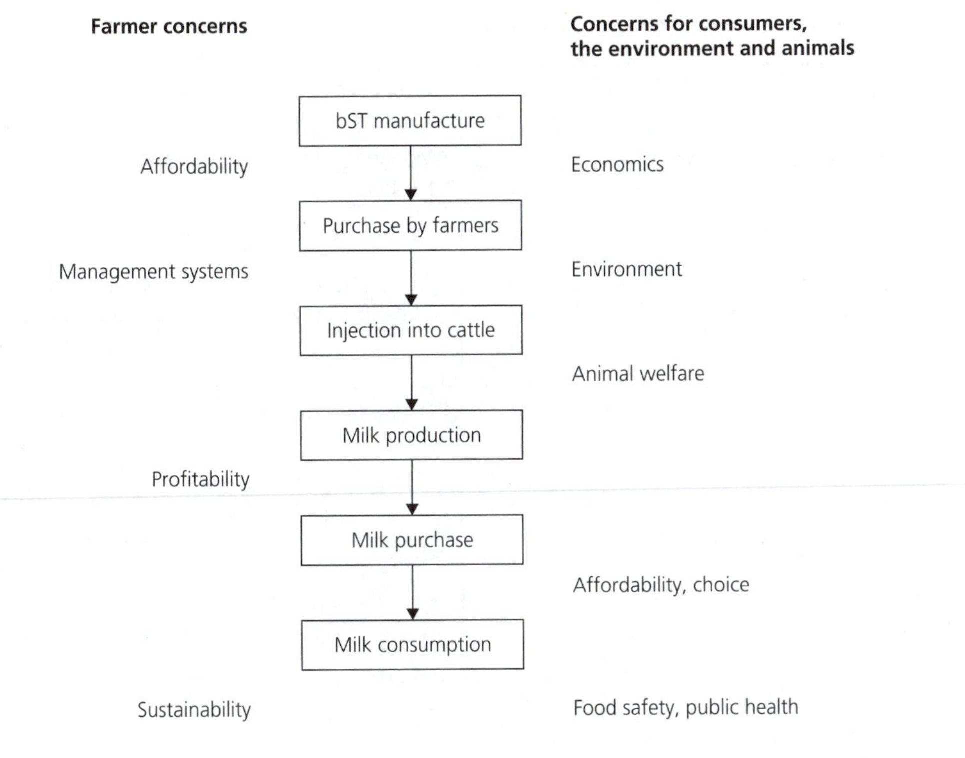

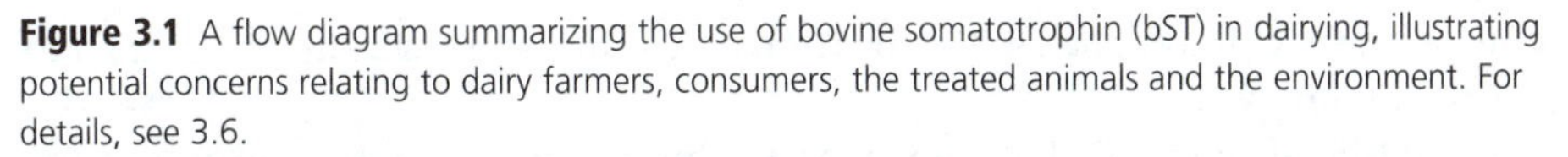
Figure 3.1 A flow diagram summarizing the use of bovine somatotrophin (bST) in dairying, illustrating potential concerns relating to dairy farmers, consumers, the treated animals and the environment. For details, see 3.6.

decisions in the USA and the European Union (EU). (A more extensive analysis is provided elsewhere.[15])

Table 3.1 shows how the use of an ethical matrix can help to summarize the ethical issues raised by this technology in a systematic way that is based on the principles that comprise the common morality. Box 3.2 describes in more detail the ways in which the different principles are specified for each of the four identified interest groups.

The first thing to appreciate is that the specifications in the cells (Table 3.1 and Box 3.2) set criteria which would be met if the principles concerned were **respected** by a proposed action. In one of the commonest ways in which the matrix is used, the impacts of the action, in this case injecting cows with bST to increase their milk yields, are compared with the conditions when bST is not used – so the *status quo* represents the baseline condition. Because some ethical impacts might be *positive* (e.g. an increase in the incomes of dairy farmers using bST, so respecting their well-being), they would be 'scored' positively (e.g. +1 on a scale of +2 to −2). On the other hand, some impacts might be negative (e.g. the cows' welfare would be **infringed** if the additional metabolic load led to more cases of lameness), so that they would be scored accordingly (e.g. −1); while some impacts might be insignificant (and so recorded as 0). A 'filled in' ethical matrix would thus show a total of 12 scores for the perceived ethical impacts of bST use

Table 3.1 The ethical matrix applied to use of bovine somatotrophin (bST) in dairy farming

Respect for:	Well-Being	Autonomy	Fairness
Dairy Farmers	Satisfactory income and working conditions	Managerial freedom of action	Fair trade laws and practices
Consumers	Food safety and acceptability. Quality of life	Democratic, informed choice, e.g. of food	Availability of affordable food
Dairy Cows	Animal welfare	Behavioural freedom	Intrinsic value
The Biota	Conservation	Biodiversity	Sustainability

Note: For details of the matrix see 3.5–3.6. For dairy farmers and consumers, both impacts and duties need to be considered, whereas for dairy cows and the biota (shaded) only impacts are involved.

on the three principles applied to the four interest groups. This assessment is referred to as an **ethical analysis**.

It should be appreciated that numerical scoring means no more than use of the adjectives *very* (for '2') and *quite* (for '1') – and if people thought numbers gave the wrong impression, those or similar words could be used instead. But some means of grading responses seems necessary.

It is also important to emphasize that it is not possible from this analysis to directly deduce the ethical acceptability, or otherwise, of bST use, for two reasons. First, the different principles will have different degrees of significance for each assessor. This might be expressed by saying the different factors carry different **weights**. The next step in the process, **ethical evaluation**, involves subjectively weighing the different impacts, allowing one to reach an ethical judgement on the acceptability of bST use.

The second reason why the analysis does not assess overall ethical acceptability is because it simply compares the impacts of two situations, neither of which might be ethically acceptable by comparison with some third option. In other words, a system adjudged marginally more ethically acceptable than another according to the analysis might still fall far short of a system which has not been investigated. For example, a case could be made for saying that a system of dairying that prioritized animal welfare, such as **organic farming**, would be a better baseline against which to compare bST use than conventional dairying systems, which already experience significant problems with animal diseases related to high productivity (7.6.4). What you get out of the matrix is totally dependent on what you put in.

BOX 3.2 MORE DETAILED SPECIFICATION OF THE PRINCIPLES IN AN ETHICAL MATRIX FOR BOVINE SOMATOTROPHIN USE (SEE TABLE 3.1)

Dairy farmers

Well-being: satisfactory incomes and working conditions for farmers and farm-workers ('satisfactory' is obviously debatable, but it is a better word than 'adequate', which might imply 'just enough to meet bare necessities')

Autonomy: allowing farmers to use their skills and judgement in making managerial decisions, e.g. in choosing a farming system

Fairness: farmers and farm-workers receiving a fair price for their work and produce, and being treated fairly by trade laws and practices

Consumers

Well-being: protection from food poisoning (and harmful agents, e.g. residues of veterinary drugs); this also refers to the quality of life citizens enjoy as a consequence of a productive and profitable farming industry

Autonomy: a good choice of foods, which are appropriately labelled, together with adequate knowledge to make wise food choices; this principle also encompasses the citizen's democratic choice of how agriculture should be practised

Fairness: an adequate supply of affordable food for all, ensuring that no one goes hungry because of poverty

Dairy cows

Well-being: prevention of animal suffering; improving animal health; avoiding risks to animal welfare

Autonomy: ability to express normal patterns of instinctive behaviour, e.g. grazing and mating

Fairness: treated with respect for their intrinsic value as sentient beings rather than just as useful possessions (instrumentally)

The biota

Well-being: protection of wildlife from harm (e.g. by pollution), with remedial measures taken when harm has been caused

Autonomy: protection of biodiversity and preservation of threatened species (and rare breeds)

Fairness: ensuring sustainability of life-supporting systems (e.g. soil and water) by responsible use of non-renewable (e.g. fossil fuels) and renewable (e.g. wood) resources; cutting greenhouse gas emissions

Many of the specifications in Table 3.1 and Box 3.2 are self-explanatory, but additional comment will be helpful in some cases.

3.6.1 Respect and infringement of principles

In theory, for any proposed action, all the principles specified in the individual cells of the matrix might either be respected (earning a positive score for those wishing to quantify the effects) or infringed (deserving a negative score). But positive and negative scores do not necessarily balance each other, even for a single specification. Thus, the duty 'not to harm' (*non-maleficence* in the terminology of Beauchamp and Childress) might be thought to be more compelling than the duty to 'do good' (*beneficence*). For example, in the case of bST use, the duty not to harm cattle is often considered much more important than the duty to improve their lot. In some cases, it might be thought preferable to maintain the distinction between between the two,[16] but this has the practical disadvantage of complicating the matrix by doubling the number of columns. So it is important to bear in mind in using the matrix, as shown in Table 3.1, that just as different principles often carry different weights, so can positive and negative effects for a single principle.

3.6.2 Clarification of consumer autonomy

Autonomy in this context is about liberty – being able to choose the sort of food you eat and how it was produced. In many respects these are *citizens'* concerns and not just consumers' concerns, because you might have legitimate views on how a food is produced irrespective of whether you consume that particular food.

But the liberty of individuals, like most things, must have limits. In a shrinking world, how free should the 'haves' in rich countries be to appropriate the Earth's resources, pollute the environment, and exploit cheap labour overseas to the detriment of the 'have-nots' (chapter 4) in both developed and developing countries? Whether or not the sense of injustice is considered a persuasive factor, it has been argued that the West's excessive concentration on liberty has led to 'a fear of freedom', which is resulting in a dangerous backlash. For example, philosopher Thomas Gauly suggests that this illiberalism is evident in the violent attacks of extremist groups on the capitalist ideology of the West (notably on 11 September 2001), in racist reactions of the political right-wing in many Western countries, and in reactions against the welfare state by those who consider that traditional standards are threatened by a debased popular culture.[17] Decisions about consumer autonomy thus concern some fairly important political issues, and are not confined to whether a food bears an informative label or not.

3.6.3 Clarification of *animals' intrinsic value*

The principle of fairness applied to farm animals (here, dairy cows) is specified as respect for *intrinsic value*, a term which needs further explanation. Some things (e.g. stethoscopes and taxis) are valuable because of their usefulness, and are said to have **instrumental value**. By contrast, **intrinsic value** is assigned where it is possessed irrespective of any usefulness; and most of us share the fundamental belief, stressed by Kant, that all people have intrinsic value. But most people sometimes (and others, often) *also* have instrumental value, so that possession of the two types of value is not mutually exclusive. For example, doctors, taxi drivers and refuse collectors all perform useful tasks, making them of instrumental value. This does not raise an ethical concern if they do their jobs by choice, and receive a fair income.

Attributing intrinsic value to dairy cows makes the assumption that in addition to their instrumental value in providing milk and dairy products, cows are also 'subjects of a life' which we have a duty to respect (2.6.4). That is, they have ethical standing (3.4). Given all we now know about the sentience, sensibilities and even 'personalities' of cows, it would be unfair to regard them *simply* as useful objects. Recent legislation gives official recognition to this concept. For example, the 1999 Treaty of Amsterdam requires that animal sentience and welfare are recognized in the implementation of EU legislation. Some governments have gone even further: for example, the Swiss Federal Constitution relating to the genetic modification of animals (and indeed, of plants and other organisms) has been amended to take into account 'the dignity and integrity of living beings' (*Würde der Kreature*).[18]

The idea that animals have *merely* instrumental value, as was commonly assumed until very recently, now seems totally discredited. Putting it starkly to emphasize the point, if someone destroyed one of their books, we might, at worst, think him a fool – but if he destroyed his healthy cat (even painlessly, by poisoning when it was asleep) we should think him depraved.

3.6.4 Conservation, biodiversity and sustainability of the biota

When the principles of well-being, autonomy and fairness are applied to the biota, a different way of assessing the impacts is adopted to that used when considering individual farm animals. The reasoning behind this is that ethical impacts on the biota are concerned with life in the wild and on the collective scale – as populations, species and breeds. Consequently, the principles need to be specified quite differently to be at all consistent with the common morality – for it is commonly recognized that Nature is not 'fair': we cannot protect the interests of prey without at the same time endangering the lives of predators. This implies that what is important is not the well-being or freedom of individual animals or plants, but rather the viability and, ultimately, survival of a population or of a species. (Some people suggest that we should intervene on welfare grounds to protect wild animals from their natural predators, and others that wild populations should be controlled by chemical contraceptives. Such views might be considered naïve.)

The point has been expressed by saying that '*Our relationship with wild animals arises out of an environmental ethic, which . . . can only be "eco-centric", that is, it must not assign value to natural beings themselves but rather to their diversity and to the ecological systems on which they depend.*'[19] This suggests that the appropriate specifications of the principles for the biota are conservation, biodiversity, and sustainability – which are all prominent environmental concerns. The rationale for translating respect for autonomy as biodiversity is that it may be seen as permitting the natural ecological interplay of the biota. Sustainability represents fairness in an *intergenerational* sense, by respecting the biotic impetus for survival. Unavoidably, there is often overlap between the specified principles, but this seems less important than risking gaps in the issues that need consideration. Such designations might, of course, be regarded as highly imaginative or figurative; but again, the more important question would seem to be 'do they address the crucial ethical concerns?' The concept of sustainability, and its relationship to the other two biotic principles, is discussed further in chapter 12.

3.6.5 The content of the cells

The factors in each cell of the ethical matrix which are relevant to performing an ethical analysis of the impacts of bST are of two major types. In some cases **evidence** is required. For example, you would need to know what increases in milk yield were obtained when bST is injected, whether any effects on the chemical nature of the milk have implications for consumer health, and whether the welfare of the injected animals is affected. For people who see science as about 'facts' the answers to such questions might appear straightforward. But as noted in 1.4, the nature of the 'facts', whether they were obtained reliably, and whether they are relevant to the question in hand, are all matters over which there is sometimes disagreement. Even the scientific theory considered by some to justify the particular data examined may be questioned, and if the source of the data is thought to be biased (e.g. if a commercial company was relied on to produce the key data supporting their own product, or if the data were produced by a pressure group known to be ideologically opposed to the product) neutral observers might suspect that the evidence was unreliable. Assessing evidence may thus entail examining different versions of the facts where there is controversy.

In contrast to factual data, other cells of the matrix require a judgement that is not dependent on the quantifiable consequences of bST use but instead concerns **values**. For example, in the pursuit of economic objectives, is it right to treat animals instrumentally by chemically altering their metabolism, or is it right to take risks with human health when appropriate scientific evidence is unavailable?

3.6.6 The ethical matrix as an ethical map

It is important to appreciate that the aim of the ethical matrix is to facilitate rational decision-making but not to determine any particular decision. Indeed, conscious of Popper's criticism of what he called 'the myth of the framework' (in which he interpreted the word framework to mean a shared ideology, such as a scientific Kuhnian paradigm, or a political philosophy like Marxism),[20] to avoid confusion it might be preferable to regard the matrix more as a *map* than a framework. It is, after all, a pluralist tool, which seeks to identify society's whole ethical terrain. (Indeed, it would be instructive to use the matrix as a computerized relief map, to show the various peaks and ravines.) So, far from constraining ethical reasoning, the matrix provides a vehicle for the expression of the full range of ethical perspectives.

The proof of this claim is that both people approving of bST use and people opposing its use can use the ethical matrix to justify their differing opinions, as has been demonstrated in workshops conducted with experts.[21] This indicates two important points about the matrix:

- it provides a means of explaining and justifying different ethical positions
- it facilitates identification of the areas of agreement and disagreement.

3.6.7 Ethical evaluations of the use of bovine somatotrophin

Box 3.3 summarizes the lines of evidence (facts and values) which have been presented for the different cells of the ethical matrix. According to different interpretations of the importance to be attached to this evidence, the governments of the USA and the EU reached opposing decisions on the acceptability of licensing bST for commercial use. Although in neither case were the decisions expressed in terms of ethical acceptability, it is clear that each would be justified, if it was requested of their supporters, in ethical terms: hardly anybody admits to acting unethically. We can thus summarize the two positions, according to the ethical criteria that have been defined.

- The ethical acceptability of bST use for those who have licensed it (e.g. the USA) would probably cite the need to respect farmers' freedom to innovate; and the economic benefits to the manufacturers of bST, to the economies of countries producing it, to the farmers using it, and, were prices to fall, to consumers of dairy products. Moreover, if its use led to reduced cow numbers it might result in marginally reduced emissions of methane. This case also rests on perceptions that the welfare of treated cows is not affected significantly (or that increased disease can be effectively treated) and that there are no risks to human safety, so that labelling is unnecessary. Job losses in the dairy industry would not be seen as an ethical issue, being merely a feature of market economies, in which competition guarantees efficient production.

- The ethical case of those who have banned bST use (e.g. the EU) would probably focus on respects in which it appears to infringe commonly accepted ethical principles. They would point to authoritative reports suggesting that bST use substantially increases the risk of pain and disease in dairy cows, and that it might present a risk to human safety through ingestion of increased IGF-I in milk. Moreover, they might consider that bST use would: reduce farmers' autonomy; undermine consumer choice if milk products from treated cattle were not labelled; jeopardize public health if rejection of dairy products followed the licensing of bST (because milk is a valuable source of dietary nutrients); and increase local pollution through the intensification of dairying.

3.6.8 Using the ethical matrix to facilitate personal and group decision-making

The above description provides a guide to identifying relevant issues in reaching a personal position on bioethical concerns. It might also suffice for using the matrix in group discussions or workshops. But employing a suitable tool for ethical analysis does

BOX 3.3 A BRIEF ANALYSIS OF BOVINE SOMATOTROPHIN USE IN DAIRYING WITH REFERENCE TO TABLE 3.1 AND BOX 3.2

Dairy farmers

Well-being: some US farmers using bST have increased their profits but economic data suggest others use it at a loss

Autonomy: farmers in the USA have an opportunity to increase productivity, but some might feel economically obliged to use bST (exemplifying the so-called 'technological treadmill')

Fairness: farmers in the USA are given the option of using a productivity-boosting technology. Farmers not using bST can label milk accordingly, but at their own expense

Consumers

Well-being: an EU report by public health experts suggested possible (but currently poorly defined) risks of consuming IGF-I (whose concentration increases in milk of treated cows); an FAO/WHO committee denied any significant health risk

Autonomy: in the USA most milk is unlabelled, denying consumers a choice on whether to purchase milk from treated cows

Fairness: there appears to be no clear evidence of an impact on milk prices

Dairy cows

Well-being: cattle suffer increased disease rates (such as mastitis, lameness, metabolic and digestive disorders), as noted on the product label, which lists 21 possible adverse side effects; the EU banned bST largely on animal welfare grounds, but the manufacturers claim the diseases are treatable by medication

Autonomy: behaviour may be adversely affected by lameness, by reduced grazing opportunities due to increased concentrate feeding, and by decreased fertility

Fairness: some people claim that the excessively instrumental use of cows is an infringement of their intrinsic value. Others claim this technology accords with accepted social norms

The biota

As quantitative data are lacking, claims are largely speculative.
Claimed **positive** features of bST use are that reduced cow numbers (because fewer cows are needed to produce the required milk yield) will lead to less environmental pollution (e.g. fertilizer use for forage growth and reduced silage run-off) and lower greenhouse gas emissions (methane is exhaled by ruminant animals).
Claimed **negative** features of bST use are that mergers in the dairy industry (as non-user farmers leave the industry), resulting in fewer but much larger dairy farms, will increase point-source pollution (e.g. excessive fertilizer use, silage run-off) and jeopardize biodiversity and sustainability by reliance on fossil fuels for fertilizer production etc. and routine veterinary medication.

Note: In the USA, bST was licensed for commercial use in 1994. In the EU in 1999, an earlier moratorium on its use was extended indefinitely.

not guarantee a genuine ethical evaluation. If users adopt a partisan position on the issue, e.g. allowing bias to influence the choice of scientific data, then the tool is unlikely to prove of value.

A conceptual device to counter this tendency, is to try *put yourself in the shoes* of each interest group in turn as the different cells specifying its interests are considered. This strategy corresponds to Rawls' veil of ignorance, behind which judgements are to be made in deciding on the just rules for society (3.3). In essence, it amounts to recognizing that ethics is concerned with caring about other beings with ethical standing that are described in the matrix. As Kemp puts it: '*ethics in its full scope aims at care of the other*',[22] and while only certain occupations are conventionally classed as 'caring professions', it is implicit in the remit of ethics that care should be exercised in relation to others (necessarily, but not exclusively, people). If someone was not prepared to admit to caring about anyone or anything other than him- or herself, it would be impossible for them to use the ethical matrix. But even if they expressed concerns for only *one* other cell of the matrix, say, *respect for farmers' profits* or, alternatively, *respect for animal welfare*, that revelation would starkly expose the value system determining their choices. In fact, experience shows that most people do ascribe some value to all cells of the ethical matrix, although the degree of value ascribed varies both with the individual and with the issue being discussed.

Of course, putting yourself in the shoes of others (developing '*an imaginative conception of others' predicaments*'[23]) is not easy, especially when the interest group concerned is non-human (although there is increasing scientific evidence, e.g. on the welfare of farmed animals, to add substance to our imaginative conceptions). But genuine ethical insight is only likely to emerge from attempts to empathize in this way. In any event, empathizing with other people, for example farmers, or people on low incomes, or in less developed countries (LDC) would seem to be within most people's imaginative capabilities.

The relative importance of the impacts recorded for each of the cells is ideally only revealed at the evaluation stage, when the separate impacts are weighed. This step is perhaps equivalent to removing Rawls' veil of ignorance, when a decision has to be arrived at by 'taking everything into consideration'. It is here that reflective equilibrium (2.4) becomes a significant factor, as one seeks the proper balance between the right and the good, and between intellect and intuition. In the words of philosopher Thomas Nagel, '*The capacity to view the world simultaneously from the point of view of one's relations to others, from the point of view of one's life extended through time,* [and] *from the point of view of everyone at once... is one of the marks of humanity.*'[24]

3.6.9 Use of the ethical matrix in public consultations and expert committees

The matrix has also been used in several other ways. For example, Matthias Kaiser and Ellen-Marie Forsberg have used it in public consultations about the future of the Norwegian fishing industry. Typically, such exercises are organized to include representatives of the major stakeholder groups, or – if the interest group is a non-human one – then those with particular expertise and/or commitment to their cause are present, such as animal welfare and environmental groups. Kaiser and Forsberg ascribe the value of the matrix to the following features. It:

- is liberal regarding the approach to be adopted, enabling it to be read equally as a utilitarian or a deontological approach
- provides substance for ethical deliberation, guiding participants so that they do not stray into irrelevant paths
- translates abstract principles into concrete issues of direct concern to participants who may have little acquaintance with, or interest in, ethical theory *per se*
- facilitates extension of ethical concerns into fields benefiting from debate, such as democratic decision-making
- captures the basic fact that because different stakeholders will be affected differently by a decision their ethical evaluations may well differ. The object is not to downplay these differences but to search for an optimal solution in the light of the conflicts.[25]

Although it is conceivable that deliberation will result in attainment of consensus, the main value of the ethical matrix in this context may be that it provides the basis for broad discussion of ethical considerations. In such circumstances it may serve as a tool in what is called **discourse ethics**, as developed by philosopher Jürgen Habermas.[26] He argues that the very diversity of views and values in the modern world requires that we continue to seek a universal justification of ethical norms, and these norms are valid if '*all those affected could agree to them as participants in a rational discourse*'. Discourse ethics thus reflects the ethical values attached to open discussion, which is an ideal of modern democratic societies.[27]

However, an important difference between the use of the ethical matrix in this and the more academic uses discussed in 3.6 is that many of the participants will have little knowledge of (and perhaps little interest in) ethical theory, although they are likely to be highly knowledgeable about the impacts of any proposed changes on their own circumstances, or those of the interest group they represent. So it is possible that the Rawlsian device by which we 'put ourselves in others' shoes' might be less important in such circumstances, where the nature of the whole group guarantees that the issues are viewed from different perspectives.

Apart from its roles in teaching and public participation exercises, the ethical matrix may serve as a useful structure for professional bodies and organizations concerned with formulating policy.[28] For example, the European Academy's investigation into the ethical and social issues associated with the development of functional foods employed the matrix to structure its deliberations,[29] while the Food Ethics Council has used the framework in several of its reports.[30] However, some people have misunderstood the aims of the matrix, assuming that it is intended to be prescriptive.[31]

3.7 A summary of aims and limitations of the ethical matrix

At its simplest, the matrix is merely a checklist of concerns, which happen to be based on ethical theory. But it can play a more important role by serving as a stimulus to ethical deliberation, and as the basis of ethical decision-making. It seeks to do this by establishing a coherent and consistent approach that gives due attention both to objective facts and to human values. It should be no more acceptable to 'fudge' an ethical valuation than it is to fabricate experimental data (15.3.1).

Above all, the aim of the ethical matrix is to encourage, in the phrase trumpeted by contemporary politicians, 'joined-up thinking'. The necessity to consider how narrowly focused interests interact with a wide range of other factors which are considered of value in society can only have beneficial effects. However, it is important to emphasize that:

- the matrix is not prescriptive: the fact that different people *weigh* the cells differently precludes its providing a definitive decision on ethical acceptability
- very few, if any, decisions that people might reach using the matrix could afford *equal* respect to all the ethical principles, so that some may need to be overridden by others, or respect for some only partially discharged
- the matrix is designed to *facilitate*, but not determine, ethical decision-making by making explicit the relevant ethical concerns and providing a reasoned justification for any decisions made
- contrary to the suspicion that the matrix necessarily complicates decision-making (with so many issues to consider), it might, depending on one's worldview, actually simplify matters, because when *all* the important factors are brought into the frame a single 'ethical' decision might suggest itself as inevitable.

Finally, it is important to state that the ethical matrix is primarily intended for use in addressing dilemmas raised by technical innovations. It is not a tool for addressing all ethical questions. Indeed, some issues that have important ethical dimensions are so straightforward as not to need any deep analysis. For example, if farm workers in LDC are paid unfairly low wages, if deceit is entailed in the marketing of a new product, or if an industrial company knowingly and wilfully pollutes the environment to cut costs - we are simply confronted with examples of injustice, dishonesty and irresponsible behaviour. It is as well to be aware that calls for an 'ethical analysis' in such cases might be cynical moves, calculated to enable the perpetrators to buy time and mount a defence of their unethical practices.

3.7.1 Use of the ethical matrix in this book

Many of the chapters in this book, especially chapters 4–12, are implicitly underpinned by the theory encapsulated in the matrix. In some cases versions of the matrix are reproduced to structure the reader's thinking. But it has been considered unnecessary (and would perhaps be tedious) to make explicit reference in every case where it might be used. Even so, in deliberating on the various issues raised in the book, the conscientious reader might find that using this tool helps to structure rational thinking – even if ideas are only (proverbially) sketched out 'on the back of an envelope'.

THE MAIN POINTS

- A framework, the ethical matrix, is described, which is designed to help ethical decision-making by clarifying the principles involved, and illustrating how they can be applied to specific cases in the biosciences and biotechnology.
- Its use is exemplified with reference to an ethical analysis of bST, an agricultural biotechnology used commercially in some countries but banned in others.
- The ethical matrix seeks to facilitate a coherent, transparent, explicit process of ethical decision-making, which allows expression of the multiplicity of views that characterizes modern democratic societies.
- The ethical matrix has been used in teaching, research, public participation exercises, and a web-based educational programme.
- The matrix can be a useful tool in aiding ethical decision-making but, like other such tools, it has several limitations.

EXERCISES

These can form the basis of essays or group discussions:

1. In view of differences between different nations, religions and cultures is it reasonable to speak of a *common morality* (3.1.1)?
2. Discuss different understandings on the term 'fair' (3.3). How should it best be interpreted in relation to: (*a*) your dealings with fellow students, (*b*) people living in other countries whom you do not know, (*c*) sentient animals, such as pigs?
3. Do you think the ethical matrix as described (3.5) omits any important ethical concerns?
4. What are the *good* points, *weak* points and *interesting* points about the ethical matrix? (3.5) Can you suggest any improvements?
5. Try using the web version of the ethical matrix (3.5) listed below in 'useful websites'.

FURTHER READING

- *Principles of Biomedical Ethics* (5th edition) by T L Beauchamp and J F Childress (2001). Oxford, Oxford University Press. Useful background reading on ethical principles.

- *Justice as Fairness: a restatement* by J Rawls (2001). Cambridge, Mass., Belknap Press. This provides a development of John Rawls' seminal ideas, first presented in *A Theory of Justice* (1971), but it is not easy reading for non-philosophers.
- *The Role of Food Ethics in Food Policy* by T Ben Mepham (2000). Proceedings of the Nutrition Society, *59*, 609–618. A more extensive treatment of the case of bST in dairying using the ethical matrix.

USEFUL WEBSITES

- **www.ethicalmatrix.net** *Compassion in World Farming* (2003): an interactive web version of the ethical matrix (by Ben Mepham and Sandra Tomkins) which explores three types of animal farming. In each case, users assess organic systems against conventional, intensive forms of production. Because the exercise is primarily designed for use in secondary schools, note that the terminology used differs slightly from that employed in this chapter, e.g. respect for autonomy is represented as 'choice'.
- **http://www.ethicsweb.ca/resources/decision-making/index.html** A collection of useful articles on ethical decision-making can be found at *EthicsWeb.ca*.
- **http://www.jcu.edu/philosophy/gensler/ethics.htm** *Gensler's Philosophy Exercises*: this is a series of exercises with multiple-choice questions.

NOTES

1. Beauchamp T L and Childress J F (1994) Principles of Biomedical Ethics (4th edition). New York and Oxford, Oxford University Press
2. Ross W D (1930) The Right and the Good. Oxford, Oxford University Press, pp. 19 *ff*
3. Gillon R (1998) Bioethics, overview. In: Encyclopedia of Applied Ethics. (Ed.) Chadwick R. San Diego, Academic Press. Vol. 1, pp. 305–317
4. Beauchamp T L and Childress J F (1994) Principles of Biomedical Ethics (4th edition). New York and Oxford, Oxford University Press, p. 206
5. Kemp P (2000) Four ethical principles in biolaw. In: Bioethics and Biolaw. (Ed.) Kemp P, Rendtorff J, and Johansen N M. Copenhagen, Rhodos International Science and Law Publishers and Centre for Ethics and Law, Copenhagen. Vol. II, pp. 13–22
6. Rawls J (1972) A Theory of Justice. Oxford, Oxford University Press
7. Ibid., p. 11
8. Ibid., p. 3
9. Rawls J (2001) Justice as Fairness: a restatement. (Ed.) Kelly E. Cambridge, Mass. and London, UK, Harvard University Press
10. Rawls (1993) Political Liberalism. New York and Chichester, Sussex, Columbia University Press
11. Mill J S (1910) [1863] Utilitarianism, Liberty and Representative Government. London, J M Dent, p. 58

12. Rawles K (1998) Biocentrism. In: Encyclopedia of Applied Ethics. (Ed.) Chadwick R. San Diego, Academic Press. Vol. 1, pp. 275–83
13. Beauchamp T L and Childress J F (1994) Principles of Biomedical Ethics (4th edition). New York and Oxford, Oxford University Press
14. Ibid.
15. Mepham T B (2000) The role of food ethics in food policy. P Nutr Soc *59*, 609–618
16. Mepham T B, Moore C J, and Crilly R E (1996) An ethical analysis of the use of xenografts in human transplant surgery. Bull Med Ethics *116*, 13–18
17. Gauly T M (1998) The Future of Liberty. London, Phoenix
18. Swiss Ethics Committee on Non-Human Gene Technology (2001) La Dignité de l'Animal. Berne, ECNH
19. Larrère C and Larrère R (2000) Animal rearing as a contract? J Agr Environ Ethic *12*, 51–58
20. Popper K R (1996) The Myth of the Framework. London, Routledge
21. Mepham T B and Millar K (2001) The ethical matrix in practice: application to the case of bovine somatotrophin. EurSafe 2001 Food Safety, Food Quality and Food Ethics. Preprints, pp. 317–319
22. Kemp P (2000) Four ethical principles in biolaw. In: Bioethics and Biolaw. (Eds) Kemp P, Rendtorff J, and Johansen N M. Copenhagen, Rhodos International Science and Law Publishers and Centre for Ethics and Law. Vol II, pp. 13–22
23. Rawls J (1951) Outline for a decision procedure for ethics. Philos Rev *60*, 177–197
24. Nagel T (1979) Mortal Questions. Cambridge, Cambridge University Press, p. 134
25. Kaiser M and Forsberg E-M (2001) Assessing fisheries – using an ethical matrix in a participatory process. J Agric Environ Ethics *14*, 191–200
26. Habermas J (2003) The Future of Human Nature. Cambridge, Polity
27. Outhwaite W (1998) Discourse ethics. In: Encyclopedia of Applied Ethics. (Ed.) Chadwick R. San Diego, Academic Press. Vol. 1, pp. 797–803
28. Schroeder D and Palmer C (2003) Technology assessment and the 'ethical matrix'. Poiesis Prax *1*, 295–307
29. Chadwick R, Henson S, Moseley B, Koenen G, Liakopoulos M, Midden C, Palou A, Rechkemmer G, Schröder D, and von Wright A (2003) Functional Foods. Berlin, Springer-Verlag
30. E.g. Food Ethics Council (2001) After FMD: aiming for a values-driven agriculture. Southwell, FEC
31. Mepham B (2004) A decade of the ethical matrix: a response to criticisms. Science, ethics and society (Preprints of 5th EURSAFE Congress). Katholic Universiteit Leuven, Belgium, pp. 271–274

PART TWO

Bioethics and human futures

O brave new world,
That has such people in't!

William Shakespeare (*circa* 1611)
The Tempest

4

The haves and the have-nots

OBJECTIVES

When you have read and discussed this chapter you should:

- appreciate the nature and scale of the differences between developed and less developed countries in terms of factors such as: population size and structure, food supply, health, wealth, education and technology
- recognize the importance of the biosciences in addressing such inequalities
- understand the different theories of global development
- be familiar with different development policies pursued by governments in developed countries
- be able to apply ethical principles to your reasoning on how to address the vast inequalities in the circumstances of the world's peoples

4.1 Introduction

It has been noted (1.2) that a theory of ethics is '*a philosophical structure, which together with some degree of empirical fact, will yield a decision procedure for moral reasoning*'. This chapter concentrates on facts, of a general and global nature, concerning the state of the human population. Taken together, they serve as a set of critical reference points for when, in later chapters, we consider the ethical implications of various issues in the biosciences.

A comprehensive account would entail an extensive review of the ways in which people live, in terms of their health, domestic circumstances, educational opportunities and material wealth. But such an approach would be far too ambitious here, so that the emphasis will be on a limited number of key indices that distinguish the circumstances of the *haves* and the *have-nots*. Even so, because this is a big subject, the treatment is necessarily brief. In order to make the task more manageable, the focus will be on key statistics for groups of countries, or sometimes for individual countries chosen to represent specific points. If at times the account seems to be too documentary, it is only necessary to recall the two principles underpinning Rawls' theory of *justice as fairness* (Box 3.1).

4.1.1 Comparing different countries

Making realistic comparisons of the circumstances in which people live in the approximately 200 separate countries of the world is far from straightforward. There is so much variation in what people consider necessary, desirable, or acceptable, that simplistic criteria such as the gross domestic product (GDP)[a] are quite inadequate. The financial wealth of different nations may give a rough indication of the overall well-being of their citizens (although there is often much inequality *within* societies), but material possessions are clearly not everything. Henry Thoreau, the nineteenth-century American advocate of the simple life, argued that one is rich in proportion not to what one *has*, but to what one can afford to do *without.*[1] Certainly, people with many material possessions do not appear to have a monopoly on happiness. On the other hand, health, education, and opportunities for personal and social development, which are almost universally regarded as basic necessities for all in civilized societies, often come at a price.

The usual distinction made between countries is that defined by the terms **developed countries** (DC) and **less developed countries** (LDC) or *developing countries.* These are so familiar that they will be employed here, even though they greatly oversimplify the issues. But in using them, it is important to note the following points.

- Not all so-called 'developing countries' are in fact developing: e.g. in many countries of sub-Saharan Africa most indices of development have been negative for some years. Hence the term LDC is used here.
- While 'development' is usually measured in numerical data, people in LDC often have strong cultural bonds and deep reserves of traditional knowledge, which represent a significant degree of development according to different criteria that cannot easily be quantified.
- There is a gradation in the circumstances of different countries such that no sharp divide between DC and LDC can be drawn.
- What are usually called *industrialized* countries (largely synonymous with DC) represent only about 20% of the global population, whereas the *third world* (a term often equated with LDC) accounts for perhaps two-thirds of the global population.

The aim in the next six sections is to summarize some key facts concerning the global human population. These are listed under the headings: demography, food, health, wealth, education, and technology.

4.2 Demography

Demography is *'the statistical study of human populations, especially with reference to size, density, and distribution'.*[2] The global population is currently (2005) more than 6.5 billion and likely to exceed 9 billion by 2050. (You can see the way the population changes second-by-second on the POP Clock website.[3]) But the increase is not because more people are being added to the world per annum (p.a.), but because fewer people are

[a] The GDP is the value of the output of all goods and services produced within a nation's borders, normally given as the total for one year. It includes the production of foreign-owned firms within the country, but excludes income from domestically owned firms located abroad.

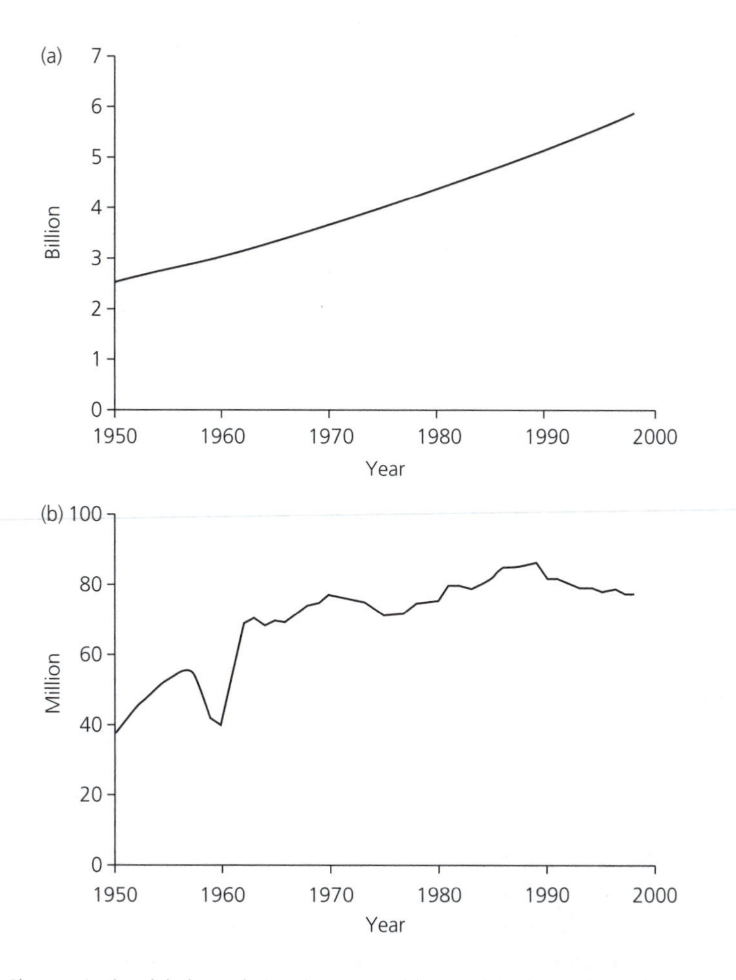

Figure 4.1 Changes in the global population since 1950. (a) Growth in the total population; (b) the annual addition to the world population. [From: Brown L, Renner M, and Halweil B (1999) Vital Signs. London, Earthscan]

Source: Census Bureau.

dying. Just under 80 million people are now added annually to the global population, but **life expectancy** is increasing steadily, and at a higher rate in LDC, where most people reside (Figure 4.1). In 2004 there were estimated to be 4.1 births and 1.8 deaths every second, giving a natural global increase of 2.3 per second.[4] There are, however, great differences between countries, in both population size and rates of growth. China and India both have populations of more than one billion, and about one billion extra people are being added to the global population every twelve years.[5] Some African countries have growth rates of more than 3% p.a., while in most European countries they are less than 1%, or populations are actually declining.[6] Overall, population growth rates are inversely related to economic development – so that, over the next ten years, about 97% of the increase in population is expected to be in LDC.

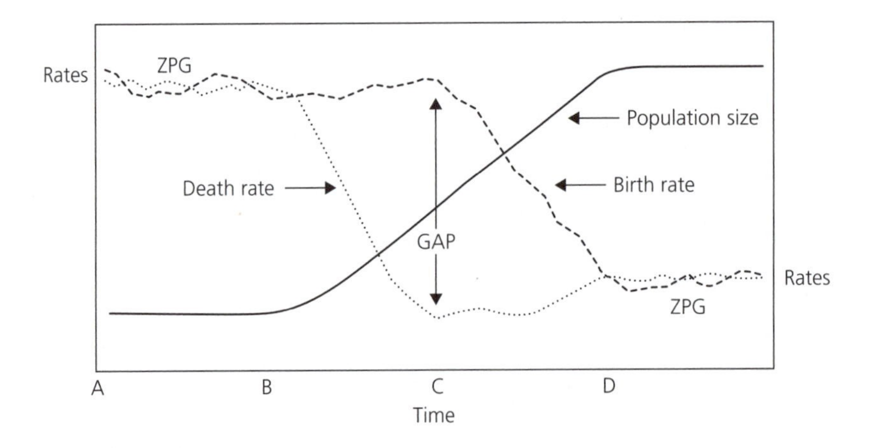

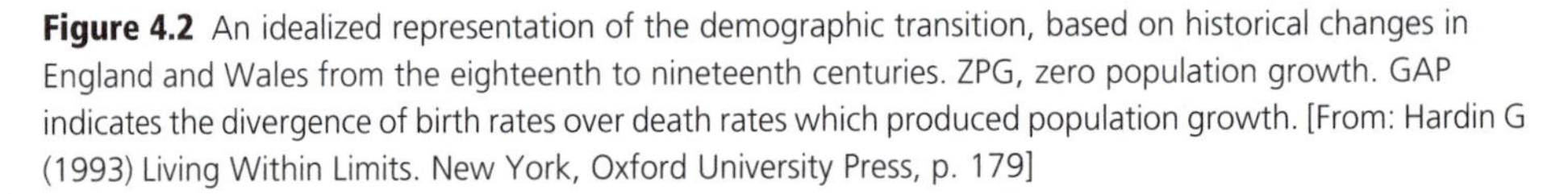
Figure 4.2 An idealized representation of the demographic transition, based on historical changes in England and Wales from the eighteenth to nineteenth centuries. ZPG, zero population growth. GAP indicates the divergence of birth rates over death rates which produced population growth. [From: Hardin G (1993) Living Within Limits. New York, Oxford University Press, p. 179]

Up to the nineteenth century, before the so-called **demographic transition**, population growth rates in DC were also very high, but they declined rapidly in the early twentieth century. In the demographic transition, the situation changes from one of high birth rates (BR) and high death rates (DR) to one of low BR and low DR (Figure 4.2). The traditional explanation for the demographic transition appeals to socio-economic factors. In such terms, the fall in DR in Britain is considered to be due to advances in the standard of living resulting from the industrial and agricultural revolutions of the eighteenth and nineteenth centuries, which meant that BR fell because parents no longer needed to have so many children. Up to then, however else they were valued, children were also often seen as a human resource necessary for economic survival - in the days before pensions or sickness benefits. (Reliable historical data are available for relatively few countries, such as England and Wales, and these tend to be taken as representative of all DC.)

Because the resulting fall in BR lagged significantly behind the fall in DR, the population (which had until then been kept in check by many early deaths) increased rapidly, leading to a veritable population explosion. Although one might suspect that DR declined as a consequence of medical advances, such as antibiotics and vaccination, these came too late to explain the reduced mortality rate. Instead, they appear to have been due largely to measures taken to improve public health, such as safer water and food supplies and improved hygiene and sanitary measures.

The idea that the demographic transition was dependent on industrialization has, however, been challenged by some demographers. They attribute it instead to cultural or educational factors, such as the process of **secularization**, as a result of which people developed a sense of autonomy which freed them from a passive acceptance of their fate or from the influence of 'other-worldly powers'.[7] This interpretation implies that reduced BR might be achieved by educational development as well as, or instead of, economic development.

While the demographic transition from high BR/high DR to low BR/low DR has long been completed in DC, many LDC are still in the early or middle stages of the transition, so that although DR have declined populations continue to grow relatively rapidly. Another informative index is the **total fertility rate** (TFR), defined as the number of children a woman has in her lifetime. The average TFR in sub-Saharan Africa exceeds 6, whereas in Europe it is less than 2 (i.e. below the sustainable rate for a stable population).[8]

Progress through the demographic transition, whatever its causes, is accompanied by marked effects on the age structure of societies. In LDC, children often form the largest sector of the population, necessarily supported financially by fewer working adults. In fact, 20% of the world's population is between 10 and 19 years old (of whom 87% live in LDC) and nearly 50% are under 25 years old.[9] On the other hand, in DC large numbers of people survive into old age: life expectancy in many DC is more than 75 years, whereas in sub-Saharan Africa as a whole it barely reaches 50 years.[10] But an **ageing population** can place an enormous financial burden on the working population - and this is likely to affect many nations in the near future.[11] Russia's population is falling by 750,000 p.a. (described by President Putin as a *'national crisis'*), and, after 2020, China's population will go into a steep decline, so that the country could lose 20–30% of its population every generation.

Population density is not just a balance between BR (a 'well-conceived topic') and DR (a 'deadly subject'): it is also determined by migration (a 'moving experience'). Migration can be a significant factor, both within countries, e.g. as people move from their villages into the cities in search of work, and internationally, as people move to wealthier neighbouring countries, such as the USA and the oil-producing states of the Middle East.

Currently, half the global population lives in cities - and soon two-thirds will.[12] Cities such as Mexico City, Sao Paulo, Tokyo, and New York have populations of between 20 and 30 million, usually associated with a vast extension of the urbanized area (*suburbanization*). While migrants can have beneficial effects in increasing the labour force, there are also serious adverse effects in many cases. A recent UN report shows that currently one billion people live in slum conditions, and in sub-Saharan Africa more than 70% of the population do so. In thirty years' time, the total number could have doubled. Such conditions are not only utterly depressing, but also often breeding grounds for disease and social unrest.[13]

4.3 Food supplies

Of all the necessities of life, food and drinking water are the most crucial - and access to them defines, more clearly than anything else, the *haves* and the *have-nots*. Although currently enough food is produced globally to feed everyone, malnutrition and undernutrition are prevalent.

4.3.1 Less developed countries

The Food and Agriculture Organization (FAO) estimates that, globally, 842 million people were undernourished during 1999–2001, of whom 798 million were in LDC. One-quarter of them were children under the age of 5 years. There are a few encouraging signs of progress, because in about twenty countries hunger rates have been reduced from the 1990–92 baseline figures. But taking LDC as a whole, the number of chronically malnourished people only declined by 19 million over the same period, and more recent analyses are even less encouraging because between 1995 and 2001 the numbers of undernourished people actually *increased* by 18 million.[14] An authoritative report in 2003 predicted that the number of malnourished pre-school children in sub-Saharan Africa is likely to increase by 2015.[15]

One critical factor is the energy available in food, still customarily reported in kilocalories (kcal) rather than the megajoules (MJ) favoured in scientific literature. In the USA and several EU countries, average daily intake exceeds 3500 kcal (in the UK it exceeds 3000 kcal), but in India it is less than 2500 kcal and in many sub-Saharan African countries it is less than 2000 kcal. Because protein can substitute to some degree for low supplies of energy in fat and carbohydrate, malnutrition is most marked when there is deficiency of all three. **Protein–energy malnutrition** often afflicts infants (i.e. less than 1 year old) and young children, when it manifests itself as **marasmus** and **kwashiorkor**, respectively. About 20% of all children have inadequate intakes of protein and energy,[16] but the severity of protein–energy malnutrition varies widely. For example, a study in India revealed that while 1.2% suffered from kwashiorkor, as many as 80% showed signs of reduced growth due to inadequate diet.[17]

Low-birth-weight babies (by definition less than 2.5 kg) who survive being born too small are likely to remain **underweight** and sick throughout childhood and adolescence. Every year, about 30 million such infants are born in LDC, their impaired growth caused by poor nutrition in the womb. In LDC as a whole, underweight, which afflicted 153 million children under the age of 5 years in 2000,[18] accounted for 3.4 million childhood deaths (mostly in Africa and Asia). Food security remains Africa's most fundamental challenge. About 200 million people on the continent are undernourished, their numbers having increased by 15% since the early 1990s, and almost doubled since the late 1960s.

Because small children have a higher *surface area to weight ratio* than adults, their dietary energy requirements are disproportionately greater – and lack of appreciation of that fact may partly explain the high rates of infant mortality in LDC. Another reason lies in infant feeding practices. Expert medical opinion asserts that the vast majority of women are able to breast-feed their children, a practice that provides optimal nutrition and immunological protection from diseases, and establishes a beneficial psychological bond between infant and mother.[19] However, the large-scale adoption of artificial (bottle) feeding in LDC deprives babies of the **advantages of breastmilk**, and the United Nations Children's Fund (UNICEF) claims that inappropriate feeding is responsible for the deaths of 1.3 million infants annually. *'The non-breast-fed child living in disease ridden and unhygienic conditions is 6–25 times more likely to die of diarrhoea and four times more likely to die of pneumonia than breastfed infants.'*[20] Because use of bottle-feeding is often due to high-pressure advertising, the resulting disease condition has been described as **commerciogenic malnutrition**.[21]

Nutrients other than energy, such as essential amino acids and fatty acids, minerals, vitamins, and micronutrients, are also vital for health. Iron deficiency is one of the most common conditions, affecting about 3.5 billion people and leading to almost 1 million deaths annually; while iodine deficiency afflicts another 2 billion. Vitamin A deficiency (VAD) currently affects more than 200 million pre-school children, leading to blindness, and sometimes death. The most severe cases are in sub-Saharan Africa and India, but the condition is also common in South America.[22]

4.3.2 Social factors

The lower peoples' incomes are, the larger the proportion they spend on food, and for the poorest people food becomes by far the largest single item of expenditure. However, even for the hungriest people, an irreducible 20% seems to be reserved for other necessities, such as clothing and soap. The poorest people spend 80% of their income to purchase 80% of the conventionally defined food energy requirement (the so-called **double eighty rule**).[23] By comparison, the average amount spent on food in the UK is less than 10%.

Apart from food supplies, food acquisition and preparation are also important considerations, because they take up many of the **time resources** of poor people. For example, a study in Ghana showed that farmers spent nearly five hours preparing a meal - made up of collecting firewood and water, and walking to the farm, market and grinding mill.[24]

Most of the widespread hunger in the world is a result of poverty, but other important factors are droughts and floods, armed conflict, and political, social and economic disruptions. The result is that about thirty nations are currently facing exceptional food emergencies, with nearly 70 million people needing emergency food aid.[25]

4.3.3 Over-consumption

Although undernutrition is rare in DC, the nutritional status of many people is far from satisfactory (11.2). This is because 300 million people are clinically obese, and of these about 500,000 people in North America and Western Europe die annually from obesity-related diseases. More generally, **over-consumption** is considered to cause several diseases, such as coronary heart disease (CHD), which is related to high blood concentrations of cholesterol, and diabetes. In the USA and most EU countries, 3–5% of the population suffer from diabetes. Annual deaths from CHD are up to 250 per 100,000 people in the USA and EU, and more than 500 per 100,000 in the former Soviet Union.[26]

4.4 Health

The World Health Organization (WHO) identifies ten leading risk factors which together account for more than one-third of all deaths worldwide. Five of these that are largely diet-related (underweight, high blood pressure, iron deficiency, high cholesterol and obesity) have been discussed in 4.3. The other identified risk factors are:

- unsafe sex
- tobacco consumption
- alcohol consumption
- unsafe water, sanitation, and hygiene
- indoor smoke from solid fuels.

There are vast inequalities in public and private expenditure on health, with the wealthiest nations, such as the USA and Switzerland, spending more than US$2500 per person p.a., whereas the poorest, such as Vietnam and Sierra Leone, spend less than US$5.[27]

Human immunodeficiency disease/aquired immune deficiency disease (HIV/AIDS) is the fourth largest cause of death, with most of the estimated 40 million cases being attributed to unsafe sexual activity. Seventy per cent of the people currently affected live in Africa, and such is the impact that life expectancy in seven sub-Saharan countries has dropped to 33 years, a fall of about 20 years over the past decade.[28] The epidemic, which kills 3 million people p.a., shows no signs of abating. In addition to Africa, the disease is also spreading rapidly in Eastern Europe and Asia.

Tobacco smoking is estimated to be responsible for 4.9 million deaths annually, most of which occur in DC, although the rate of increase is now most marked in LDC. A report from the British Medical Association (BMA) in the UK has stressed the effects of smoking on reproductive fertility. Men who smoke have an increased likelihood of impotence, lower sperm counts, more malformed sperm, and a poorer response to fertility treatment. Women smokers are more likely to have painful or irregular periods, pre-cancerous changes in the cervix, and an earlier menopause. They also take longer to conceive and have a poorer response to fertility treatment. Smoking during pregnancy puts both the mother's and baby's health at risk.[29]

Half the world's population is also exposed to a different form of indoor air pollution, resulting from burning solid fuels for cooking and heating; the toxic fumes are estimated to kill 2.2 million people p.a.

Global alcohol consumption has increased in recent decades, with the largest increase in LDC. Worldwide, alcohol leads to 1.8 million deaths p.a., and is estimated to cause 20–30% of the cases of oesophageal cancer, liver disease, epilepsy, motor vehicle accidents, murders and other intentional injuries.

4.4.1 Water and sanitation

Unsafe water and sanitation are responsible for the deaths of the approximately 17 million people p.a. who suffer from the curable infectious and parasitic diseases which cause diarrhoea, measles, malaria, and tuberculosis. Water supplies, both in terms of quantity and quality, are highly variable between countries. More than 1 billion people have no access to clean drinking water, while 2.4 billion (about 40% of the global population) lack safe and hygienic sanitation.[30]

But the inadequacy of water supplies seems likely to increase in future: sixteen African countries experienced droughts in 1995. Countries of North Africa experience the worst level of scarcity (each person receiving less than 1000 m^3 p.a.), but India, China, and most EU countries are described as being 'on the verge of water stress'.

4.5 Wealth

We live in a world in which there are vast disparities in income. In some countries the average person earns more than US$40,000 per year, but at the same time 1.2 billion live on less than US$1 per day and almost 3 billion (about half the world's population) on less than US$2 per day.[31] The 20% of the global population who live in the highest income countries account for 86% of total private consumption expenditure, whereas the poorest 20% account for only 1.3%.[32] Such inequalities in wealth translate into stark differences in consumption. The richest 20%, compared with the poorest 20% (values in brackets):

- consume 58% of total energy (compared with 4%)
- consume 45% of all meat and fish (compared with 5%)
- consume 84% of all paper (compared with just over 1%)
- own 87% of all vehicles (compared with less than 1%)

In consequence, one child born in the industrialized world adds more to consumption and pollution than do 30 to 50 children born in LDC.[33] In view of such differences, the common assertion that the people in LDC have 'too many children' has a rather hollow ring. The enormity of the wealth of the ultra-rich is evident from the data in Box 4.1.[34]

4.5.1 Conspicuous consumption

It is clear that the gulf between the haves and the have-nots (or at least between the very rich and the very poor) is not just a problem of the poor not having enough: it is also a problem of others having too much. In his classic *The Theory of the Leisure Class* (1899) the American sociologist Thorsten Veblen analysed consumption as a social phenomenon, and identified the ways in which, through its purchasing patterns, a relatively small leisure class extended its influence over virtually the whole of society.[35] Veblen described how each social group sought to copy the group above it, so that what he called **conspicuous consumption** (effectively, keeping up with the Jones's, through flaunting

BOX 4.1 INEQUALITIES IN WEALTH

- The three richest people in the world together have assets that exceed the combined GDP of the 48 least developed countries.
- The 15 richest people together have assets that exceed the total GDP of sub-Saharan Africa.
- The wealth of the 32 richest people exceeds the total GDP of South Asia.
- Of the richest 225 people, the USA has the most (60 with a combined wealth of US$611 billion).
- The income of the world's richest 5% is 114 times that of the poorest 5%.

your possessions) permeated American society. The insights have even greater force in the twenty-first century, when advertising is so much more persuasive and pervasive.

Conspicuous consumption is ethically problematical for various reasons. It deprives the 'have-nots' of their just deserts, pollutes the global environment, will adversely affect the options of future generations, and places unsustainable stresses on the environment. There would seem to be a strong case for taking Thoreau's philosophy (4.1.1) more seriously, and finding fulfilment in activities that do not depend disproportionately on material possessions or high levels of consumption.

4.5.2 Progress in less developed countries

However, despite the vast disparities between the average citizen of LDC and DC, the situation for many in LDC has improved, with a marked increase in their consumption expenditures over recent decades. This has been especially prominent in the high-growth economies of Asia and Latin America. For example, by the mid-1980s, China had become the largest manufacturer of television sets; and in India by the mid-1990s more than 70% of rural households owned a portable radio, a bicycle, and wristwatches, and more than 20% had refrigerators. Even so, a number of problems are associated with this growth in consumption. For example:

- the expansion in consumption is badly distributed, with perhaps 20% left out
- the consumption patterns are both socially divisive and environmentally damaging
- the rights to information and product safety are difficult to maintain in a global market.[36]

Clearly, the crucial feature of population size (4.2) is not its density in terms of people per square mile, but the average resource depletion and environmental pollution that the environment can withstand. Hong Kong, with a density of more than 14,000 people per square mile, and the Netherlands, with more than 1000 per square mile, are flourishing societies, whereas Africa, many parts of which are extremely poor, has a population density of only 55 per square mile.

4.6 Education

In the modern world, where education confers many advantages, illiteracy is a severe impediment to development. But whereas in most DC almost all the adult population is literate, the proportion in several LDC (particularly in Africa) is very low, often less than 50% of the population. More than 870 million people in LDC are illiterate;[37] and it should be noted that the official definition of literacy only amounts to the ability to read and write a simple sentence.

Moreover, the gender gap is very pronounced in many LDC, with less than 50% of girls receiving primary education and less than 10% enrolled for secondary education. This is reflected in women's participation in government, which is often less than 10% that of men in the least developed countries, even though their *economic activity rate* is often equivalent to that of men.[38]

Of course, education should not be seen as simply a matter of being able to sign your name and read simple documents. If properly developed as part of a broad curriculum, literacy is the route to enfranchising people previously oppressed by the inability to understand or question those who govern their lives, and it has the potential to open up a whole world of ideas, allowing them, probably for the first time, to participate in the dialogue on which the democratic process depends.

One area where education is crucially important is in critically assessing the claims of advertisements, which can often be unreliable or deceptive to a gullible audience. Global advertising is now a US$435 billion business, but if all forms of marketing are included the figure is probably of the order of US$1 trillion.[39]

4.7 Technology

It is a truism that in DC we live in a 'technological society'. Almost the whole of our lives is supported by a technological infrastructure that supplies virtually all our needs and wants - both in terms of material resources and through its impacts on educational, cultural and recreational activities. Even those who seek to 'get away from it all' often fly to their destination - and, armed with patent medicines to ward off illness, maintain contact with civilization by mobile phone, while recording it all on their digital cameras.

For most people in LDC such opportunities are non-existent. They might happily forgo the mobile phone and the camera if technology could be harnessed to supply their basic needs, but the prospects of that happening are remote. The suggestion is often made that the vast differences between DC and LDC are due to climatic and ecological factors, but this appears to be only a part of the reason. In 1820, at the start of the modern growth era, the average income per capita in the tropical world was about 70% of that in the temperate world. By 1992, it was only 25%. Five possible explanations for this have been advanced:

- technologies for human development (e.g. for health and agriculture) are **ecologically specific** and cannot be transferred easily across different zones
- temperate zone countries had a **head start** because by 1820 their technologies were more productive and integrated into international markets
- technological innovation offers increasing **returns to scale**. Richer populations increase market demand, amplifying the gap between the temperate and tropical zones
- **social dynamics**: urbanization and the demographic transition (4.2) further fuelled economic growth in the temperate zone
- temperate zone countries exerted **geopolitical dominance** over countries in the tropical zone through colonialism, which entailed suppression of local industry and neglect of education and healthcare.[40]

The result was the emergence of the divide between what we now call the DC and LDC, which is reflected in the disparities discussed above (4.2–4.6). If it is assumed that technology is a critical 'engine of change', altering this situation will require a major reallocation of global resources. Data on scientific research funding illustrate the point. In 1994, research and development (R&D) spending in North America was US$178.1 billion and in Western Europe US$135.5 billion, but in Africa it was just US$2.3 billion. Only 2.2% of GDP was spent on R&D in Africa compared with 22.2% each for Western Europe and North America.

In consequence, Africa is almost completely dependent on imported technology. This raises two major problems. First, Africa is too poor to afford anything approaching the quality and quantity of technology needed – because for the DC companies who develop it commercial, rather than humanitarian, objectives are the priority. Second, the imported technology contributes to a significant net outflow of resources from the continent: about 75% of R&D project budgets go towards salaries of the foreign experts involved.[41] On the other hand, it is important to realize that LDC are not all the same. For example, both China and India *'have world class capacity in a number of scientific and technological areas... including biotechnology and pharmaceuticals'*.[42]

4.7.1 Intellectual property rights

Intellectual property rights (IPR) is a generic term (e.g. it includes **patents** and copyrights) which seeks to provide legal protection for creators, such as authors, artists, and inventors, from someone copying or using their work without permission (Box 4.2).

Some people argue that IPR are necessary to stimulate economic growth, which will itself lead to poverty reduction. Thus, by stimulating development of new technologies, agricultural and industrial production will increase, thereby facilitating domestic and foreign investment, technology transfer, and the availability of new medicines. They consider that a system that works for DC could also be effective in LDC.

Others, however, dispute this. They believe that IPR do not stimulate invention in LDC because they lack the necessary human and technical capacities. Moreover, poor people would be unable to afford the products even if they were developed. Instead foreign firms are able to use IPR to obtain patent protection to drive out domestic competition, and to increase the costs of essential medicines and agricultural inputs.[43] The TRIPS (Trade-Related Aspects of Intellectual Property Rights) Agreement is one of the three pillars of the World Trade Organization (WTO) (the others being trade in goods and trade in services) which is legally binding for all WTO members (although the least developed countries have until 2006 to comply, and may seek extension of that deadline). These legal issues are difficult to summarize briefly, but a UK Government Commission expressed concerns over the current global intellectual property regime and stated: *'We need to make sure that the IP* [intellectual property] *programme facilitates, rather than hinders, the application of the rapid advances in science and technology for the benefit of developing countries.'*[44]

4.7.2 Biopiracy

Biopiracy has been defined as the process through which the rights of indigenous cultures to genetic resources and knowledge are erased and replaced by those who have exploited this knowledge and biodiversity. Large numbers of patents are claimed to have

BOX 4.2 INTELLECTUAL PROPERTY RIGHTS: SOME DEFINITIONS

Patent: an exclusive right awarded to an inventor to prevent others from making, selling, importing or using their invention, without licence or authorization, for a fixed period of time. In return, society requires the patentee to disclose the invention to the public.

The invention must be:

- novel
- inventive
- capable of industrial application
- sufficiently described
- not inherently unpatentable, which would be the case e.g. if it was:
  - not a discovery
  - an immoral invention
  - essentially a biological process
  - a plant or animal variety

The Agreement on Trade Related Aspects of Intellectual Property Rights (TRIPS): sets minimum requirements for IPR in eight areas which must be enacted international law by members of the **WTO**.

Features are:

- enforcement mechanisms are required
- built-in review mechanism of agreement
- monitored by a council of WTO members

been granted on genetic resources and knowledge from LDC without the consent of the possessors.

A classic case of alleged biopiracy is the use by multinational corporations of the neem tree in India. The tree is characterized by its great versatility, having uses in medicine, contraception, as a construction material, as a fuel, and in agriculture as an animal feed, fertilizer and insecticide. The many virtues are apparently largely due to its containing a number of potent compounds, notably azadirachtin. Several extracts of neem have been patented by US biotechnology companies, one of which, W R Grace, argued that although traditional knowledge inspired the research, the patented compositions and processes were *'considered sufficiently novel from the original product of nature and the traditional method of use to be patentable'*. Such claims have been vigorously contested by the Indian scientist and rights activist Vandana Shiva.[45]

4.8 Addressing the inequalities

With such a rapidly growing global population and the marked differences between DC and LDC discussed above (that are largely a consequence of the 'luck of the draw'), we are all faced with the prime ethical question 'How should I act?' Mistakes may have been made in the past, and doubtless by our own society and its political leaders, but what matters now is deciding on the right course of action in future. All the classical theories of ethics (such as utilitarianism, deontology, and virtue ethics - see chapter 2) would seem to require us to act in ways that will improve matters, even though they appeal to different forms of justification (such as maximizing happiness, acting out of duty, or in accordance with human rights).[46] But what *precisely* needs to be done?

A simplistic answer is that the rich countries of the world should provide more assistance to the poor, which because of the vast differences between them might allow the poor to be rescued from their dire plight without the rich being unduly penalized. But if that were the case, we might ask why it hasn't been done already. The answer lies, at least in part, in different analyses of the root causes of the problem of poverty, and hence in the different solutions proposed to solve it. Almost certainly, it also lies partly in ignorance and partly in indifference.

As we have noted, the most basic requirement is for people to have access to enough nutritious food to lead healthy lives; but in principle, as population increases it might reach a level at which it exceeds the Earth's **carrying capacity**. There are, of course, issues concerning the way in which available supplies are distributed, but the carrying capacity represents the ultimate physical measure of the **limits to growth**.

4.8.1 Neo-Malthusians

That population growth would soon outstrip the Earth's ability to produce adequate food was perhaps most famously argued by the Reverend Thomas Malthus, a Cambridge economist, who wrote *An Essay on the Principle of Population* in 1798.[47] His argument, which became significantly revised in subsequent editions over the next 25 years, is readily summarized in the following quotations:

> *Population, when unchecked, increases in a geometrical* [i.e. exponential] *ratio. Subsistence* [i.e. food] *increases only in an arithmetical ratio* [i.e. linearly]. *By that law of Nature that makes food necessary to the life of man, the effects of these two unequal powers must be kept equal.*

Malthus, who considered that '*the passion between the sexes is necessary and will remain nearly in its present state*', deduced that a stable population could only be achieved by an increase in the death rate, '*because gigantic inevitable famine stalks in the rear, and with one mighty blow levels the population with the food in the world*'.

In later editions of his book, Malthus placed more emphasis on reducing BR, which in terms of the sexual standards of the time was to be achieved by delaying the age of marriage. But the central message remained essentially unchanged - that food supplies place a limit on population growth because '*the power of population is indefinitely greater than the power in the earth to produce subsistence for man*'.

It is clear that Malthus's pessimistic forecast has not materialized. In fact, since he first wrote, at the end of the eighteenth century, the global population has increased almost eightfold. There appear to be several explanations for this. For example, in Britain, they range from the increased productivity of the land due to the agricultural revolution coupled with increased food imports, to lower BR due to contraceptive technologies and the desire for smaller family sizes that has accompanied economic development. Elsewhere, new areas of land have also been opened up for cultivation.

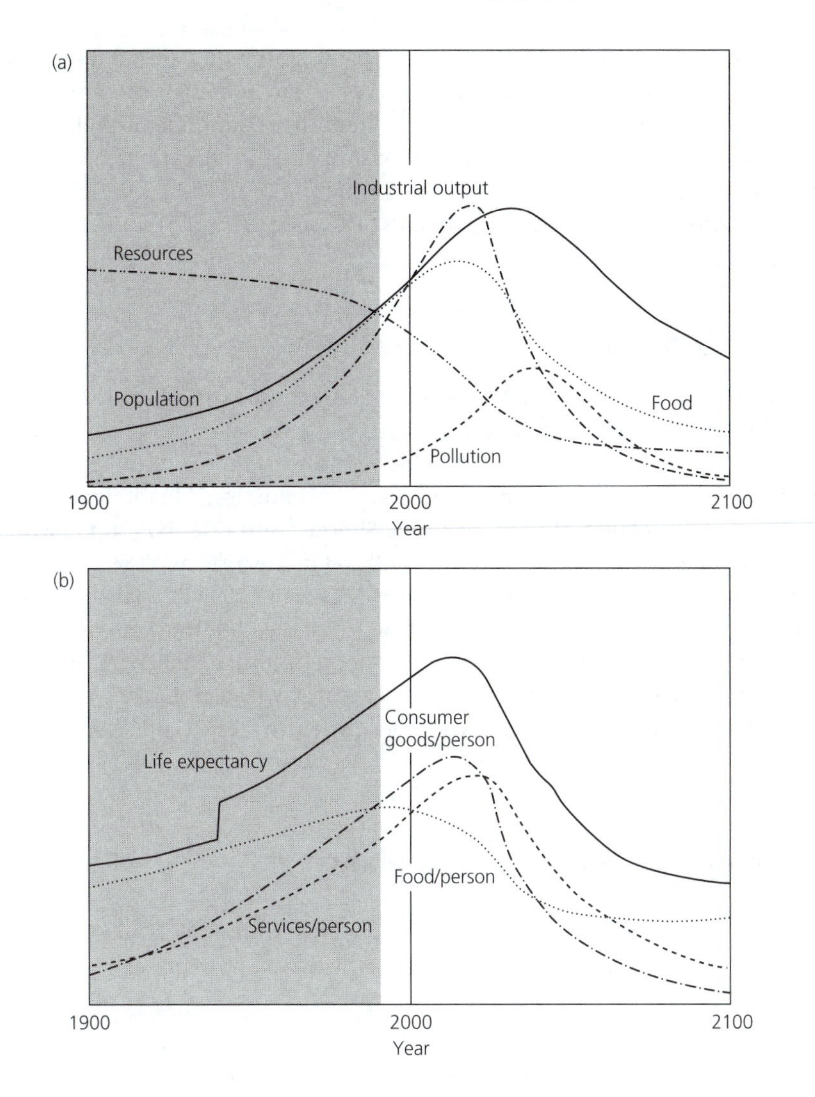

Figure 4.3 Scenario 1: The Standard Run. A computer prediction of several global variables based on the assumption that world society proceeds along its historical path as long as possible without major policy change. (a) State of the world; (b) material standard of living. [From: Meadows D H, Meadows D L, and Randers J (1992) Beyond the Limits. London, Earthscan, p. 192]

But this has not stopped demographers appealing to the Malthusian argument many times in the last 200 years.[48] And of course, however much the so-called **Malthusian solution** (mass starvation) has been delayed, it is still possible that, sooner or later, Malthus's 'power of population' will be unleashed to exert its devastating effects. Among prominent modern **neo-Malthusian** writers are Lester Brown[49] and Paul and Anne Ehrlich.[50] They point to the positive feedback loops that seem certain to bring humanity to the very limits of survival in the near future (Figure 4.3). In fact, notable computer-model predictions (12.6) suggested that food supplies are but one constraint. Others include the rapidly diminishing reserves of raw materials on which the consumption patterns of modern society depend, and the rising tide of pollution which threatens to outstrip the Earth's capacity to absorb or neutralize its effects.[51] The latter point is emphasized by the vastly greater consumption and pollution characteristics of life in DC (4.5), and the fact that as populous LDC, such as China, industrialize, they are adopting similar lifestyles.

4.8.2 Anti-Malthusians

Neo-Malthusian arguments have been countered by those (**anti-Malthusians**) who see a large and growing population as both a sign, and a cause, of prosperity. This optimistic perspective can be traced back to the eighteenth-century Scottish philosopher Adam Smith, who laid the foundations of modern economics.[52] In modern times it has been promoted by economists such as Julian Simon[53] and Ester Boserup,[54] who claim that population growth has stimulated agricultural innovation and we can confidently rely on human skills to satisfy future needs as they have done to a large degree in the past. This view is often coupled with a belief that neo-Malthusians exaggerate the severity of the current problems and the likelihood of future challenges. It is an interesting fact that many neo-Malthusians are scientists, who emphasize ecological constraints, while many anti-Malthusians are social scientists, especially economists.[55]

4.9 Alternative ethical action plans

These different assumptions may be said to give rise to four types of action plan to address the perceived population problem, which can be represented as a simple typology:

	Neo-Malthusian	Anti-Malthusian
Altruistic	NMA	AMA
Egoistic	NME	AME

Neo-Malthusian altruists (NMA) consider that time is rapidly running out, and that the ethical strategy required to ameliorate the problems of the needy will involve DC governments urgently taking a wide range of pro-active measures. These will aim to increase material help through aid programmes, help to reduce BR through family-planning programmes, improve educational opportunities, and implement fair-trade arrangements. But some NMA also recognize that economic policies currently pursued by DC serve to undermine the attempts of LDC to make any progress. This is because of the phenomenon of **zero-sum**,[56] which describes the fact that when resources are limited, their appropriation by people in DC denies most people in LDC adequate access to them. If this claim is valid, promoting economic development in LDC will also entail DC reducing their levels of consumption and pollution to enable LDC to have more opportunities.

Neo-Malthusian egoists (NME) are even more pessimistic than NMA, taking such a gloomy view of the future that they have sometimes been labelled 'prophets of doom'. According to this assessment, things have already gone too far, with much of humanity being portrayed as living in a **demographic trap**, in which more births simply lead to more deaths. In such circumstances, medical scientist Maurice King has questioned whether public health measures to save life should be introduced at all.[57] Similarly, the ecologist Garret Hardin argues that we are in a 'lifeboat situation', in which those people in LDC who can escape the threat of imminent starvation are attempting to clamber aboard the DC lifeboat as the only realistic means of survival. But Hardin argues that if DC take on board all the 'refugees' the boat will sink. On this basis, his egoistic **lifeboat ethics** seems to demand that people in DC take measures (e.g. strict immigration control) to ensure their own future welfare, and that of their descendants, rather than risk jeopardizing human life as a whole.[58]

Anti-Malthusian altruists (AMA) are optimists. They point to the fact that science and technology have had such a significant effect on society that past predictions (from Malthus onwards) have been confounded. Their claims are often couched in terms such as 'biotechnology will feed the world' (see 11.7) and, in general, they are strong advocates of increasing investment in new technologies, such as genetic modification, as ways of addressing social problems. They may concede that, to date, these technologies have not been used to significant degree in the interests of LDC, but suggest that political and economic changes might remedy this situation in future. A prominent advocate of this approach is statistician Bjorn Lomborg, who has an altogether more optimistic view of our global future than NMA.[59]

Anti-Malthusian egoists (AME) are sceptical of the worst predictions of the NMA and NME, but cautious of what they might regard as the scientistic view (2.5.2) that technology has all the answers. They regard themselves as realists in believing that social development is primarily driven by political and economic factors. Since 'charity begins at home', the most appropriate course of action is to let people in individual countries decide how to cope with future challenges through the political and economic mechanisms which operate in liberal democratic societies, i.e. at the ballot box and by their purchase decisions in a free market.

Of course, in reality, people might occupy intermediate positions in this scheme. Some individuals might, for example, combine aspects of NMA and AME in their personal philosophy. But such an analysis helps in identifying the rationales people adopt in defending their ethical positions. Clarifying our reasoning might lead us, taking everything into consideration, to either change our views or to discover stronger justification for them.

4.10 Development policies

The types of bioethical issue discussed in this chapter have major policy implications. People's ethical analyses may well affect personal decisions about a choice of job, financial investments, donations to charities, or voting intentions. But the most widespread effects are likely to be implemented as a result of decisions made by governments and by international governmental bodies. In the latter context, bioscientists may be said to have particular responsibilities in their role as advisers, because biological understanding, for example in relation to food requirements, reproductive physiology, and population dynamics, has a crucial bearing on political decisions.

4.10.1 Acts of charity

An instinctive reaction to crises in LDC, such as outbreaks of famine, is that we ought to play our part as individuals by making donations to charity, or becoming involved in fund-raising events, so pursuing the NMA approach. Philosopher Peter Singer enunciates the following principle to address such circumstances: *'if it is in our power to prevent something very bad happening, without thereby sacrificing anything of comparable moral significance, we ought to do it'.*[60] We might gain some sense of satisfaction from so doing, believing ourselves to be acting with generosity: after all no one is compelling us to do so.

But for philosopher Thomas Nagel, charity should be rejected as a satisfactory solution to the problem, for two main reasons. First, it is left to the donors to decide when they have been *generous enough*; and second, it assumes that what the donors decide to donate is *rightfully theirs* in the first place. The latter must be seriously questioned in the face of such radical inequalities as are evident between people born into LDC and DC. For Nagel, there is something wrong with an international market economy in which many people are malnourished while others live well, when there is enough to feed everyone adequately. For he says: *'even if it doesn't involve anyone's doing anything wrong, the system that permits this outcome is still morally objectionable'.*[61] Those persuaded by this argument may feel that directing their efforts to political change is an ethical obligation.

4.10.2 Governmental action

Official programmes, especially those of UN agencies, also adopt the altruistic positions conforming to NMA or AMA. The fundamental justification is contained in the UN Declaration of Human Rights, first formulated in 1948, key articles of which in the present context are listed in Box 4.3.

Over the last fifty years, there have been many attempts to build on these articles to make them relevant to prevailing circumstances, and currently the critical role of the biosciences could hardly be more evident. For example, at the UN Millennium Summit in 2000, agreement was reached on a number of Millennium Development Goals, many of which would be addressed by improved nutrition. Thus, good

BOX 4.3 KEY ARTICLES OF THE UNITED NATIONS UNIVERSAL DECLARATION OF HUMAN RIGHTS

- Article 1: *All human beings are born free and equal in dignity and rights*
- Articles 25: *Everyone has the right to a standard of living adequate for the health and well-being of himself and his family, including food, clothing, housing and medical care and necessary social services, and the right to security in the event of unemployment, sickness, disability, widowhood, old age, or other lack of livelihood in circumstances beyond his control*
- Article 26: *Everyone has the right to education* (which) *shall be free at least in the elementary and fundamental stages*

nutritional status:

- reduces poverty by boosting productivity (Goal 1)
- leads to improved educational outcomes, because malnutrition reduces mental capacity (Goal 2)
- empowers women, because girls who are better nourished stay at school longer and have more control in future choices (Goal 3)
- reduces child mortality, over half of which is due to malnutrition (Goal 4)
- improves maternal health, as malnutrition is a major risk factor in maternal mortality (Goal 5)
- slows the onset of AIDS in HIV-positive people, and reduces the incidence of malaria (Goal 6).[62]

In 2005, a UK Government Commission made a number of recommendations to alleviate the problems faced by Africa.[63] These included: increased aid; writing off the debts of sub-Saharan countries to the World Bank; ending subsidies and other trade-distorting support for DC agriculture (see 4.10.5); and introducing reforms to eliminate corruption and ensure that aid provided is used for 'developing economic growth and reducing poverty'.

In contrast to such altruistic programmes, the egoistical positions (NME and AME), by adopting rigorously utilitarian reasoning, disregard the ethical significance of the Declaration of Human Rights, and of the duties incumbent on others to respect them. But, as Nagel has observed, *'We should be suspicious of a result that coincides so perfectly with our economic self interest.'*[64]

4.10.3 **Contrasting population control policies**

One strategy for development lays emphasis on population control, and in recent times, the People's Republic of China has become its most prominent example. Believing that population was increasing far too rapidly, in 1979 the Government introduced the **one child policy** which aimed to *'promote happiness, national prosperity, and social progress'* by a strict regime of control. Couples agreeing to only have one child are allowed access to better schools, housing, and health services. The policy has certainly been effective in reducing fertility rates, but many claim that its sudden and strict implementation has had serious drawbacks.[65] For example, the preference for male babies has resulted in widespread use of abortion and infanticide to achieve this outcome, which not only causes much physical and mental suffering, but may also result in a significant gender imbalance. The rapidly reduced numbers of young people will in future have to support a largely ageing population (4.2).

If couples break their undertaking to limit family size to one child, pressure is put on the women to undergo abortion, or submit to sterilization, and families lose financial benefits previously obtained. According to Amnesty International, violence has been used by officials in the process of securing abortions and sterilizations and the payment of birth control fines, but the Chinese government claims that family planning is pursued in accordance with the relevant principles on human rights.[66]

It is often thought that there is no alternative to stringent birth control policies, but it is important to examine the experience of others whose policies might be as effective but less coercive. A prime example is the Indian state of Kerala, in which women have traditionally held a high social status. Here, there are high levels of both **women's education** and **empowerment**, which are thought to have been key factors in the highly effective family planning campaigns, and which are coupled with a very high literacy rate (over 90%) and low rate of infant mortality. The result is that the TFR is less than 2.0.[67] Primary education is compulsory, and provision of health services relatively high, in many cases providing free treatment. This holistic, egalitarian approach to population control contrasts with medical/surgical strategies adopted by others. It may provide a model for LDC population policies in future.

4.10.4 Aid agencies

Various UN agencies have programmes that aim to promote development in LDC. For example, the UN Development Programme (UNDP) has targets to:

- halve, between 1990 and 2015, the proportion of people suffering from hunger
- reduce by two-thirds, between 1990 and 2015, the mortality rate in under-5-year-olds
- halt, and begin to reverse, the spread of HIV/AIDS by 2015.[68]

Comparable targets form the UN Millennium Development Goals (4.10.2), while the UN Population Fund (UNFPA) aims to ensure universal access to reproductive health, including family planning and sexual health, and to support population and development strategies in LDC.[69] Most DC governments also have international development programmes. For example, in the UK, the overall aim of the Department for International Development (DFID) is *'the elimination of poverty in poorer countries'*, in pursuit of which the Government spends over £3.5 billion annually.[70]

The chief objectives of the International Bank for Reconstruction and Development, usually known as the World Bank, are *'to assist in the reconstruction and development of territories of members by facilitating the investment of capital for productive purposes'*. The bank, which has more than 180 members, grants loans to member nations, for the purpose of financing specific projects.[71]

Various non-governmental organizations (NGO) are also prominent advocates of the NMA approach. Charities such as Oxfam, Christian Aid, and Action Aid all pursue development and famine relief programmes, and they also have policy departments that make recommendations for government action.

4.10.5 Different approaches to economic development

However, problems encountered by LDC may be exacerbated, at least in part, by practices and policies that DC pursue in their own interests, and/or those actually aimed at helping LDC. So an ethical assessment of aid policies entails examining whether the stated altruistic intents of DC are indeed effective. Part of the difficulty lies in the concept of development.

The widely accepted view over the last fifty years was that development is about socio-economic change, with the two main concerns being those of enabling LDC to catch up with DC, and of tackling absolute poverty – represented by malnutrition, disease, and low life-expectancy (see 4.2–4.7). Within a broadly egalitarian framework, there are three approaches to economic development, which adopt different strategies. **Trickle-down** sees the aim as increased incomes for the poor as they benefit from a growing economy. **Equitable growth** is also concerned with increasing income, but places emphasis on the way it is distributed. **Basic needs** approaches deny that increased income is desirable in itself: what matters is satisfying the needs for food, clothing, healthcare, schooling, etc. – which it cannot be assumed will be provided by increased incomes.[72] But trickle-down is still a prominent strategy in encouraging development.

Despite some progress, many LDC remain in seriously disadvantaged positions which often show little sign of real improvement. For example, DC operate high tariffs and steep agricultural subsidies, with those rich countries belonging to the Organization for Economic Cooperation and Development (OECD) providing their farmers with more than US$300 billion p.a. in agricultural subsidies. Moreover, the heavy burden of debt repayments of many LDC has been compounded recently by reduced development assistance, which has fallen from 0.34% of the donor countries' gross national product (GNP)[b] in 1990 to 0.23% in 2002. There are sharply divided opinions on whether the trade liberalization agenda of the WTO will benefit LDC or not.[73]

However, the dominant economic development paradigm (1.5.2) has more recently been called into question, by, for example, Nobel-prize-winning economist Amartya Sen.[74] He claims that development is a much richer concept than can be measured by indices such as GNP, and meeting the economic criteria may not only fail to promote development, but may actually make some matters *worse*, e.g. by reducing access to basic needs. Since 1990, the UNDP has paid attention to a whole range of non-monetary indices in assessing LDC progress. So, in pursuing its programme, UNDP now employs the **Human Development Index** (HDI), devised by Sen, which takes into account adult

[b] The GNP is the total value of all goods and services owned by a country. It is the GDP plus income from abroad, minus income earned by foreign investors within the country.

literacy and life expectancy, in addition to GNP per person; and it also expresses the mean of these on a scale of 0.0 to 1.0. Norway and Sweden top the table with HDI of 0.95, whereas at the bottom, Niger and Sierra Leone have indices of about 0.27.[75]

4.10.6 Aid and trade

Food aid and trade are two, at first sight contrasting, strategies for assisting LDC. Trade, being a largely self-serving activity, might appear to be less virtuous than aid, which sounds altruistic. But as economist John Marsh points out, the distinction is not always so clear cut.[76] When DC governments 'dump' food on LDC (calling it *aid*) it is, in reality, simply a means of disposing of their food surpluses, which by undermining the efforts of LDC farmers to make a living may do more harm than good. In some cases food aid might lead to dependency, with LDC governments having little incentive to invest in agriculture if food is being provided free. On the other hand, trade might stimulate economic activity in LDC and have far-reaching benefits for sustainable development. However, there are clearly some situations, such as famine conditions, when providing food aid is a moral obligation. Neither trade nor aid should be dismissed out of hand, and both would seem to have roles to play.

4.10.7 Limits to affluence

An important element of one version of the NMA position discussed above (4.9) is that relieving the condition of the *have-nots* will entail reduced consumption and pollution by the *haves*. But the assumed need for growth is a fundamental tenet of capitalist economics. In order to survive in a competitive capitalist economy, surplus profits are reinvested in new production, so that capital, like living organisms, tends to reproduce in an exponential fashion.[77]

It is obvious that this is an unsustainable process. Sooner or later, the sources of the raw materials for production of goods (fossil fuels, minerals, etc.) and the sinks to absorb the wastes (greenhouse gases, toxic chemicals, etc.) will be exhausted – unless there is a radical change in the throughput of the entire system. So an important question is 'How long will it be before such change becomes urgent?' This crucial issue is discussed in 12.6.

4.11 Ethical reasoning in formulating development policies

The facts presented in paragraphs 4.2–4.7 can hardly be disputed, but perhaps not surprisingly, we have seen that people advance different philosophical arguments, and give credence to different interpretations of the empirical facts in support of their ethical reasoning. However, this does not necessarily lead to different practical outcomes – and some of the differences might prove be more matters of emphasis than fundamental disagreement. A radical difference of opinion would be one that rejected all concerns for others in less fortunate circumstances, on the basis that it is none of our business. 'What has it got to do with me?' is a not uncommon (rhetorical) question from people who have already decided that the answer is 'nothing'.

How should one respond? According to biochemist N W Pirie, there are three types of valid response. First, people in DC should act out of **sympathy**, because if we had the ill-luck many others endure, *we* would be experiencing the stark injustices suffered by many in LDC. Second, we should act from a motive of **restitution**, since much of the material prosperity enjoyed by DC has been built on the exploitation of people in LDC in the days of Empire; many would claim that little has changed now, as powerful Western interests continue to dictate the terms of trade to their own advantage. But a third reason conforms to Hobbes' notion of **egoistic prudence** (1.8), because living in a world which is so unequal in wealth, health and education can be seen as a recipe for social unrest. Thus, it has been argued that the (unjustifiable) terrorism that now so dominates public consciousness might well have its roots in the (understandable) resentment of the *have-nots* in the face of the conspicuous privileges enjoyed by the *haves*. If so, there might be both deontological and utilitarian justifications for aiming for a more equal world.

A more formal way of examining the ethical case for action is to employ an ethical matrix (3.5) to explore how different policies would impact on the worlds' peoples, now and in future, and on the wider environment. For this purpose, it would be necessary to modify the version illustrated in Figure 3.1, e.g. by listing LDC and DC as separate interest groups. But however it was used, the matrix would be a reminder that our actions should be guided by *prima facie* respect for the principles of well-being, autonomy and fairness. And if 'respect for the well-being of DC societies' was considered to over-rule 'respect for the well-being of LDC societies', there would need to be some very powerful arguments to justify that view.

THE MAIN POINTS

- There are vast differences between the lives of people in the developed fifth of the world (DC) and the rest of the world's 6.5 billion people, most of whom live in LDC.
- Life expectancy, nutritional standards, health, wealth, education and technological opportunities are all far lower in LDC, and in many countries of sub-Saharan Africa the conditions are deteriorating.
- Some people claim that in such circumstances people in DC should not seek to remedy the disparity, both because attempts to help LDC are doomed to failure and because this could harm DC.
- Others argue that people in DC have an ethical duty to assist in LDC development, on the grounds of compassion, restitution and prudence.
- The potential role of the biosciences in development is critical, through their influence on policies for family planning, health, nutrition, education, agriculture and environmental management.

EXERCISES

These can form the basis of essays or group discussions:

1. Discuss the ethics of the *lifeboat situation* described by Garret Hardin (4.9).
2. Peter Singer argues that we (e.g. in DC) have a moral duty to help people in LDC if we can do so 'without sacrificing anything of comparable moral significance' (4.10.1). How far should that help go?
3. The Chinese government's birth control policies (4.10.3) are often criticized for paying too little attention to human rights. Use an ethical matrix (see 3.5) to compare these policies with those in the Indian state of Kerala (4.10.3).
4. Discuss the suggestion (4.10.7) that relieving the conditions of the *have-nots* will entail reduced consumption and pollution by the *haves.*
5. Which of Pirie's three arguments (4.11) for DC giving assistance to LDC (if any) do you consider the most persuasive? Give the reasons for your preference.

FURTHER READING

Textbooks on global development issues tend to get out of date quickly in terms of accurate documentation but the analyses provided often remain valuable. Among the more useful are:

- *Population and Food: global trends and future prospects* by Tim Dyson (1996). London, Routledge. A fairly optimistic analysis of the prospects for feeding the growing global population.
- *The World Food Problem* by David Grigg (2nd edition) (1993). Oxford, Blackwell. The book describes the complex interactions between natural resources, human needs, human skills and population pressures.
- *Beyond the Limits* by Donella Meadows, Dennis Meadows, and Jorgen Randers (1992). London, Earthscan. Based on computer program predictions, the authors argue that, while a sustainable society is technically feasible, if current trends continue we face virtually certain global collapse within the lifetime of children currently alive.

USEFUL WEBSITES

Up-to-date policy information and statistical data are available on numerous agency websites. For example:

- **http://www.undp.org/mdg/Poverty.pdf** *United Nations Development Program* (UNDP). Poverty and hunger: provides access to the annual Human Developments reports.
- **http://www.unfpa.org/swp/** *United Nations Population Fund* (UNFPA): a source of useful reports.
- **http://fao.org/** *United Nations Food and Agriculture Organization* (FAO): a source of useful reports.

NOTES

1. Thoreau H D (1910) Walden. Letchworth, Dent
2. Longmans Dictionary of the English Language (1984) Harlow, Longmans
3. World POPClock (2004) http://www.ibiblio.org/lunarbin/worldpop
4. Pop Clock (2004) Population counter. http://www.census.gov/ipc/www/world.html
5. Dyson T (1996) Population and Food. London, Routledge, p. 13
6. Ibid., p. 32
7. Coale A J and Watkins S C (Eds) (1986) The Decline of Fertility in Europe. Princeton, Princeton University Press
8. Dyson T (1996) Population and Food. London, Routledge, p. 32
9. United Nations Population Fund (UNFPA) (2003) http://www.unfpa.org/swp/2003
10. Dyson T (1996) Population and Food. London, Routledge, p. 32
11. Wattenberg B J (2004) Fewer: how the new demography of population will shape our future. Chicago, I R Dee
12. Reader J (2004) Cities. London, Heinemann
13. UN Human Settlements Programme (2003) The challenge of the slums. http://www.unhabitat.org/global-report.asp
14. The State of Food Insecurity in the World (2003) http://www.fao.org/docrep/006
15. International Food Policy Research Institute (2003) Global hunger predicted to persist through 2050. http://www.ifpri.org
16. United Nations Development Programme (1998) Human Development Report. New York, Oxford University Press, p. 50
17. Grigg D (1993) The World Food Problem (2nd edition). Oxford, Blackwell, p. 11
18. United Nations Development Programme (2004) Poverty and hunger [amended data 30 October 2003] http://www.undp.org/mdg/Poverty.pdf
19. Mepham T B (1987) The Physiology of Lactation. Milton Keynes, Open University Press
20. United Nations Childrens Fund (2004) http://www.unicef.org/nutrition/index_breastfeeding.html
21. Jellife D B (1972) Commerciogenic malnutrition? Time for a dialogue. Nutr Rev *30*, 199–202
22. Millstone E and Lang T (2003) The Atlas of Food. London, Earthscan, pp. 22–23
23. Pacey A and Payne P (1985) Agricultural Development and Nutrition. London, Hutchinson, p. 82
24. United Nations Development Programme (1998) Human Development Report. New York, Oxford University Press, p. 53
25. United Nations Food and Agriculture Organization (2002) Progress in reducing hunger has virtually halted. http://fao.org/newsroom/news
26. Millstone E and Lang T (2003) The Atlas of Food. London, Earthscan, pp. 24–25
27. United Nations Development Programme (1998) Human Development Report. New York, Oxford University Press, p. 51
28. United Nations Development Programme (2004) Human Development Report. http://hdr.undp.org/overview
29. British Medical Association (2004) Smoking and Reproductive Life. London, BMA
30. Mepham D (2003) Clean Water, Safe Sanitation. London, Institute for Public Policy Research
31. United Nations Development Programme (2003) Human Development Report. http://hdr.undp.org/overview, p. 5

32. United Nations Development Programme (1998) Human Development Report. New York, Oxford University Press, p. 2
33. Ibid., p. 4
34. Ibid., p. 30
35. Veblen T (1994) [1899] The Theory of the Leisure Class. New York, Dover
36. United Nations Development Program (1998) Human Development Report. New York, Oxford University Press, p. 47
37. United Nations Development Program (2003) Human Development Report. http://hdr.undp.org/overview, p. 6
38. United Nations Development Program (1998) Human Development Report. New York, Oxford University Press, pp. 154–155
39. Ibid., p. 63
40. United Nations Development Program (2001) Human Development Report. http://hdr.undp.org/overview. P. 96
41. Kinoti G K (2000) In: Livestock, Ethics and the Quality of Life. (Eds) Hodges J and Han I K. Wallingford, CAB International, pp. 221–242
42. Commission on Intellectual Property Rights (2002) Integrating Intellectual Property Rights and Development Policy. London, CIPR, p. 2
43. Ibid., p. 1
44. Ibid., p. 9
45. Shiva V (2004) The neem tree: a case study of biopiracy. http: www.twnside.org.sg/title/pir-ch.htm
46. Dower N (1996) Global hunger: moral dilemmas. In: Food Ethics. (Ed.) Mepham B. London. Routledge, pp. 1–17
47. Malthus T R (1910) [1798] An Essay on the Principle of Population as it Affects the Improvement of Society. London, J Johnson
48. Arnold D (1988) Famine: social crisis and historical change. Oxford, Blackwell
49. Brown L R (1976) In the Human Interest. Oxford, Pergamon
50. Ehrlich P R and Ehrlich A H (1990) The Population Explosion. London, Hutchinson
51. Meadows D H, Meadows D L, and Randers J (1992) Beyond the Limits. London, Earthscan
52. Smith A (1925) [1776] An Enquiry into the Nature and Causes of the Wealth of Nations. London, Routledge
53. Simon J (1992) Population and Development in Poor Countries. Princeton, Princeton University Press
54. Boserup E (1990) Economic and Demographic Relationships in Development. Baltimore, Johns Hopkins University Press
55. Atkins P and Bowler I (2001) Food in Society: economy, geography, culture. New York, Arnold, p. 110
56. Thurow L C (1971) The Zero-Sum Society. New York, Basic Books
57. King M (1990) Health is a sustainable state. Lancet *336*, 664–667
58. Hardin G (1993) Living within Limits. New York, Oxford University Press
59. Lomborg B (2001) The Skeptical Environmentalist. Cambridge, Cambridge University Press
60. Singer P (1979) Practical Ethics. Cambridge, Cambridge University Press, p. 108
61. Nagel T (1977) In: Food Policy. (Eds) Brown P G and Shue H. New York, Free Press, pp. 54–62
62. UN Standing Committee on Nutrition (2004) Nutrition for improved outcomes. Geneva, SCN, p. iii

63. Commission for Africa (2005) Our common interest. http://213.225.140.43/english/report/introduction.html
64. Nagel T (1977) In: Food Policy. (Eds) Brown P G and Shue H. New York, Free Press, pp. 54–62
65. China's one child policy (2002) http://axe.acadiau.ca/
66. Amnesty International (2003) www.amnesty.org
67. National Commission on Population (2004) http://populationcommission.nic.in/cont-en.htm
68. United Nations Development Programme (2002) Annual Report
69. The United Nations Population Fund (2004) http://www.unfpa.org/about/index.htm
70. Department for International Development (2004) http://www.dfid.gov.uk/
71. World Bank (2004) http://web.worldbank.org/
72. Segal J (1991) In: Ethics and Agriculture. (Ed.) Blatz C V. Moscow, University of Idaho Press, pp. 600–605
73. International Monetary Fund (2002) Globalization: threat or opportunity? http://www.imf.org/external/np/exr/
74. Sen A (1999) Development as Freedom. Oxford, Oxford University Press
75. United Nations Development Programme (2004) http://hdr.undp.org/overview Human Development Report
76. Marsh J S (1996) Food Aid and Trade. In: Food Ethics. (Ed.) Mepham B. London, Routledge, pp. 18–34
77. Hoogendijk W (1991) The Economic Revolution. Utrecht, Green Print

5

A time to be born?

OBJECTIVES

When you have read and discussed this chapter you should:

- appreciate the advantages and limitations of the range of birth control techniques available, and the ways in which birth rates are affected by social pressures
- understand the applications of various existing and prospective assisted reproductive technologies (ART) and the ethical questions they raise
- be aware of legal controls on the use of ART
- be able to perform a simple ethical analysis of the use of specific ART, using the ethical matrix
- be aware of the political background to the legalisation of embryo research

5.1 Introduction

Ethics is nothing if not about choices, and choosing whether to bring a new person into the world, and what sort of person, are among the most important choices people can make. In many ways, the choices have become more complicated in recent years because our options, and our ability to influence events, have grown much greater. But whatever new technologies are, or will soon be, at our disposal, currently the children we can have are still largely a consequence of the sexual partners we choose (determining the genetic heritage of our children – their so-called *nature*) and the way they are brought up (their *nurture*). The relative significance of nature and nurture is unclear; if the word 'voluntary' means anything (2.1.2), some aspects of our behaviour must ultimately elude analysis in terms of the two factors, or the interactions between them.

This chapter considers the ethical implications of reproductive choices, particularly as they are influenced by modern biotechnologies. But we begin with a consideration of methods used to *prevent* birth, usually referred to as *birth control* because they also allow parents to plan the number and spacing of their children.

Table 5.1 Birth-control methods

Technology	Reliability	Risks/problems	Comments/examples
Steroidal Ovulation Inhibitors			
Steroidal ovulation inhibitors: *the pill*	+ +	Cardiovascular diseases; hypertension; cancer	Many formulations: e.g. *combination pill* and *minipill*
Subdermal implants	+ + + +	Difficult; painful removal. Expensive	*Norplant*: silastic capsules, 5-year protection
Injectables	+ + + +	Not reversible	*Depo-Provera*, 3-month–2-year protection
Inhibitors of Fertilization			
Male condom (m)	+ +	Unreliable	Cheap, accessible. X–D
Female condom	+	Uncomfortable; expensive	
Diaphragm	+ +	Cumbersome	Needs spermicide application
Cervical cap	+ +	Uncomfortable; infections	Needs spermicide application
Spermicides	+	Unreliable; allergic reactions	Foams, creams, gels. X–D
Interceptives and Abortifacients			
Intrauterine devices (IUD)	+ + +	Pelvic inflammatory disease	*Progestacert*
Postcoital contraception	+	Vaginal bleeding; nausea	Morning-after pill: *Mifepristone, Methotrexate*
Menstrual extraction (pre-emptive abortion etc.)	?	Psychological trauma	Vacuum aspiration of uterine lining
Surgical Sterilization			
Vasectomy (m)	+ + + +	Generally irreversible	
Tubal ligation	+ + + +	Generally irreversible	

Notes: Apart from those marked (m), all technologies are applied to women. X–D indicates methods which also provide effective protection against sexually transmitted diseases, such as HIV/AIDS. Typical chances of pregnancy: + + + +, <0.4%; + + +, <5.0%; + +, <20%; +, >20%. Compiled from information in Knight (1998).[1]

5.2 Birth control

During the twentieth century, a multiplicity of birth-control techniques was devised, which are summarized in Table 5.1. WHO defines the ideal **contraceptive** (an agent preventing fertilization) as one that is: *'highly effective, easy to apply, readily reversible, inexpensive, easily distributed, has no serious side effects or risks, and is acceptable in the light of the religious, ethical, and cultural background of the user'*.[2] But it is clear that no existing technique fulfils all these criteria. Some are associated with significant health risks, and some are unreliable, inconvenient or expensive. And those who draw a clear distinction between contraception and birth control (preventing the development of an embryo) face a significant additional problem, because *breakthrough ovulation* can occur in women taking an oral contraceptive. So although the contraceptive may successfully prevent pregnancy, this might be because abortion is induced (the contraceptive acting as an **abortifacient)** rather than ovulation inhibited. Moreover, in the current climate of sexual freedom in Western societies, a drawback of most contraceptives is that they provide no effective protection against sexually transmitted diseases.

In essence, birth-control practices seek to dissociate **sexuality** from **fertility**. Contraceptive technologies provide several ways of doing this, but given their limitations it is perhaps not surprising that many people fail to use them. The *acceptability* of contraceptives depends on factors such as: their financial cost and convenience of use; cultural and religious norms; the educational level of potential users; and perceptions of the risks and benefits entailed.

5.2.1 Abortion

As noted (5.2), some contraceptives may lead to abortion of the early embryo. But intentional abortion of the fetus is also a common form of birth control. **Abortion** is defined as the termination of pregnancy at any point between conception and birth, resulting in the death of the fetus (a term applied to all stages of development *in utero*).[3] Technically, abortions can be induced by procedures such as suction evacuation, prostaglandin administration, or use of abortifacients, which prevent implantation of the fertilized egg. Under the UK Abortion Act (1967), abortion was permitted up to 28 weeks of gestation, a time when the fetus was thought capable of being born alive. But an amendment in the Human Fertilisation and Embryology Act (1990) lowered the age to 24 weeks because medical advances have enabled the survival of some children born at that age, and extended the opportunities for aborting a fetus likely to develop into a disabled child. Figure 5.1 illustrates the appearance of the fetus at different stages of its early development.

5.2.2 Social and cultural factors

Major religions have advocated large families (**pronatalism**) and condemned artificial means of contraception. For example, the Christian Book of Genesis urges adherents to *'Be fruitful and multiply, and replenish the Earth'*, although it is with reference to natural law (2.3) that the Roman Catholic Church prohibits all forms of artificial contraception (and especially abortion), and allows only 'natural' forms of contraception that limit intercourse to the woman's safe period. Although about 1 billion people belong to the Catholic Church (mostly in LDC), **birth rates** (BR) in several Catholic countries have declined significantly, so that it seems many people ignore their Church's teaching.

BR are also strongly influenced by legal, political, and economic factors. For example, a low minimum age of marriage (as low as 12 or 13 years old for girls in many countries) may increase BR, while government policies on child benefits and female education

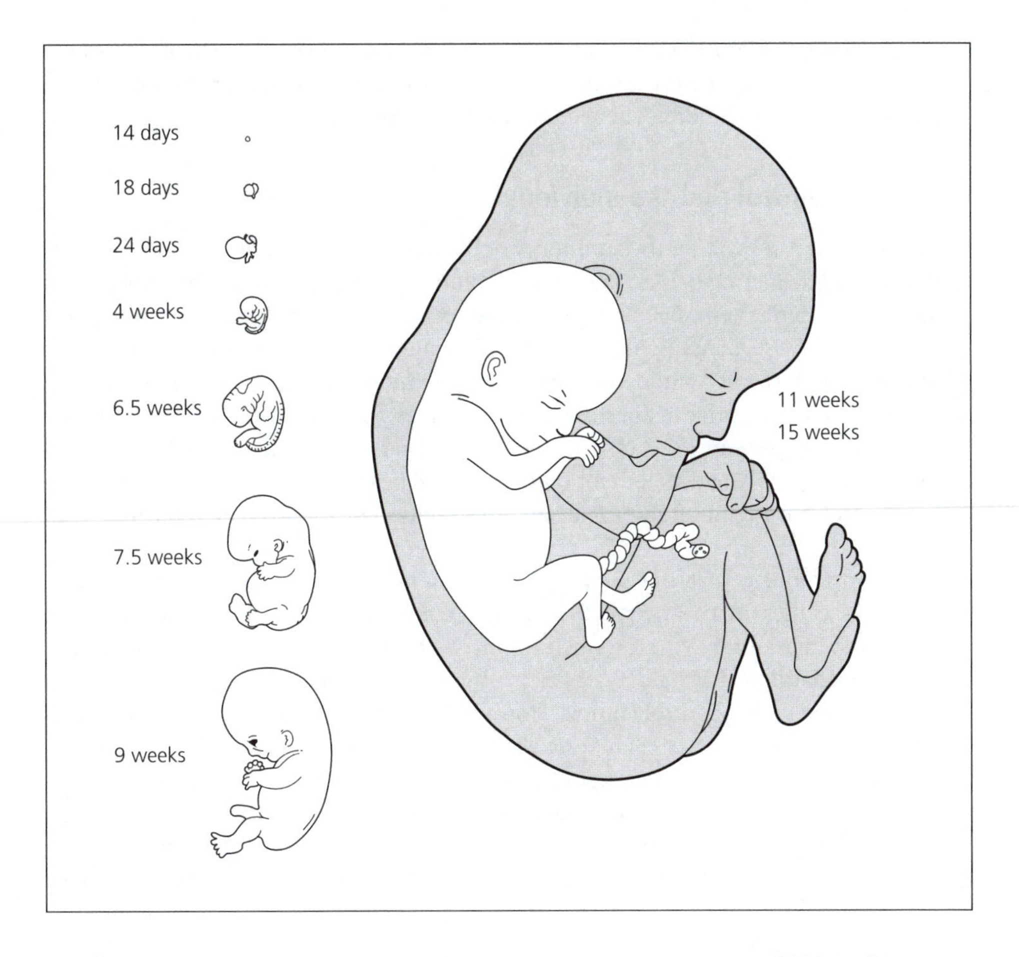

Figure 5.1 Early stages of human fetal development. Diagrams are about 75% actual (life) size. [From: Arey L B (1965) Developmental Anatomy. London and Philadelphia, W B Saunders]

may, depending on their level, increase or decrease BR. In LDC, children often add to the family workforce and provide security for parents in their old age – which offer strong reasons for pronatalism. But economic development generally seems to counter this effect. For example, BR tend to fall when:

- women receive formal education
- women increasingly enter waged employment
- state welfare provisions are increased
- (potential) parents invest more time and money in material possessions, personal advancement and leisure activities.

Moreover, human **creativity** (as expressed in art, music, science, and culture in general) may well be linked psychologically to **procreativity**, so that procreation might be viewed as but one of the channels through which sexuality can be expressed.

Ethical issues are most evident at two levels: the level of the individual, especially women; and in relation to social policy, when governments consider the health and stability of their populations. These are discussed in turn below.

5.2.3 Birth control and the individual

A critical factor is respect for the autonomy of the women who bear the responsibility of gestating and delivering babies, and whose voluntary participation in this process is a basic human right. Responsibility for contraception also falls disproportionately on women (Table 5.1). Consequently, few people would deny that women have a right to prevent the birth of a child conceived unwillingly (most offensively as a result of rape) by use of interceptive or abortifacient techniques, or by surgical abortion. Whether this right also applies in cases of negligence, by either party, when engaging in intercourse, is more problematical. But for many, it would be regarded as the *child's right* to be 'wanted', and this might often provide an ethical rationale for post-coital birth-control measures.

It is worth noting that the sexual freedom many people now enjoy is relatively recent, and earlier in the twentieth century, contraception, and indeed all sexual activity that did not have reproductive *intentions*, was considered obscene. Marie Stopes in England and Margaret Sanger in the USA were vilified for their pioneering attempts to open birth-control clinics.[4] However, recently attitudes are changing again, so that in the USA there has been *'a steady shift away from the short-lived "sexual revolution" to a more and more strident opposition to sex education'*.[5] People holding more traditional attitudes, who believe that the embryo attains ethical standing at the time of conception, might consider that the doubts over whether oral contraceptives act as inhibitors of ovulation or as abortifacients (5.2) are grounds for rejecting their use. This may partly account for the fact that in the USA female sterilization is the commonest means of birth control.[6]

Much of the ethical debate over induced (elective) abortion focuses on the *ethical standing of the fetus* (3.4). The extreme *conservative* view (although this is not necessarily one that is held by political Conservatives, or *vice versa*), considers that the fetus has the same ethical standing as an adult, and so should be treated morally according to the same principles: it follows that it has a right not to be harmed. This is the view promoted, perhaps most prominently, by the Roman Catholic Church. The opposite, extreme *liberal* view (with the same caveats as above) considers that although the fetus is in some sense human, it has so few human attributes (and at its earlier stages, is more akin to a mass of tissue) that not only is abortion ethically permissible but, according to utilitarian reasoning, the fetus may be used instrumentally (3.6.3) for the benefit of humans if the possibility arises, e.g. in experiments (5.7) or medical therapies. An intermediate position considers that the ethical standing of the fetus increases as it develops, so that the decision as to how it is morally permissible to treat it depends largely on the emergence of some distinctive character. Normally, this relates to its ability to experience sensations, and in particular to feel pain.

5.2.4 Political influences

In the USA, abortion was legalized in 1973; and currently more than 50% of all pregnancies are unplanned, of which more than half are terminated by elective abortion.[7] But recently there has been a strengthening of the anti-abortion lobby, and an intensifying backlash (sometimes involving violence, and even murder) associated with the growth of a hard-line political movement. Those campaigning in favour of abortion are often called the **pro-choice** movement and those opposed the **pro-life** movement. Although similarly polarized groups also exist in the UK, the debate is far less heated.

Globally, government population policies have been devised with different aims. States have often adopted pronatalist (5.2.2) measures as components of supremacist doctrines, e.g. the Italian Fascists in the 1920s and the National Socialists (Nazis) in Germany in the 1930s. Other states have introduced **antinatalist** policies to curb population growth in the interests of improving, or maintaining, standards of living – on the assumption that 'more lives mean more misery'. But this presents something of a dilemma for utilitarians who adopt the so-called **hedonistic total principle** (see 2.5). This states that, other things being equal, we should attempt to maximize the total sum of happiness.

For example, imagine two scenarios, one ('A' in Figure 5.2) in which a small number of people in a society have, on average, a pleasant and fulfilling standard of living, and another (B) in which, because of a higher BR, there are more people experiencing, on average, a lower standard of living. However, despite the differences in average well-being, the total sum of 'happiness' is greater in B than in A. As philosopher Derek Parfit

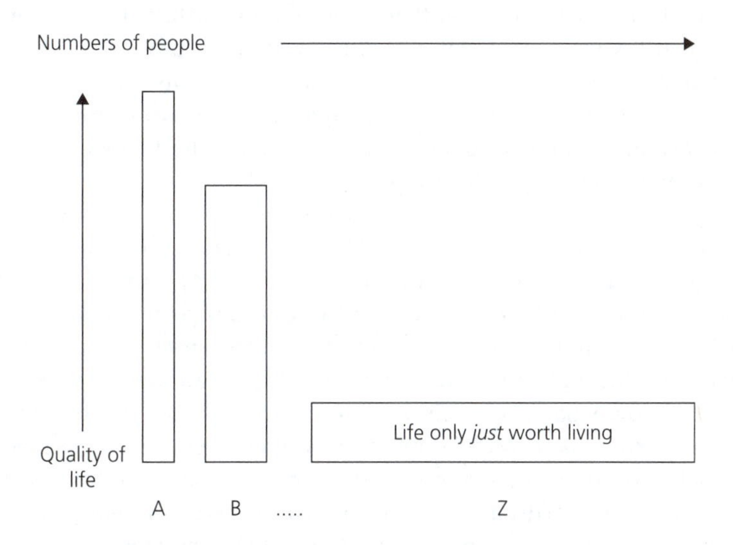

Figure 5.2 The consequences of rigorously applying the *hedonistic total principle*. The total sum of 'human happiness' of a large number of people represented by Z could exceed that of smaller numbers represented by A and B, even though the lives of people in Z were 'only just worth living'. Derek Parfit calls this the *repugnant conclusion*.

points out, this line of reasoning could, logically, produce what might seem a 'repugnant conclusion' – that a very large number of people (Z) might, despite living lives of such misery that they only just exceeded the level at which life would *not* be worth living, still experience a total sum of happiness greater than smaller numbers of 'happy people' in situation A.[8] This is, of course, a thought experiment. But it forces us to ask 'how many people should there be?' and that entails consideration of what makes life worth living, and how Parfit's 'repugnant conclusion' can be avoided. (Indeed, there must be real concern that current trends are moving towards a situation in which 'A' might apply to DC and Z to certain parts of LDC.)

Different population control policies were discussed in 4.10.3. Often, as with the Chinese government, both incentives and disincentives are used to encourage reduced BR. Some people argue that the Chinese approach offends human rights, but it has been claimed that despite its apparent harshness it is at least egalitarian – the policy is applied fairly across society. Sometimes, governments have resorted to other problematical incentives, such as India's policy of 1975–76, when transistor radios were offered to men willing to consent to vasectomy on the spot. This, at the time attractive, inducement denied the man the possibility of careful reflection or discussion with his wife.[9] Clearly, in all such cases, utilitarian justifications need to be balanced (and questioned) against impacts on human rights.

5.3 Infertility

Contraception is, of course, only one side of the birth-control coin, because for many people the problem they face is not that of preventing birth, but a condition of sub-fertility or infertility which makes it very difficult, if not impossible, for them to have children naturally. From one perspective, infertility is a disease, and so is strictly speaking a medical issue. But not everyone agrees. So it is worth considering 'What counts as a disease?', because the question has important ethical implications for several reproductive and genetic technologies to be examined.

At what point, it might be asked, is it reasonable for any of us to question the hand that we have been dealt in life's health lottery and seek to improve matters? We can easily recognize the contrasting states of people at the extreme ends of the scale, such as the Olympic athlete and the accident victim existing in a *permanent vegetative state*. But because there is no sharp dividing line in the spectrum of conditions, the definitions of health and disease must be, to a degree, arbitrary. Moreover, as it is impossible to know what it is really like to be someone else, we can never define objectively what states of 'dis-ease' other people suffer. Even so, most people's perceptions of health and disease are almost certain to be socially conditioned to a large degree; and when the medical conditions are considered sufficiently severe (microbiologically or mentally), governments enforce legally binding restrictions on people's liberty.

Much, then, depends on what society (or, more realistically, the experts to whom politicians defer) considers acceptable. But there are dangers in relying on experts who have a vested interest in promoting a particular view. For example, from the best of motives, some doctors might consider that pre-emptive medical treatment is the best means of maintaining patients' health. Or, less worthily, some doctors might be tempted to try to increase their status and incomes by declaring people 'ill' who had previously believed themselves healthy.

The effects of our changed attitudes to health in modern industrialized society have been profound – resulting in a process of **medicalization**, which philosopher Ivan Illich characterized as an *'internalised colonisation of liberty by affluence'*.[10] That concise definition might be unpacked by saying that, as people become better off, they increasingly purchase medical 'improvements' to their health, to levels above those previously thought quite acceptable. The classic example is cosmetic surgery, whereby people who are in no sense ill employ doctors to beautify their faces and bodies. (Of course, in some cases, physical disfigurement is so severe as to *require* surgery.) A related phenomenon is **commodification**, which is the transformation of what was previously free, or at least commonplace, into a commercial product. A prime example is baby milk, where mothers' breasts, which for millennia have served amply to feed their infants, have now for many babies been supplanted by a modified form of cows' milk (*infant formula*) delivered from a feeding bottle.[11]

5.3.1 Medicalization of birth

In view of the above discussion, it might be asked whether infertility should be classed as a disease. For some people, it is *not* a disease because the prospective parents are not ill in the usual sense of the word – and perhaps rather less disabled than many, such as those who are blind or deaf but cope satisfactorily without medical intervention. Or it might be argued that infertility is really a *socially constructed* disease, a consequence of social pressures that encourage the belief that childbearing and motherhood are necessary aspects of being a *real* woman. This is sometimes associated with the view that we need to recognize that we all have limitations, many of which we can learn to accept with dignity, especially when medical expertise is needed for far more pressing concerns.

Such claims have, however, been challenged strongly, e.g. by the leading fertility expert Robert Winston, who argues that *'infertility is actually a terrible disease affecting our sexuality and our well-being'*.[12] This finds support in a form of natural law claim (2.3), namely, that the desire for children is a fundamental, biologically determined, aspect of human nature. Consequently, it is often argued, people have a *right* to reproduce – irrespective of whether they are biologically equipped to do so. And if that is so, it would seem to follow that others (usually in the form of governments) have a duty to provide the means whereby this right can be asserted.

The above debate provides sufficient reason for addressing these issues in a book not specifically concerned with medical questions. But further justification is surely provided by the fact that some of the reproductive technologies to be discussed (chapter 6) aim to avoid certain illnesses or to *enhance* certain characteristics in the children to be born, and thus revive the ethical concerns raised about eugenics in earlier times (1.4.3).

5.4 An outline of reproductive technologies

Many of the procedures to be discussed are so-called 'new' **assisted reproductive technologies (ART)**, but they are often combined with practices that have been in use for much longer, such as the use of donor sperm in **artificial insemination** (AI), and the employment of **surrogate mothers** to carry babies during pregnancy.

5.4.1 Artificial insemination

Several ART now used by people were first employed in farm animals. For example, AI with frozen sperm, which was first routinely used in the breeding of dairy cattle, has been available for infertile couples for fifty years. It has allowed men to bank their sperm before undergoing chemo- or radiotherapy (processes that might adversely affect gametogenesis), or going into dangerous military operations. It also allows some couples to have children when sexual intercourse is physically impossible, or for women to be inseminated with a donor's sperm (DI).

In a notable case, Diane Blood was inseminated with sperm removed from her husband when in a coma before he died, although, because he left no written consent, she went to Belgium for treatment because DI was illegal in such circumstances in the UK.[13] AI is accomplished using a fine-bore tube to deposit sperm in the neck of the woman's uterus.

5.4.2 *In vitro* fertilization

However, use of ART in humans is often considered to date from 1978, with the birth of Louise Brown, who became known as the first **test-tube baby**. The technique of ***in vitro* fertilization** (IVF) was pioneered in England by Robert Edwards and Patrick Steptoe, as a means of enabling couples who were infertile or subfertile to have children.[14]

Since then, several developments have modified the practice of IVF and other ART. For example, endoscopic surgery, video endoscopy, and sophisticated ultrasound devices have, collectively, allowed less invasive ways of monitoring egg development and the receptivity of endometrium. Developments in robotics and precision-controlled tools for gamete and embryo micromanipulation, together with the production of recombinant forms of hormones and of monoclonal antibodies, have facilitated manipulation of the various steps that culminate in embryo implantation in the uterine wall (Box 5.1).[15] In the UK in 2003, an average cycle of IVF cost £1800; and only 10% of women are treated on the NHS. However, despite these technological developments, the success rate of IVF is not high. In the UK, the latest available figures (2001) show that the overall percentage of live births per treatment cycle was 21.8% (and 25.1% for women less than 38 years old).[16]

Problems can arise at a number of steps. Apart from the inability of the ovaries to produce follicles or the inaccessibility of the follicles, eggs may fail to become fertilized, or, if so, the embryos may fail to develop. But the commonest cause of IVF failure is the inability of the embryo to implant in the uterine wall. More seriously, the welfare of the woman and her children can be adversely affected. For example, the use of drugs to

BOX 5.1 *IN VITRO* **FERTILIZATION (IVF) PROCEDURES**

IVF entails the following steps:

- Treating the woman with hormones to stimulate superovulation in the next cycle (since success rates are increased by the subsequent transfer of more than one embryo).
- Careful monitoring of the woman's hormone levels to detect the precise moment of egg ripening.
- Egg collection, either by ultrasound guidance or laparoscopy. In the former method, with the woman sedated, a fine needle is passed through her bladder or vagina, and following ultrasound visualization of each ovarian follicle the eggs are sucked into tubes containing culture medium. Laparoscopy is a more invasive process and requires general anaesthesia.
- Sperm is provided by the man, on the same day as egg collection, by masturbation: a previously collected sample will have been used to check the viability of the sperm. Alternatively, frozen sperm can be used.
- Eggs, graded according to their state of maturity, are incubated with fresh culture medium. To achieve insemination, sperm are added to the medium and the Petri dish returned to the incubator until the following day.
- Eggs are inspected 12 to 18 hours after insemination, and if fertilization has not occurred re-insemination may be attempted.
- Transfer of the embryos may occur immediately, or after a further 24 hours incubation (when the first cleavage of the embryos will have occurred). In the latter case, embryos are graded according to quality, and the best two (or, exceptionally, three) selected for transfer into the woman, while others may be frozen for possible future use, or used in research.
- Embryo transfer involves flushing the embryos, suspended in incubation medium in a fine flexible catheter, into the uterus via the cervix. The process is painless, but sometimes additional hormone treatment is given to improve the chances of implantation.
- Pregnancy testing can be applied 14 days after embryo transfer.

cause superovulation sometimes results in a potentially life-threatening condition called **ovarian hyperstimulation syndrome**; while the use of several eggs increases the changes of multiple births, and the consequent risks of delivering low-birth-weight babies, who suffer higher incidences of health complications.[17]

5.4.3 Developments in the *in vitro* fertilization procedure

The basic IVF procedure is now subject to a number of variations. In brief, these are:

- Gamete intrafallopian transfer (GIFT), in which sperm and eggs are mixed together and transferred to one or both of the woman's fallopian tubes, where fertilization occurs
- Zygote intrafallopian transfer (ZIFT), in which a putative zygote is transferred into a fallopian tube, following a period of incubation of the eggs with sperm
- Various micromanipulation techniques are employed to overcome the barriers to incorporation of seminal DNA into the oocyte, when the sperm has poor motility or morphology. Examples are: partial zonal drilling (PZD), subzonal insemination (SUZI), and intracytoplasmic sperm injection (ICSI).[18]

5.4.4 The perspective of compassion

Of course, matters of birth, health and death affect people deeply. And what might seem to be merely academic discussions about whether infertility is a disease or not tend to recede into the background when faced with real-life problems and their demonstrable medical solutions. Undoubtedly, many people's lives have been greatly enriched by the use of ART – although some have had their hopes raised to no avail. The development of ART can thus be seen as an act motivated by compassion. Perhaps what matters in future is the *ethical management* of ART, and setting appropriate rules for how prospective new ART should be regulated, or possibly banned. This has been a job for governments (5.5).

5.5 The Human Fertilisation and Embryology Authority

As noted ironically by Ed Yoxen, ways have now been found of loosening the association between those two old partners, sex and reproduction, which had been inextricably linked together for aeons. *'Contraception permits sex without reproduction; new reproductive technologies...permit reproduction without sex.'*[19] But the introduction of ART has also signalled the abandonment of the old socially approved restrictions that have traditionally governed who is allowed to beget children, and with whom. Such developments raise some important ethical concerns.

In the UK, regulations controlling ART were introduced by the Human Fertilisation and Embryology (HFE) Act (1990), following the publication in 1984 of the report of an enquiry chaired by philosopher Mary Warnock.[20] In 1991, the Human Fertilisation and Embryology Authority (HFEA) was set up, implementing recommendations of the report. It aims to ensure *'that all UK treatment clinics offering IVF or donor insemination, or storing eggs, sperm, or embryos, meet high medical and professional standards and are inspected regularly'*. The HFEA, one of the few statutory bodies of its kind in the world, also licenses and monitors all human embryo research (5.7). Policy and licensing decisions are taken by the Authority's 21 members, who are appointed on the basis of their personal knowledge and expertise. To ensure a balance of viewpoints, more than half of the membership must come from disciplines other than medicine or human embryo research.[21]

Infertility and subfertility can result from a number of causes. For example, in some cases the man produces poor-quality sperm or has a low sperm count, or the woman may suffer from damaged fallopian tubes or fail to ovulate. Complete infertility, a rare condition, might be due to a premature menopause or azoospermia (absence of sperm in the semen). Which particular ART is used obviously depends on the nature of the problems. Table 5.2 indicates the range of techniques that can be applied; in the following discussion, coordinates are cited to identify the different people and techniques involved.

5.5.1 Third, fourth and fifth parties

If the standard techniques described prove inadequate for a couple to conceive a child, they may resort to a number of strategies using gametes provided by others. In the UK, the HFE Act licenses sperm donors – a misnomer, because the men are paid up to £15 plus expenses for their services. DI (5.4.1) involves the use of sperm from an anonymous donor (Table 5.2: A5:D1), who is aged 18–55 years and has undergone screening for conditions such as HIV and hepatitis, and an attempt is made to match physical characteristics such as body build and eye colour with the *legal prospective father* ('husband' in Table 5.2).[22] In some clinics, this service is also available to single women and lesbian couples. By contrast, in the USA a fully commercial system operates, and 'donors' advertise through agencies and on the internet.[23]

The other main technique used is IVF (5.4.2) in which the *legal prospective mother* ('wife' in Table 5.2) incubates an embryo(s) resulting from fertilization, with the husband's sperm, of eggs supplied by a donor (Table 5.2: B6:E2:A8), who, under the HFE Act, is not allowed payment in the UK.[24] (However, in the USA there is a developed

Table 5.2 Summary of procedures in begetting, gestating and caring for a child

Interest Group	Sperm 1	Egg 2	Gestate 3	Care 4	AI 5	IVF 6	Screen 7	Implant 8	Dispose 9
A Wife									
B Husband									
C Embryo/fetus/ child									
D Donor m									
E Donor f									
F Surrogate f									
G (m/f) Adopter I									
H (m/f) Adopter II									

Notes: The terms 'husband' and 'wife' are used here to cover recognized, socially accepted relationships, even when not legally certified. The lightly shaded area indicates the traditional way in which children are conceived, born, and cared for, whereas the darker shaded areas indicate alternative procedures that have been employed. The shaded cells identified (lighter or darker shaded) show the direct involvement of the interest groups specified, e.g. G and H allow for adopters to be heterosexual or homosexual couples, or a single person. In the text, issues are identified by the coordinates. For example, the claimed right of a child to know the identity of its genetic father, when born as a result of the mother's insemination with donor sperm, is identified by C:D1; the putative parents' decision over whether to destroy or implant an embryo produced by IVF, following results of pre-implantation genetic diagnosis (PGD) are identified by A,B:C7–C9. Many permutations are possible.

AI, artificial insemination; IVF, *in vitro* fertilization; m, male; f, female. Screen refers to all forms of prenatal testing, e.g. including PGD (see 6.3.3) and analysis of samples from amniocentesis or chorionic villus sampling (see 6.1). Dispose refers to disposal of embryos following storage and/or embryo experimentation, or abortion of fetuses.

market in human eggs, with women charging *'$3000–6000 for a clutch of ten eggs'.*[25]) In the UK, egg donors must be 18–35 years old, have been offered full counselling, and be screened for infections such as hepatitis and HIV. Many donors are women undergoing sterilization or who have surplus eggs after their own fertility treatment; but some are friends, relatives, or others acting out of altruism (**non-patient volunteers**). As for sperm donors, detailed records are kept, but until 2004, these details remained confidential. However, a change in the law will now allow children to know the identity of their biological parents when they reach the age of 18 years, and it is speculated that this may reduce the number of people willing to donate sperm or eggs.

A third possibility arises when the wife is unable or unwilling to carry the fetus during pregnancy. If she and her husband produce an embryo through IVF, it can be transferred to a **surrogate mother**, who will carry the baby to term (Table 5.2: A,B6:F3). In a notable case in 2004, a woman gave birth to twins who were her own grandchildren, and who had been conceived by IVF.[26] Or if donors provide sperm or eggs other outcomes result (such as D6,E6:F3; or D6,A6:F3; or E6,B6:F3 in Table 5.2). Under the Surrogacy Arrangements Act (1985) in the UK, surrogacy services are permitted, but profits are prohibited: only justifiable expenses for the pregnancy are allowed. Although surrogate mothers receive no fee, in the UK it is estimated that an IVF surrogacy costs about £10,000, while in the USA the figure is about US$50,000.[27]

It is clear that some highly complex relationships can arise, which result in rather complicated definitions, especially when IVF is performed with both sperm and eggs provided by donors and gestation is carried out by a surrogate mother. For example, Box 5.2, taken from the HFE Act (1990, now amended), defines the term *father* with respect to these ART.[28] On occasion, accidental mix-ups occur, and the wrong sperm is used to fertilize eggs; which becomes quite obvious when people of different skin colour are involved.

A helpful scheme is to identify: the **intentional parents**, who wish to have a child; the **genetic parents**, who provide the gametes; the **gestational mother**, who carries and gives birth to the baby; and the **nurturing parents**, who raise and care for the child.[29] Traditionally, the same woman performed all four roles: now, in some cases, four different women might be involved (and three, or more, different men).

5.5.2 Less common and proposed arrangements

Since ART were first introduced, and legal provisions permitted novel genealogical relationships, the traditional boundaries of acceptable practice have been subject to continuous challenge. The following are notable examples.

Older mothers. The upper age limit for fertility has until recently been set by the menopause, occurring around about 50 years in most women. But ART can subvert this age constraint by several strategies. Insemination of frozen superovulated eggs allows pregnancy to occur up to age 60 years or more, with the eggs provided by the woman herself or by a donor. Concerns have been expressed that such older mothers may unfairly deprive the children of the normal expectations of a parent able to fulfil the physical and emotional demands which accompany rearing a child. But the same might also be said to apply to elderly fathers, who generally remain fertile to a much later age.

BOX 5.2 MEANING OF 'FATHER'

(Section 28 the HFEA Act, 1990)

(1) This section applies in the case of a child who is being or has been carried by a woman as the result of the placing in her of an embryo or of sperm and eggs or her artificial insemination.

(2) If –
 (a) at the time of the placing in her of the embryo or the sperm and eggs or of her insemination, the woman was a party to a marriage, and
 (b) the creation of the embryo carried by her was not brought about with the sperm of the other party to the marriage,then, subject to subsection (5) below, the other party to the marriage shall be treated as the father of the child unless it is shown that he did not consent to the placing in her of the embryo or the sperm and eggs or to her insemination (as the case may be).

(3) If no man is treated, by virtue of subsection (2) above, as the father of the child but –
 (a) the embryo or the sperm and eggs were placed in the woman, or she was artificially inseminated, in the course of treatment services provided for her and a man together by a person to whom a licence applies, and
 (b) the creation of the embryo carried by her was not brought about with the sperm of that man,then, subject to subsection (5) below, that man shall be treated as the father of the child.

(4) Where a person is treated as the father of the child by virtue of subsection (2) or (3) above, no other person is to be treated as the father of the child.

(5) Subsections (2) and (3) above do not apply –
 (a) in relation to England and Wales and Northern Ireland, to any child who, by virtue of the rules of common law, is treated as the legitimate child of the parties to a marriage,
 (b) in relation to Scotland, to any child who, by virtue of any enactment or other rule of law, is treated as the child of the parties to a marriage, or
 (c) to any child to the extent that the child is treated by virtue of adoption as not being the child of any person other than the adopter or adopters.

(6) Where –
 (a) the sperm of a man who had given such consent as is required by paragraph 5 of Schedule 3 to this Act was used for a purpose for which such consent was required, or
 (b) the sperm of a man, or any embryo the creation of which was brought about with his sperm, was used after his death, he is not to be treated as the father of the child.*

(7) The references in subsection (2) above to the parties to a marriage at the time there referred to –
 (a) are to the parties to a marriage subsisting at that time, unless a judicial separation was then in force, but
 (b) include the parties to a void marriage if either or both of them reasonably believed at that time that the marriage was valid; and for the purposes of this subsection it shall be presumed, unless the contrary is shown, that one of them reasonably believed at that time that the marriage was valid.

(8) This section applies whether the woman was in the United Kingdom or elsewhere at the time of the placing in her of the embryo or the sperm and eggs or her artificial insemination.

(9) In subsection (7)(a) above, 'judicial separation' includes a legal separation obtained in a country outside the British Islands and recognised in the United Kingdom.

* Following the Diane Blood case (see 5.3.1), in 2000, paragraph 6(b) was amended to allow a deceased man to be recognized as the father, when his sperm was used posthumously.

Contract pregnancies. Surrogacy entails impregnating one woman to gestate a baby who will be raised by another – the intentional parent (5.5.1). Generally in surrogacy arrangements, the intentional father's sperm fertilizes the surrogate's egg – so she provides both genetic material and gestation. Bioethicist Laura Purdy argues that in such cases *'it would make more sense to use a more accurate and objective term like contract pregnancy'*.[30] Often such arrangements appear satisfactory at the personal level, in that parents are pleased and the children unharmed. But there have been less happy outcomes, e.g. when the gestating woman has fought to retain custody, or when women have engaged in the practice for monetary reward, unaware of the emotional burden it might entail. More generally, it has been argued that the practice reinforces undesirable stereotypes of women as 'breeders'.

Some gay couples have arranged for surrogate mothers to gestate embryos produced by semen from one of the partners (Table 5.2: D6:F3: G,H4). A gay couple in the USA have between them fathered five children by IVF (in one case, quadruplets), born to the same surrogate mother – a single woman with three other children.[31]

Aborted fetuses. A possible ART entails the use of tissue from aborted fetuses, as a plentiful source of gametes which are in short supply. Research reported in 1993 suggested that egg-producing follicles could be grown in ovarian tissue taken from human fetuses. In 1994, the HFEA conducted a public consultation on the issue, and on the strength of the strong public reaction the procedure was banned in the UK.[32] But renewed interest, and the ease of international travel (dubbed **reproductive tourism**),[33] suggest that the technique might soon be available elsewhere. There is even a possibility that by using embryonic stem cells that can be transformed into sperm (6.5), two men might become the biological parents of a child which was then gestated by a surrogate mother. Eggs might be produced in the laboratory by the same process.[34]

5.5.3 Assisted reproductive technologies on the horizon?

Cloning by somatic cell nuclear transfer. One of the highest-profile prospects is the possibility of producing babies by cloning, using the technique of somatic cell nuclear transfer (SCNT) first employed successfully in sheep with the birth of Dolly (9.3.2). That process involved fusing the nucleus of a somatic cell from an adult sheep with the enucleated egg cell of a second sheep by exposing them to an electric current.[35] The employment of the technique in many other species of mammal implies that this is technically feasible in humans, albeit a serious risk to the child's health because of its low success rate. Another major concern is that in the aged DNA of an adult cell nucleus about one-quarter of the genes are likely to have accumulated errors in their code. According to geneticist David Galton: *'It would be outrageous to handicap a child from birth with a multitude of genetic errors and subsequent deformities due to the deteriorated condition of the adult nucleus used for transfer.'* [36]

A variation on the cloning theme is illustrated in Figure 5.3, where a possible future procedure is compared with standard IVF.[37] In this novel procedure the fertilized eggs are allowed to go through a number of cell divisions before removing their nuclei and transferring them to enucleated donor eggs, with which they are fused. The claimed

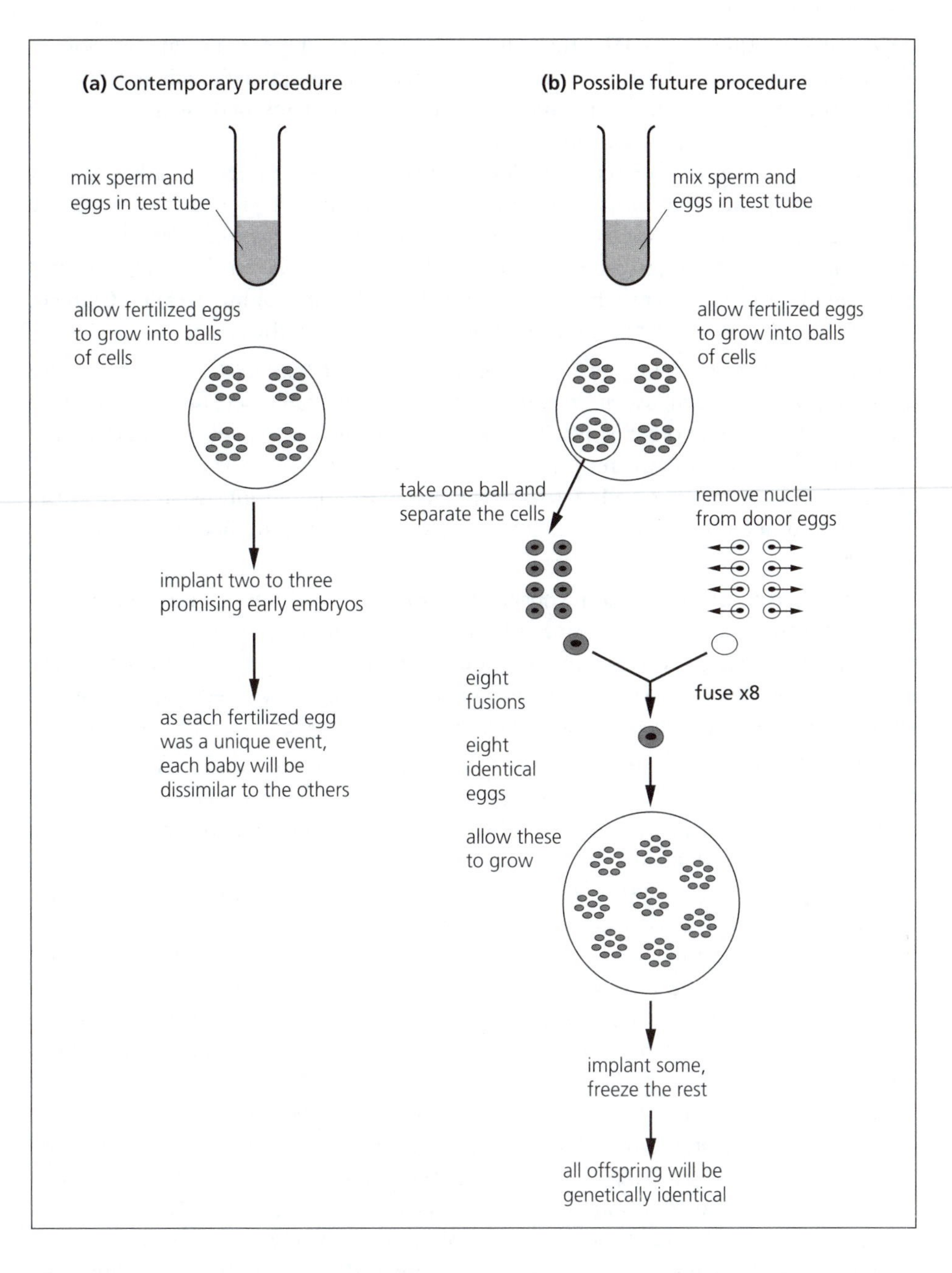

Figure 5.3 The prospective treatment of infertility by use of a limited type of cloning that uses cells arising from a fertilized egg (b) compared with the current *in vitro* fertilization procedure (a).
[From: Curran B (2003) A Terrible Beauty is Born. London, Taylor and Francis, p. 136]

advantage is that it avoids the multiple hormone treatments many women experience – and the 'donor' is, in fact, a cell rather than a unique individual. However, all offspring will be genetically identical, so that 'identical twins' of different ages could be born if embryos were frozen and used years later.

But most people strongly reject the idea of reproductive cloning.[38] In 2001, the UK Government passed the Human Reproductive Cloning Act, which makes it an offence to *'place in a woman a human embryo which has been created otherwise than by fertilization'*, and similar legislation has been enacted in other countries. Even so, periodic claims are made that human babies have been cloned, e.g. by the company called Clonaid, although to date no evidence has been produced supporting the claim.

Parthenogenesis. In 2004, Japanese scientists reported *'the development of a viable parthenogenetic mouse individual from a reconstructed oocyte containing two haploid sets of maternal genome'*, or in other words, by combining the nuclei of two mouse eggs a mouse had been created with two mothers.[39] The single viable mouse was a result of 460 attempts to grow embryos, and there appear to have been significant welfare risks. So the likelihood of men becoming redundant partners in reproduction does not seem to be imminent!

Uterine transplantation. The prospect of uterine transplantation into women who have a diseased or damaged uterus has been given credibility by successful operations, and subsequent pregnancies, in animals. Interest has also been shown in the possibility of men being able to gestate embryos by this means, or even ectopically (in the abdominal cavity), which occurs spontaneously, but successfully, in a very few women.

Ectogenesis. Perhaps the ultimate ART would be ectogenesis, a procedure by which babies would be produced entirely outside the body. The sinister possibilities of ectogenesis were described in Aldous Huxley's famous science-fiction novel *Brave New World,* written in 1932, in which he envisaged a society in which cloned babies, produced by IVF on a conveyor belt system, were designed for specific tasks.[40] Superior eggs were inseminated with superior sperm to produce Alphas and Alpha Pluses, and provided with the best prenatal treatment. In contrast, biologically inferior sperm and eggs produced embryos (subjected prenatally to alcohol and other protein poisons), which became almost sub-humans, capable only of performing unskilled work and kept happy by frequent access to the opposite sex, gratuitous entertainment, and daily doses of a sedative drug.

However, some scientists now see the prospect of ectogenesis as a welcome advance. Thus, reproductive physiologist Roger Gosden argues: *'Artificial wombs would put an end to all the medical and social difficulties of surrogate motherhood. And conventional maternity would become unnecessary for anyone. ... There would be advantages for the baby too. Safe in its bottle, it would be spared any harm arising through ignorance or carelessness of the parents'.*[41] But it seems likely that many people will find the prospect of ectogenesis abhorrent and/or highly irresponsible, in that it would expose the growing fetus to undoubted physical and psychological risks.

Italian resistance. Despite the marked liberalization of attitudes evident from the preceding paragraphs, not all Western governments have followed the same route. In Italy in 2004, the Senate overwhelmingly approved a law banning the use of donor sperm, eggs, or surrogate mothers. It also limits AI to *'heterosexual couples in stable relationships'* and excludes gay couples and single women.[42]

5.6 Ethical aspects of assisted reproductive technologies

Two important issues raised by ART are those concerning the **right to reproduce** and the distinction between **public** and **private morality**.[43]

5.6.1 Reproductive rights

Article 16 of the UN Declaration of Human Rights, formulated in 1948, states that *'Men and women of full age, without any limitation due to race, nationality, or religion, have a right to marry and found a family.'* But how should we interpret these words more than 55 years later, when social norms are so different and, certainly in Western countries, there is widespread public acceptance of unmarried people living together in intimate sexual relationships, whether as heterosexual or homosexual couples? A critical question concerns reproductive rights. Does the right assigned in the Universal Declaration to married couples now extend to non-married couples, homosexual couples, and single people? If people want to have children, is it anybody else's business? And if ART enable people to have children who could not otherwise do so, is there an ethical problem? We need to explore the implications of the claimed rights.

Rights have been defined as *'advantageous positions conferred on some possessor by law, morals, rules, or other norms'*[44]: they are often divided into **positive rights** and **negative rights**. Exercise of the former seeks to ensure people receive something desirable (such as a fair wage for work done), whereas the latter seeks to protect them from suffering from something undesirable (such as being imprisoned for no good reason). Reproductive rights have been considered to be both positive and negative rights, because it is argued that:

- *positive*: once the technology exists to assist conception in subfertile or infertile people, it is right for it to be made available to all who want children
- *negative*: it would be an infringement of people's liberty and privacy to interfere with their right to have children.

A feature of rights is that they entail (*correlative*) duties on others to observe them. For example, if a couple had a right to have a child by IVF, it would follow that someone (or usually the government) had a duty to provide the means to do so. But this raises some problems. If an individual had a *right* to have a child, e.g. by IVF using a donor's gamete, this would claim more than is physiologically possible, since two sexually active people are normally necessary. So asserting this right entails claiming that something that only occurred as a result of human ingenuity is now a *necessary* provision for everyone. (Pushing this to an extreme, does this suggest everyone has a right to go to the moon because a few people have done so?)

Another problem is that even if it is technically possible to employ ART successfully, the fact that resources are used (e.g. doctors' time and hospital facilities), that could be employed differently, poses questions about priorities. How serious a condition is infertility? And how valid is a claim on resources made by single or gay people in order to have a child by IVF?

Then there are the important questions concerning the negative rights of others who might be affected by the application of ART. These include, for example:

- the infant who may remain ignorant of its parentage, or discover they were gamete 'donors'
- the surrogate mother who might suffer distress when she has to sever a strong bond with the child she has carried
- the husband/partner who might subsequently regret the decision to agree to his wife bearing a child to whom he is not genetically related.

It is not only rights that might be problematical, because there are also some significant health and financial risks. For example, the woman undergoing treatment (the wife or egg donor) might experience ill health, particularly if she suffers from the hyper-stimulation syndrome, while the low success rate of the procedures (5.4.2) can add significant financial costs to the emotional and physical burdens experienced.

5.6.2 **Public morality and private morality**

It seems clear that the claim (5.6.1) that reproduction is a private matter is difficult to maintain once the bedroom door has been left ajar. Even considering techniques which involve the husband and wife's gametes, the employment of medical staff and hospital/clinic facilities makes this a matter of public concern because of the alternative uses to which resources might have been put. (Of course, the routine hospital delivery of babies is also now extensively medicalized.) But when ART are involved, when regulatory authorities like the HFEA have to be established, and when extensive legal arrangements need to be implemented, then the issues clearly entail public policies.

But public concerns do not stop at the level of resource allocation. There are also big questions about public morality in the broadest sense. In the twenty-first century many who live in Western society have become used to the diversity of our social life, to the highly varied appearances, customs, and lifestyles of their fellow students and neighbours. But older people often find it difficult to adjust to the rapid change from a society in which people shared similar values and 'knew their place' – to a modern society where it seems 'anything goes'. It is not long ago that parentage outside marriage was so unusual, and disapproved of, that the following dialogue would have been regarded as amusingly risqué:

Doctor: I've got good news for you, Mrs Brown
Patient: It's Miss Brown, actually
Doctor: I've got bad news for you, Miss Brown[45]

But despite the sexual revolution of recent decades, for many people a child is still properly the outcome of two people's mutual love, so that purchasing the ejaculate of an anonymous sperm donor offends a basic norm of acceptable behaviour; and the fact that it is impossible to prevent the practice legally adds nothing to its ethical acceptability. On the other hand, even for critics of DI, the resulting child, who is totally innocent of its origins, merits the fullest ethical respect.

No one, it seems, is exempt from the need to consider the future - especially the younger generation, on whom it will have the greatest impact. Perhaps an important question here concerns the *direction* of social change and whether the impetus is so great that an apparently commendable tolerance and respect for individual choice might end up producing a society which no one would have wanted if they had foreseen the eventual outcome. But another important question concerns the extent to which the majority in society is entitled to impose its moral code on others.

Philosopher H L A Hart argued that moral problems exist on two levels, the *primary* and *critical*. Primary questions are concerned with whether some practice is right or wrong, but critical questions are concerned with whether, if the law intervened (say by imposing a ban on a practice), the infringement of liberty would itself be wrong, not least because enforcing the ban would entail an unjustified intrusion on privacy.[46] Generally speaking, in Western societies, liberty and privacy are accorded high priority. It is as though liberal democracies now underwrite the famous lines attributed to eighteenth-century French philosopher Voltaire: *'I disapprove of what you say, but I will defend to the death your right to say it.'*

5.6.3 **Analysing the desire for children**

An ethical analysis of people's desire for children using ART is facilitated by dissecting the concept into its component elements to clarify the grounds for support or objection. Philosopher Ruth Chadwick has suggested that it is helpful to bear in mind *'the distinction between the concepts of begetting, rearing, and bearing'*,[47] and this plan will be adopted in the following discussion. Indeed, it has already been used in essence in Table 5.2, where reference is made to begetting, gestating and caring for children.

The desire to beget a child is often interpreted in masculine terms, as to *sire* a child, but the motivation to be a genetic parent is surely no less strong in women. When DI or egg donation are used the desire is clearly frustrated. And if either gamete is donated the desire to have a child *with a particular person*, so that it is the unique outcome of two people's genetic influences, is at best diluted. Some forms of motivation might be based on social pressures, in that a child is wanted to 'keep up with the Jones's, or for economic or legal reasons in order to provide an heir. Such considerations will influence the acceptability of procedures using donated gametes.

Gestating a fetus for nine months might seem to provide compensation for any dilution of the genetic inheritance when donor gametes are used. Where a relative donates a gamete the parents might find the process even more acceptable (although for others it could have overtones of incest). But surrogacy adds a further barrier to the sense that it is the parents' *own* child, and where this occurs with donated gametes the *intentional parents'* (5.5.1) actual biological connection with the child might be slight. It is almost as though the child had been adopted when still an embryo.

Gestating a fetus normally culminates in the delivery of the baby, but the increasing use of anaesthetics and resort to Caesarean section means fewer women now experience 'natural birth'. This is, of course, a very subjective experience, but it has been claimed that, at least in some cases, mothers *'become conscious of a sense of exaltation and incomparable happiness as they watch the arrival of their child'.*[48]

If the main element of the desire for children is that of caring for them, it might be thought that ART are unnecessary because of the possibility of legally adopting a child.[49] In practice, the desire might not always be readily satisfied because of restrictions on the availability of suitable children. The ease of access to abortion means that fewer children are available for adoption than used to be the case, and many are children who have suffered hardship or abuse from their earlier carers, which may dissuade some potential adopters from embarking on what might seem a very demanding venture. Even so, many find adoption a mutually rewarding practice for parents and children alike.

5.6.4 Ethical analysis of assisted reproductive technologies

To clarify thinking on the acceptability of the various ART, a form of the ethical matrix (Table 5.3) might be used. The relevant interest groups will obviously include those listed in Table 5.2, but ideally attention should also be given to close relatives, demands on the whole health service, and society at large.

Table 5.3 An ethical matrix specifying respect for the three principles (well-being, autonomy, and fairness) in a case where eggs are supplied by an egg donor

Respect for:	Well-Being	Autonomy	Fairness
Mother	Fulfilment/health	Choice	Rights/affordability
Father	Fulfilment	Choice	Rights/affordability
Child/Fetus	Health	Respect for identity	Non-discrimination
Egg Donor	Emotional reward/health	Choice	Adequate expenses
Society	Social harmony	Democratic choice	Fair resource allocation

Notes: For more details on use of the matrix see 3.5. The matrix can be used to help identify the extent to which each of the specified principles is respected, and to facilitate deliberation on how positive and negative impacts should be weighed.

In using Table 5.3, specification of the principle of respect for well-being will not only take account of the sense of fulfilment of people previously unable to produce children, but also the risks to health (physical and mental) of all involved. Respect for autonomy will take account of the people's freedom of choice to employ ART (which has been discussed above in terms of their reproductive rights), but also consider impacts on others, including, most importantly, the child. Respect for fairness will assess: the impacts on the prospective parents (which might be regarded as positive for some but negative for those unable to afford the treatment); on the child (who might suffer from a lack of identity if donor gametes were used); and on people in society, who might consider that ART threaten the sense of stability that has traditionally underlain public order and respect, or that they represent an unfair use of resources given other pressing medical priorities.

More generally, it is worth considering the likely trajectories of technological change. For example, fears have been expressed that on the basis of 'protecting the unborn child', IVF may be used increasingly by people who are not infertile, because the natural procreative practices are deemed 'too risky'.[50] Or it may be used, possibly using frozen embryos, to select a more convenient time for parenthood in a woman's career plans. Unusual possibilities, such ectogenesis and 'two male genetic parents', provide useful thought experiments in clarifying one's reasoning - although they might become a reality sooner than we expect.

For some people, philosopher Lawrence Hinman's observation may prove decisive: *'The danger we face with the development of reproductive technologies is that (the) tendency to turn everything into a commodity will increase. We are moving at least dangerously close - some would say we have already crossed the line - to buying and selling sperm, eggs, and even the use of wombs.'*[51]

5.7 Embryo experimentation

One of the consequences of IVF is that a supply of eggs surplus to requirements is usually obtained as a result of superovulation. Eggs not used in IVF, or stored for subsequent use, can (if permission is given) become available for experimental purposes. This section provides a brief historical account of how an ethically contentious issue was addressed politically and in legislation, which serves as a useful case study of the ways in which ethical reasoning has been applied to public policy.

5.7.1 Research on embryos

Although it was only after 1978 that a plentiful supply of early embryos became available, research on embryos and fetuses was undertaken much earlier. The 1967 UK Abortion Act made it legal to perform abortions *'to preserve the physical or mental health of a pregnant woman or to avoid the birth of a seriously handicapped child'*, which led to an immediate rise in the number of abortions. This, in turn, resulted in a rapid fall in the numbers of children offered for adoption; a change that seemed to justify the IVF programme (5.4.2), which aimed to improve the chances of subfertile couples having their own children.[52]

The ethics of embryo research came to prominence in the 1980s, following publication of the Warnock Report (5.5). The report argued that, subject to several conditions, experiments should be permitted on human embryos up to fourteen days after fertilization. It was reasoned that before fourteen days the primitive streak (the initial band of cells from which the embryo begins to develop) is not fully formed and there is a possibility that the embryo may divide to form identical twins. But if only *individuals* can be said to have ethical standing (3.4), it would seem to follow that it would not be unethical to perform destructive experiments on embryos up to this stage. However, the report recommended that it should be *'a criminal offence to handle or use as a research subject any live human embryo derived from IVF beyond that limit'.*[53] Although, in the words of the report, *'a majority of us agreed that research on human embryos should continue'*, in fact, seven of the committee's sixteen members dissented from this view, and the report includes their statements to this effect.

Initially, the Warnock Report was not well received, either in Parliament or by the general public, and embryo experimentation proved highly contentious for several years. Many of the debates in the mid-1980s echoed the abortion debates twenty years earlier, with traditionalists (mostly represented in Parliament by Conservatives) considering that legalization of IVF would lead to a breakdown of family values, while the use of embryos in experiments would undermine the rights of the 'unborn child' and open the door to 'irresponsible' actions by scientists. The Conservative government of the day considered that this was a matter of conscience, and adopted no official party line. But in 1985 a Private Member's Bill to *prevent* any use of IVF embryos for experimental purposes was passed by a majority of 238 votes to 66.

5.7.2 The dynamics of political change

However, this initial opposition was turned round by a concerted campaign by a group of 'progressive' Conservatives, who were prominently represented in the Cabinet, supported by most Labour Members of Parliament (MPs). When the HFE Bill was finally put before Parliament in 1990, the clause approving continuation of embryo research was *carried* by 364 votes to 93. The key factors in the success of the pro-research lobby were threefold:

- a claimed moral legitimacy for embryo research by referring to the early embryo as a **pre-embryo**, devoid of any recognizably human or individual characteristics
- an emphasis on the possibility of controlling genetic diseases by using pre-implantation genetic diagnosis (PGD; see 6.3.3) and only implanting healthy embryos
- replacement of the *rhetoric of fear* by a *rhetoric of hope*, which fostered the anticipation of better things to come for society at large.[54]

Sociologist Michael Mulkay has shown that scientists played a key role in persuading politicians to change their minds, as illustrated by the strong advocacy of embryo research by leading professional bodies, such as the Royal Society, BMA, and Medical Research Council (MRC; see 14.4), and by the influential science journals *Nature* and *New Scientist*. So it is important to examine the arguments they advanced and their underlying (if only implicit) ethical reasoning.

The justification for embryo experiments is essentially utilitarian (2.5), appealing to the claimed benefits for prospective parents. The case was perhaps argued most strongly by Robert (later Lord) Winston, who announced just before the Commons vote in 1990 that, using PGD to ensure that they had girls, three women were then pregnant with healthy babies, who would otherwise (because of a sex-linked disorder) have risked having deformed male babies. This announcement is believed to have been highly influential in persuading many MPs to support the Bill. More generally, research on human embryos was considered essential, because although animal experiments provide useful background data, they cannot supply sufficiently reliable information.[55]

Several key arguments opposing experimentation were put by clinical neurologist John Marshall, a member of the Warnock Committee. He argued that since the embryo from the time of fertilization is a *potential* human being, there is no point in time at which it is ethical to treat it as non-human. He also claimed that the term 'pre-embryo' was inappropriately introduced to imply that the experts themselves opposed embryo experimentation, but that experimenting on the pre-embryo was somehow qualitatively different. Moreover, he argued that the claimed benefits of such experimentation were exaggerated, *'So that it seems that the scientists have rather abandoned their strictly scientific attitude and joined the public relations exercise.'*[56] Indeed, Mulkay claims that the scientists in high-profile London clinics acted as 'gatekeepers', selectively controlling news releases *'in accordance with their own interests and their own definition of the situation'.*[57]

Mulkay's sociological analysis suggests that a fundamental limitation of much debate is that ART are regarded as an independent variable, so that ethical deliberation is often confined to examining possible legal responses that might be adopted to control their use. But, he claims, this approach fails to recognize that ART are themselves **social products**, which are closely bound up with other social factors extending way beyond the fields of scientific and medical expertise.[58] Indeed, according to political scientist Langdon Winner, rather than requiring legislation, such is its power to shape society that *'technology* **is** *legislation'*.[59] It is worth considering whether such an analysis, if it were more widely appreciated, might provide members of society with grounds for challenging the dominant scientific view that endorses the ethical acceptability of embryo experimentation.

THE MAIN POINTS

- Most contraceptive technologies available (chemical, barrier and surgical) have drawbacks and many people resort to abortion. Socio-economic development, e.g. by increasing education and income, is often a more effective and ethically acceptable way of achieving birth control.
- Infertile and subfertile people are now often able to have babies by technologies such as AI, IVF, and gamete and zygote intrafallopian transfer – procedures that are regulated in the UK by the HFEA.
- Such techniques challenge traditional boundaries, especially when they permit single or gay people (of either sex) or postmenopausal women to produce children, sometimes (when both donor gametes are involved) with no genetic link to the 'parents'.
- Justification for ART is advanced in terms of both positive and negative human rights, and identifies the often competing claims of private and public morality.
- Debates over human embryo research in the 1980s illustrate the political context in which such decisions are made and the critical role played by technology in shaping social norms.

EXERCISES

These can form the basis of essays or group discussions:

1. Discuss the ethical justifications advanced for abortion (5.2). What are the consequences of countries permitting 'abortion on demand'?
2. Consider the contrasting views on whether infertility is a disease, and what the government's duties are with respect to infertile couples (5.3).
3. Should gay people (singly or as a couple) be allowed to produce genetically related children, using donated eggs, sperm and/or a surrogate mother? (5.5)
4. Discuss the scientific and ethical issues surrounding the potential use of eggs from aborted fetuses in assisted reproduction programmes (5.5).
5. When, and on what grounds, do you consider the human embryo attains ethical status? What implications does your view have for whether embryos should be used in research? (5.7.1).

FURTHER READING

- *Ethics, Reproduction and Genetic Control* (revised edition) edited by Ruth Chadwick (1992). London, Routledge. A valuable collection of essays.
- *Making Babies: a personal view of IVF treatment* by Robert Winston (1996). London, BBC Consumer Publishing. By a pioneer of ART.
- *Designer Babies* by Roger Gosden (1999). An upbeat account of reproductive technologies by a leading reproductive scientist.

USEFUL WEBSITES

- **http://www.hfea.gov.uk/Home** *Human Fertilisation and Embryology Authority*: the HFEA licenses all UK clinics offering ART services and has a key role in formulating policy on embryo experimentation.
- **http://ethics.acusd.edu/LMH/Papers** *Reproductive technologies and surrogacy* (L M Hinman): a useful source of ideas in applied ethics, maintained by a philosopher.
- **http://www.ccels.cardiff.ac.uk/launch/gardnerpaper.html** *Therapeutic and Reproductive Cloning – a scientific perspective* (R L Gardner, 2004): a lecture by a leading scientist in this field.

NOTES

1. Knight J W (1998) Birth control technology. In: Encyclopedia of Applied Ethics. (Ed.) Chadwick R. San Diego, Academic Press. Vol. 1, pp. 353–368
2. Ibid.
3. Gibson S (1998) Abortion. Encyclopedia of Applied Ethics. (Ed.) Chadwick R. San Diego, Academic Press, *1*, pp. 1–8
4. Soloway R A (1982) Birth control and the population question in England 1877–1930. London, University of North Carolina Press
5. Callahan J C (1998) Birth-control ethics. In: Encyclopedia of Applied Ethics. (Ed.) Chadwick R. San Diego, Academic Press. Vol. 1, pp. 335–351
6. Knight J W (1998) Birth control technology. In: Encyclopedia of Applied Ethics. (Ed.) Chadwick R. San Diego, Academic Press. Vol. 1, pp. 353–368
7. Knight J W (1998) Birth control technology. In: Encyclopedia of Applied Ethics. (Ed.) Chadwick R. San Diego, Academic Press. Vol. 1, pp. 353–368
8. Parfit D (1986) Overpopulation and the quality of life. In: Applied Ethics. (Ed.) Singer P. Oxford, Oxford University Press, pp. 145–164
9. Battin F P (1998) Population issues. In: A Companion to Bioethics. (Eds) Kuhse H and Singer P. Oxford, Blackwell, pp. 149–162
10. Illich I (1977) Limits to Medicine. London, Penguin
11. Mepham T B (1987) The Physiology of Lactation. Milton Keynes, Open University Press
12. Cited by Frith L (2001) Reproductive technologies, overview. In: The Concise Encyclopedia of the Ethics of New Technologies. (Ed.) Chadwick, R. San Diego, Academic Press, pp. 351–362
13. Gosden R (1999) Designer Babies: the brave new world of reproductive technology. Phoenix, London, p. 216
14. Edwards R (1990) In: Experiments on Embryos. (Eds) Dyson A and Harris J. Routledge, London, pp. 42–54
15. Baggott la Velle L (1997) Human Reproduction. Cambridge, Cambridge University Press
16. Human Fertilisation and Embryology Authority (2004) http://www.hfea.gov.uk/

17. Baggott la Velle L (2002) Starting human life: the new reproductive technologies. In: Bioethics for Scientists. (Eds) Bryant J, Baggott la Velle L, and Searle J. London, J Wiley, pp. 201–231
18. Ibid., pp. 217–220
19. Yoxen E (1986) Unnatural Selection? London, Heinemann
20. Warnock M (1984) A question of life: the Warnock report. Oxford, Blackwell
21. Human Fertilisation and Embryology Authority (2004) http://www.hfea.gov.uk/
22. Ibid.
23. Gosden R (1999) Designer Babies: the brave new world of reproductive technology. London, Phoenix
24. Human Fertilisation and Embryology Authority (2004) http://www.hfea.gov.uk/
25. Gosden R (1999) Designer Babies: the brave new world of reproductive technology. Phoenix, London
26. Wright O (2004) Woman gives birth to her own grandchildren. Times. (30 January 2004) http://timesonline.co.uk/
27. Gosden R (1999) Designer Babies: the brave new world of reproductive technology. London, Phoenix
28. Human Fertilisation and Embryology Act (1990) http://www.hmso.gov.uk/acts/1990/
29. Hinman L M (2004) Reproductive technologies and surrogacy. http://ethics.acusd.edu/LMH/Papers
30. Purdy L M (1998) Assisted reproduction. In: A Companion to Bioethics. (Eds) Kuhse H and Singer P. Oxford, Blackwell, pp. 163–172
31. Bioedge, 24 April 2004 www.australasianbioethics.org
32. Human Fertilisation and Embryology Authority (1994) Donated ovarian tissue in embryo research and assisted conception. Public consultation document. London, HFEA
33. Pennings G (2002) Reproductive tourism as moral pluralism in motion. J Med Ethics *28*, 337–341
34. Dennis C (2003) Developmental biology: synthetic sex cells. Nature *424*, 364–366
35. Wilmut I, Schneike A E, McWhir J, Kind A J, and Campbell K H S (1997) Viable offspring derived from fetal and adult mammalian cells. Nature *385*, 810–813
36. Galton D (2001) In Our Own Image: eugenics and the genetic modification of people. London, Little, Brown and Co., p. 65
37. Curran B (2003) A Terrible Beauty is Born: genes, clones and the future of mankind. London, Taylor and Francis, p. 136
38. Kaebnik G E and Murray T H (2001) Cloning. In: The Concise Encyclopedia of the Ethics of New Technologies. (Ed.) Chadwick R. San Diego, Academic Press, pp. 51–64
39. Kono T, Obata Y, Wu Q, Niwa K, Yamamoto Y, Park E S, Seo J-S, and Ogawa H (2004) Birth of parthenogenetic mice that can develop to adulthood. Nature *428*, 860–864
40. Huxley A (1932) Brave New World. London, Chatto and Windus
41. Gosden R (1999) Designer Babies. London, Phoenix, pp. 177–183
42. BBC News (2004) Italy bans donor sperm and eggs. http://newsvote.bbc.co.uk/ (8 January 2004)
43. Chadwick R (1992) Having children. In: Ethics, Reproduction and Genetic Control (revised edition). (Ed.) Chadwick R. London, Routledge, pp. 3–43
44. Cambridge Dictionary of Philosophy (1999) (2nd edition). (Ed.) Audi R. Cambridge, Cambridge University Press, p. 796

45. Gilbert G N and Mulkay M (1984) Opening Pandora's Box: a sociological analysis of scientists' discourse. Cambridge, Cambridge University Press, p. 184
46. Hart H L A (1963) Law, Liberty and Morality. Oxford, Oxford University Press
47. Chadwick R (1992) Having children. In: Ethics, Reproduction and Genetic Control (revised edition). (Ed.) Chadwick R. London, Routledge, p. 9
48. Dick-Read G (1960) Childbirth Without Fear. London, Pan, p. 25
49. Humphrey M and Humphrey H (1988) Families with a Difference. London, Routledge
50. Steinberg D L (1990) In: The new reproductive technologies. (Eds) McNeil M, Varcoe I, and Yearley S. Basingstoke, Macmillan, p. 114
51. Hinman L M (2004) Reproductive technologies and surrogacy. http://ethics.acusd.edu/LMH/Papers
52. Mulkay M (1997) The Embryo Research Debate. Cambridge, Cambridge University Press, pp. 8–9
53. Warnock M (1984) A Question of Life: the Warnock report. Oxford, Blackwell, p. 66
54. Mulkay M (1997) The Embryo Research Debate. Cambridge, Cambridge University Press
55. Trounson A (1993) In: Embryo Experimentation: ethical, legal and social issues. (Eds) Singer P, Kuhse H, Buckle S, Dawson K, and Kasimba P. Cambridge, Cambridge University Press, pp. 14–25
56. Marshall J (1990) In: Experiments on Embryos. (Eds) Dyson A and Harris J. London, Routledge, pp. 55–64
57. Mulkay M (1997) The embryo research debate. Cambridge, Cambridge University Press
58. Ibid.
59. Winner L (1977) Autonomous Technology: technics-out-of-control as a theme in political thought. Cambridge, Mass., MIT Press, p. 323

6

Reproductive choices

OBJECTIVES

When you have read and discussed this chapter you should:

- be aware of the various tests available and being developed to identify genes liable to cause certain diseases and the strategies used to cope with them
- understand ethical issues concerning the management of personal genetic information
- appreciate the potential of stem cells in medical therapies, and the ethical implications of therapeutic cloning, *designer babies* and *saviour siblings*
- understand arguments advanced for *utopian eugenics*
- be able to perform ethical analyses of novel gene technologies using the ethical matrix

6.1 Introduction

Quite apart from the issue of how to control birth rates (chapter 5), society is constantly being presented with challenging new questions about the *sort* of children people might want to have. It is clear that the combination of upbringing and environment (nurture) is a powerful influence in moulding human character, but our increasing knowledge of genomics now confirms that an individual's genetic inheritance (nature) also plays a critical role.

Until recently, a person's choice of sexual partner has been essentially the only way to influence the child's genotype, although in some cultures, even now, a preference for boys has often led to infanticide of baby girls. But since the introduction of **amniocentesis**, allowing sex determination by examining fetal cells in the amniotic fluid in the uterus at about 16 weeks of pregnancy, it has also been possible to abort fetuses (5.2.1) which are not of the desired sex. For example, in the late 1970s, of 750 women attending one Indian hospital, 450 were told their fetus was female, and of these 430 were aborted.[1]

More recently, the introduction of several methods for prenatal diagnosis of chromosomal and genetic diseases, in addition to amniocentesis, has permitted elective abortions when the fetus is carrying genes for a disabling condition. These methods include **chorionic villus** sampling (which entails analysis of a biopsy of the membrane surrounding the fetus in the uterus from about 11 weeks of pregnancy) and fetal ultrasound scanning. Deciding to abort a fetus because of the likelihood of the child suffering a debilitating disease is an example of **negative eugenics**, which many people might consider an unequivocally compassionate act. But, as with many seemingly straightforward issues, closer inspection reveals complications, the ethical implications of which deserve close scrutiny.

6.2 Prenatal screening

Prenatal screening aims to determine the prevalence of a disease-related gene in a population, when there is no reason to believe that any *particular individual* carries the gene in question. The discussion here is limited to screening of the embryo and fetus, although some similar and some different ethical issues are raised by genetic screening of children and adults. Hundreds of DNA-based gene tests are now available, including those shown in Box 6.1.

However, the fact that a test is available does not necessarily mean that it should be used in screening programmes. In 1993, a Nuffield Council on Bioethics (hereafter 'Nuffield') report on genetic screening concluded that the following six factors need to be taken into account before introducing a screening programme for a particular gene:

- the predictive power and accuracy of the test
- the benefits of informed personal choice in reproductive decisions
- the psychological impact of the outcome of the screening on individuals and families
- therapeutic possibilities
- possible social and economic disadvantage relating, for example, to insurance and stigma
- the resource costs and relative priority, in view of limited resources, of establishing a screening programme.[2]

BOX 6.1 EXAMPLES OF DISEASES FOR WHICH DNA TESTS ARE AVAILABLE

- Alzheimer's disease: an early onset form of senile dementia
- cystic fibrosis: affecting principally the lungs and pancreas
- Huntington's disease: a progressive, fatal, degenerative neurological disease
- phenylketonuria (PKU): mental retardation due to a missing enzyme
- Tay–Sachs disease (TS): a fatal neurological disease of childhood
- Duchennne muscular dystrophy: caused by a recessive gene on the X chromosome
- inherited breast and ovarian cancers

In short, since it would be too expensive to test everyone for everything, priority should be given to the ability to make a difference to people's quality of life, without the risk of causing unnecessary distress or hardship. If the effect on any of the six criteria listed in Box 6.1 were to be negative, the value of the test might be considered negligible. However, one major motivation for screening might well be that of developing better drugs, based on more accurate understanding of individual responsiveness and susceptibility to side effects (**pharmacogenomics**). For similar reasons, food companies developing **nutraceuticals** are also interested in the results of genetic screening (11.6.5).

6.3 Reproductive decision-making

Potential parents who know they have, or might be carriers of, genetic disorders are faced with difficult decisions about whether to have children of their own. Typically, the options in such cases are:

- to proceed with a planned pregnancy
- not to have children
- to adopt children
- to use donor eggs, sperm, or embryos (see 5.4, 5.5) to avoid the risk of the child being born with the genetic disorder.

6.3.1 Prenatal genetic diagnosis

However, another option in certain cases is to proceed with the pregnancy but also resort to **prenatal genetic diagnosis** (PND), which is a form of test that is both person- and condition-specific. Some types of PND are invasive (such as amniocentesis and chorionic villus sampling, see 6.1) and consequently they carry a small risk of inducing an abortion. In 2002–2003 about 2600 such tests were performed in the UK to determine whether the baby had a genetic abnormality for conditions such as cystic fibrosis (Figure 6.1) and Duchenne muscular dystrophy. According to the Human Genetics Commission: *'If the baby is found to be affected, people can then use this information either to prepare for the birth of a child with a genetic disorder or to terminate the pregnancy.'*[3] Although there are some conditions, such as spinal bifida, where advance warning of the disease may be relevant to the method of delivery and care of the newborn, in practice *'in most cases, the reason for prenatal diagnosis is to prevent the birth of a child with a disability'*.[4] Clearly, this decision raises important ethical questions.

6.3.2 The decision to abort a fetus

The extreme ends of the scale in Box 6.1 are perhaps represented by Tay–Sachs disease (TS) and phenylketonuria (PKU). In TS, an autosomal recessive disease in which there is an inability to produce the enzyme *hexosaminidase,* the baby appears normal at

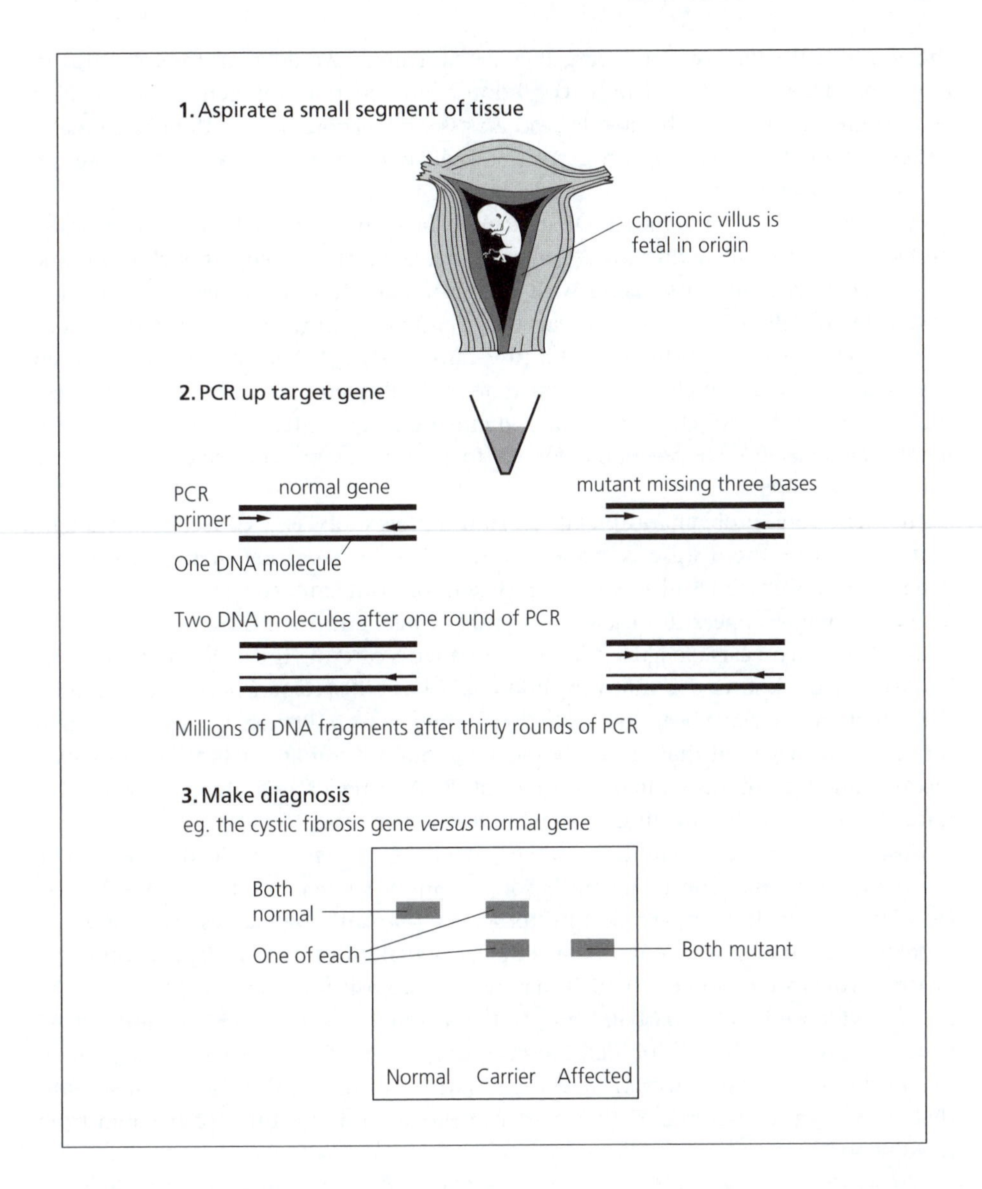

Figure 6.1 Screening the fetus for possible gene defects. Use of polymerase chain reaction (PCR) analysis allows identification of the cystic fibrosis gene. [From: Curran B (2003) A Terrible Beauty is Born. London, Taylor and Francis, p. 94]

birth, but after three months it develops mental retardation, and then goes blind and becomes paralysed. There is no known cure and death usually occurs at about 2 years of age. But in PKU the situation is quite different. Failure to produce the liver enzyme *phenylalanine hydroxylase* (which catalyses the conversion of phenylalanine to tyrosine) means that phenylalanine accumulates in the blood (and is excreted in the urine), causing brain damage. However, by consuming a diet low in phenylalanine, the effects of the disease can be largely avoided. Parents would seem to be entitled to have prenatal knowledge of both conditions, but on quite different grounds: PKU because the child should be able to lead an essentially normal life; TS because many parents might feel that it is wrong, when avoidable, to produce a child whose life is destined to be so short and full of misery.

Between these two extremes are many conditions which are perhaps more ethically problematical. Down's syndrome is due to the presence of three copies of chromosome 21, which is generally associated with mental retardation and heart abnormalities. About 50% of sufferers die before 5 years of age and most of the remainder in their teens or twenties. But people who survive those early years with Down's syndrome often lead happy lives, in which they demonstrate sociability and affection, and can sometimes attend ordinary school and find employment in sheltered work. The Government's target is for all pregnant women to be offered prenatal screening for the condition by 2005.

Another more problematical case is Huntington's disease. Caused by a single dominant allele, the disease is characterized by involuntary movements and mental deterioration, sometimes distressingly mistaken for drunkenness. However, symptoms do not normally appear until about 40 years of age, and then death usually occurs within five to ten years. So a decision to abort a fetus carrying the gene will necessarily be taken in the knowledge that the child could have 40 years of normal life, and by then a cure might have been found. In 2004, a report from Beverly Davidson's laboratory in Iowa suggested that currently incurable brain disorders such as Huntington's disease might soon be treatable by use of **RNA interference therapy**, which in mice is able to switch off a disease gene but leave others unaffected.[5]

Most of the clinical conditions discussed above are monogenic diseases where the presence of the gene is diagnostic of the onset, sooner or later, of the disease. However, many diseases are multifactorial, so that although a test may indicate a predisposition to the condition, whether it actually occurs will depend on other factors, avoidance of some of which may reduce the risk considerably. Moreover, the genetic evidence is often partial. Thus, tests for mutations in the BRCA1 and BRCA2 genes suggest that their combined presence only accounts for a small percentage of the incidence of breast cancer in large families with a history of the disease. This means that even negative results for these tests can provide no guarantee of freedom from the disease.[6]

6.3.3 Pre-implantation genetic diagnosis

For people using IVF (5.4.2), **pre-implantation genetic diagnosis** (PGD) provides another option for avoiding the risk of passing on a serious genetic disorder to their children. PGD allows prospective parents to avoid the birth of a child with disabilities by discarding an embryo rather than having it implanted.[7] PGD is used at the six- to ten-cell stage, when one or two cells are removed without apparently harming the embryo or its subsequent ability to implant. Cellular DNA is then tested for monogenic diseases using the **polymerase chain reaction** (PCR) and for chromosomal aberrations using **fluorescent *in situ* hybridization** (FISH). The number of recognized,

BOX 6.2 THE HUMAN GENOME PROJECT

With the completion of the human genome sequence in 2003, geneticists hope to compile a map identifying and locating each of the approximately 30,000 genes in the complete genome, and use it to catalogue the genetic differences between individuals in the population. The human genome consists of about 3.2 billion base pairs, which are located on the 46 chromosomes (23 each from mother and father).

All humans are 99.9% genetically identical, so the difference between any two randomly selected individuals is only 3 million base pairs, and between a mother and her biological child, about 1.5 million base pairs. But close genetic identities are not confined to humans. We share 98% of our genes with chimpanzees, 90% with rats and mice, and even 36% with fruit flies.

One of the commonest metaphors is that the genome is the 'book of life'. Thus, assuming 3000 words per page, it would take 1000 volumes, each of 1000 pages, to catalogue an individual's genome. Another metaphor represents the genome as the 'blueprint' for an organism: in the case of humans it results in 10^{13} cells, all made according to a single set of instructions.

Ethical, legal and social implications

About 5% of the funding of the Human Genome Project (HGP) was initially allocated to exploring ethical, legal and social implications (ELSI). This programme had the following components:

- **Fairness in the use of genetic information** by insurers, employers, courts, schools, adoption agencies and the military, among others
- **Privacy and confidentiality** of genetic information
- **Psychological impact and stigmatization** due to an individual's genetic differences
- **Reproductive issues** including adequate informed consent for complex and potentially controversial procedures, use of genetic information in reproductive decision-making, and reproductive rights
- **Clinical issues** including the education of doctors and other health service providers, patients, and the general public in genetic capabilities, scientific limitations, and social risks; and implementation of standards and quality-control measures in testing procedures
- **Uncertainties** associated with genetic tests for susceptibilities and complex conditions (e.g. heart disease) linked to multiple genes and gene–environment interactions
- **Conceptual and philosophical implications** regarding human responsibility, free will *versus* genetic determinism, and concepts of health and disease
- **Health and environmental issues** concerning GM foods and microbes
- **Commercialization of products** including property rights (patents, copyrights, and trade secrets) and accessibility of data and materials.

potentially disabling genetic conditions is growing rapidly as a result of data revealed by the Human Genome Project (HGP, Box 6.2), so that prospective parents are being faced with increasingly difficult choices. Figure 6.2 illustrates the technique used to identify blastocysts free of the gene causing cystic fibrosis.

The first successful use of PGD was in 1990, when two sets of twins were born to parents at risk of passing on a serious sex-linked disorder if they had had sons.

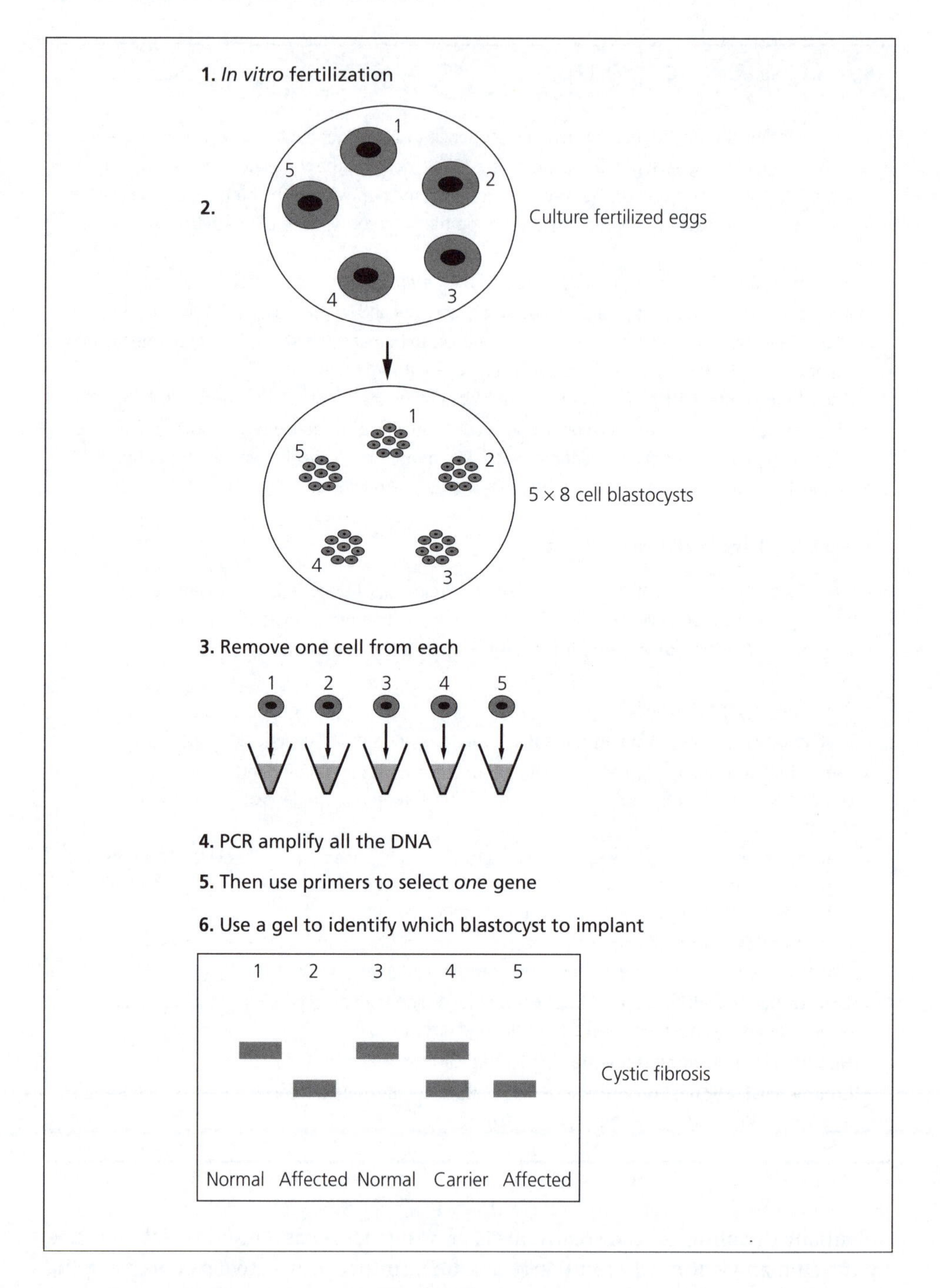

Figure 6.2 A combination of *in vitro* fertilization and pre-implantation genetic diagnosis (using polymerase chain reaction (PCR)) can avoid implantation of a blastocyst carrying the cystic fibrosis gene. [From: Curran B (2003) A Terrible Beauty is Born. London, Taylor and Francis, p. 95]

These cases proved to be important in the debates over the ethics of embryo experimentation (5.7.2). In the UK all cases of PGD must be approved by the HFEA, and since 2001 there have been about fifty live births using this technique. The HFEA also now licenses use of the technique of **aneuploidy** to screen for chromosomal abnormalities in embryos produced by IVF, thus allowing parents to decide against implantation of specific embryos.[8]

6.4 Ethical issues in managing genetic information

Several ethical issues concern the way genetic information is managed by society. These relate to questions such as *access, consent, confidentiality* and *discrimination*. Information about your genes is different from many other forms of medical information. When you have a consultation with your doctor it is taken for granted that what is said is confidential, and that you alone have the right to decide whether to share it with anyone else. But genetic information is almost certainly also a concern for your close family relatives, your sexual partner, and for your (prospective) children. It thus becomes an important question as to whether relatives have a *right* to be informed.

Even if people are assumed to have a right to know about their own, and their close relatives', genetic condition, how easy is it to assert the right *not* to know?[9] From one perspective, genetic screening is a public health measure which, like vaccination, aims to reduce the incidence of disease in the community. But the results of screening may overrule an individual's autonomous decisions 'to take life as it comes', and not feel condemned by their genetic fate. It is all very well saying that patients can choose to remain ignorant of test results, but social pressures will almost certainly make that difficult. Moreover, as technological advances such as the **gene chip** allow simultaneous multiple screening for several conditions, it is possible that instead of personal autonomy being enhanced by increased knowledge, it will become undermined. Too much information about possible risks affecting yourself, your children and your wider family might induce a state of acute anxiety.

But although individuals might have reservations about 'knowing too much', this is not a view shared by others, such as drug companies (6.2), employers and insurance companies. Imagine, for example, a company dealing with industrial chemicals which was able to screen potential employees for genetic susceptibility to certain of its toxic products. This might reduce the need to make otherwise necessary (and costly) improvements in the safety of the workplace, but it would discriminate against some people because they happened to have the 'wrong genes'.

Insurance companies also clearly have an interest in assessing accurately the risk that they might have to pay out large amounts for expensive medical treatments for clients whose medical condition could have been predicted from genetic screening. But should people be penalized for having a particular genetic make-up, rather than for self-inflicted risks due to habits like excessive drinking and smoking? Practicalities might also lead to injustices. Nuffield foresaw difficulties of assessing evidence based on *'what may be slender evidence on genetic susceptibility to develop polygenic and multifactorial diseases'* and, recognizing *'the possibility of abuse'*, recommended that British insurance companies should not require genetic tests as a prerequisite of obtaining insurance.[10]

6.4.1 Personal genetic information

So important are these issues that they were the subject of the first major enquiry undertaken by the UK Human Genetics Commission (HGC), which in 2002 made a number of recommendations about the handling of personal genetic information. Their report *Inside Information* noted that because genetic knowledge brought people into a special moral relationship with others, it was necessary to define a basic concept promoting the common good. This concept of **genetic solidarity and altruism** (developed from Article 17 of the Universal Declaration on the Human Genome and Human Rights[11]) was explained as follows:

We all share the same basic human genome, although there are individual variations which distinguish us from other people. Most of our genetic characteristics will be present in others. This sharing of our genetic constitution not only gives rise to opportunities to help others but it also highlights our common interest in the fruits of medically based genetic research.

Building on this basic concept, the HGC defined the major principle informing their work in the following terms: *'Respect for persons affirms the equal value, dignity, and moral rights of each individual. Each individual is entitled to lead a life in which genetic characteristics will not be the basis of unjust discrimination or unfair human treatment.'*[12] From this they derived four secondary principles, concerning **privacy, consent, confidentiality** and **non-discrimination**. In each case, individuals were to be accorded respect for the principle, subject to exceptions for *'the weightiest of reasons'*, such as for particular forensic reasons or in consideration of public health.

The HGC also sought to clarify what information is **sensitive** and what **private**. Genetic information comes in different forms, and not all of it *can* be private. For example, eye colour is difficult to disguise, even if hair colour is not; but some information, like one's blood group, is rarely considered sensitive. On the other hand, genetic information indicating the likelihood of a late-onset debilitating disease is almost certain to be considered both private and sensitive. In the present context, the distinctions made by the HGC (Figure 6.3) have important implications for people's reproductive choices.

Protecting the individual from invasion of genetic privacy is clearly a matter of major concern. But the secure storage of data in computerized databases is intrinsically difficult, and concerns are often expressed that opportunities for abuse will be increased when access is allowed to commercial organizations engaged in health-related research. The HGC noted *'some public disquiet'* but considered that because development of medicines and treatments was largely a commercial undertaking, denial of access would have serious adverse consequences. As a safeguard, it was recommended that all personal genetic information should be *anonymized,* although the encryption code should be capable of being reversed when necessary.

In addition, it was recommended that: *'Non-consensual or deceitful obtaining of personal genetic information for non-medical purposes should be made a criminal offence.'*

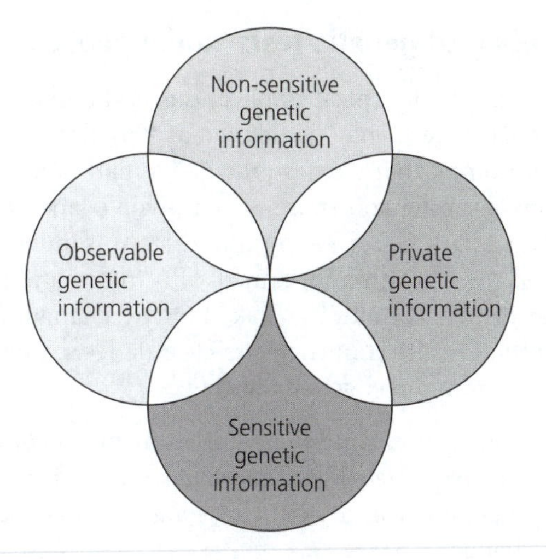

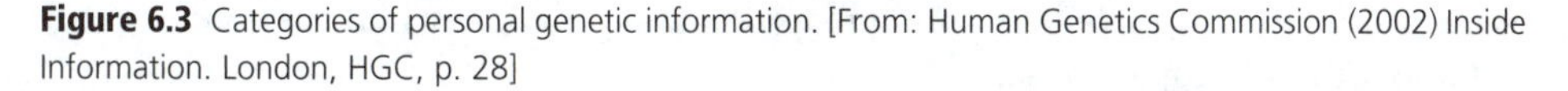

Figure 6.3 Categories of personal genetic information. [From: Human Genetics Commission (2002) Inside Information. London, HGC, p. 28]

The potential for abuse of genetic information would certainly seem to be a very serious concern. Consider the ease of genetic profiling, secretly and on people who have not given consent, by analysis of saliva on a beer glass, or of a stray hair removed from clothing. In future, it seems it might be almost impossible to protect yourself from an invasion of your genetic privacy by someone intent on accessing your profile. And from there it could be only a short step to genetic discrimination, or blackmail.

6.4.2 The UK Biobank

Analysis of genetic information will be an important role of the UK Biobank, which was established in 2003, with the aims of supporting research to improve the prevention, diagnosis, and treatment of illness and the promotion of health. Backed by the MRC, Wellcome Trust and Department of Health, from 2005 the project will follow the health of up to 500,000 volunteers over many years, collecting data on environmental and lifestyle factors, and linking them to medical records and to biochemical and genetic analyses of blood samples.[13]

Although Biobank is a long-term project, concerns have already been expressed about the confidentiality of the data, and about obtaining consent for use over a long timescale – when new opportunities might arise, but it could be very expensive to constantly seek consent on highly technical matters. An ethics council has been established to draw up strict guidelines on data use. Pharmaceutical companies will be allowed access to the information on the database, but all data will be anonymized. Insurance companies will have access to neither individual nor anonymized data, although the police may gain access under the terms of a court order.

6.4.3 Direct marketing of genetic tests to the public

A development with important ethical implications is the marketing of genetic tests directly to the public. Such tests may be defined as 'any tests to detect differences in DNA, genes or chromosomes, that are not provided as part of a medical consultation'. These cover tests done at home and those which involve laboratory tests on samples collected at home. This might be seen as a positive step, as individuals take increasing responsibility for their own health and, armed with information about their susceptibilities to particular disease conditions, make lifestyle changes (e.g. through diet or exercise) to promote their health. But there are clear dangers. The HGC identified two broad potential 'harms' from direct genetic testing:

- the impact on individuals of misinterpreted or erroneous predictive health information that overstates the role of genetics in causing common diseases, and which might result in delays in proper medical advice being sought, or in expensive and unproven dietary or lifestyle changes
- the possibility of people performing inappropriate genetic tests on children or other adults without proper consent.[14]

Consequently, the HGC recommended that there should be strict controls on such genetic testing, and that most tests providing predictive health information should *not* be offered in this way. Instead, they suggested that there should be a well-funded NHS genetics service, supported by a genetically literate primary-care workforce. There was particular concern about the possible misuse and abuse of tests performed at home. For example, for complex disorders such as high blood pressure, genetic factors may only account for a few percent of the risk, whereas changing habits in relation to diet, exercise, smoking and anti-hypertensive drug treatment could reduce the risk of a heart attack by up to 80%.

6.4.4 The Government White Paper

In 2003, the UK Government's Genetics White Paper claimed that *'genetics promises a more personalised approach to healthcare with interventions tailored to each person's genetic profile'*, and announced that £50 million was to be allocated over the following three years to develop specialized genetics services. Beginning with wider antenatal screening, one possible goal is for *all babies* to have their genetic profiles recorded at birth, to identify individual inherited risks of diseases developing later in life.[15] However, in 2005 a joint report of the HGC and the National Screening Committee argued that the costs of such a programme would be prohibitive. At current prices it could cost up to £10,000 to screen one infant for all genetically determined diseases, and with 60,000 children born in the UK annually this would place an impossibly large financial burden on the NHS.[16] The project was also questioned by those who argue that people would not be protected from genetic discrimination, and that in the absence of legal safeguards to prevent employers and insurers abusing genetic test information, the increasing use of such tests could lead to the emergence of a 'genetic underclass', excluded from insurance and employment.

Apart from such concerns, doubts have been expressed as to the current capability of genetic profiling to yield valuable information, partly because there have been few well-designed epidemiological studies and clinical evaluations of recommended interventions,[17] but also because most molecular geneticists lack the statistical know-how to analyse properly the wealth of data they obtain from DNA microarrays.[18] It seems likely, however, that genetic testing will increase markedly in future. This places information providers under an obligation to also supply advice on how to use the information wisely. The process involved, **genetic counselling**, is a complex and skilled activity,[19] but it would seem to be essential if people are to benefit from, and not be harmed by, the results of tests. The results of the Biobank project (6.4.2) might ultimately help in achieving this objective.

6.4.5 Geneticization

An important question affecting society's whole approach to health (and the implications for employment and insurance) is whether genetic explanations of disease are always justified by the scientific evidence. Some people claim that the role of genes is usually exaggerated, for which the term **geneticization** has been coined. By analogy with the phenomenon of medicalization (5.3.1), *'Geneticisation refers to an ongoing process by which differences between individuals are reduced to their DNA codes, with most disorders, behaviours, and physiological variations defined, at least in part, as genetic in origin.'*[20] This might be taken to imply that disease can best be countered by eliminating it at the beginning (by abortion or discarding IVF embryos) or, possibly, by gene therapy (6.4.7).

The concept of geneticization has been explored by several authors, at more popular[21] and more academic levels.[22] In essence, it is argued that, while increased knowledge about DNA will throw light on the structure and function of different proteins, it is far less likely to help us understand the network of metabolic relationships underlying most diseases and disabilities, because they depend on many factors not influenced by genes. These include dynamic genetic processes such as **epigenesis** (gene interactions) and **pleiotropy** (the multiple effects of a single gene or protein).[23] Despite this, DNA is frequently discussed as if it was the 'be-all and end-all' of biology.

Such misperceptions sometimes result from the way science is reported by the news media (13.6.2). For example, a news item claiming *'Scientists have cloned a gene that helps brain cells communicate, a step that may lead to improved drugs for schizophrenia'* may not be inaccurate, so much as far-fetched. If the message was deconstructed, it might mean nothing more than the fact that a DNA sequence implicated in the synthesis of the dopamine receptor had been identified, and since this is active in brain tissue it is likely to be affected, in one way or another, in conditions such as schizophrenia, a condition most people know to be distressing. Negative results would not be newsworthy, but the cumulative effect of selective reporting of positive associations, albeit often tentative, could introduce an unconscious bias, particularly with the lay public.

6.4.6 Disability

One perspective usually missing from debates on the use of genetic information in addressing disability is that of disabled people themselves, and this at a time when anti-discrimination legislation has been introduced and barriers to their access to public facilities are increasingly removed. It is a diverting coincidence that the most famous lines in English literature, Hamlet's *'To be or not to be, that is the question',* have acquired poignant significance in the debate over the ethics of whether to abort a human fetus or discard an IVF embryo – and that a present-day namesake of the Bard is prominent in the debate. Academic and disability rights campaigner Tom Shakespeare claims that many disabled people *'deeply resent the hyperbole surrounding discoveries in the new genetics, as well as the negative language that is often used about disability'* – frequently couched in terms such as 'the burden of disability' and 'the horrors of genetic disease'.[24] He argues that many impairments are not necessarily disabling if the right social supports are provided. It might be seen as ironic that moves to prevent the birth of disabled people are being made when medical advances have resulted in their greatly increased lifespans, reduced illness, improved mobility, and enhanced abilities to communicate with others. The ability to overcome physical disability is amply demonstrated by the thousands of disabled athletes who participate in the Paralympic Games (most recently in Athens in 2004).[25] Indeed, some people consider the label 'disabled' is quite inappropriate, regarding their status as one of 'cultural identity' (6.5.4).

Perhaps we need to look again at what disability means. There is a case for saying that most of us are, at times, disabled: inevitably when infants, but often when ill, and usually in old age. Another perspective on this question concerns dependency. If the alleged 'problem' with disabled people is that they are perceived to be dependent on society, and hence a burden, it should be recognized that *most* of us are, to a large degree, dependent on others, e.g. for food, housing, health, education and recreation; and the wealthier people are, the more dependent they tend to become, as they buy services which poorer people perform for themselves.

Disabled people are, in fact, often better described as 'differently abled', their limitations in some respects being more than compensated for by other, possibly heightened, abilities. Disability has not prevented some people becoming distinguished in several fields, such as science and the arts: just think of Stephen Hawking and Beethoven. The point is emphasized by consulting the insightful website of the *International Center for Bioethics, Culture and Disability,*[26] which is run by the Canadian biochemist and bioethicist Gregor Wolbring, himself a 'thalidomider' (see 8.3.3). (The website contains an inspiring poem about Dr Wolbring.) So the view that elimination of disability is the best 'solution' is likely to encourage an attitude of intolerance more generally, much to society's cost. Indeed, it might be said that a measure of a society's moral development is revealed in the way it treats those who, for whatever reason, do not conform to the norms of the majority. Moreover, if the number of people with inherited disability diminishes, disability may become stigmatized, with parents who have decided against abortion coming to be regarded as 'irresponsible'.

Opposition to eugenic, and indeed all, abortions is promoted by organizations such as the UK Society for the Protection of the Unborn Child (SPUC)[27] and by disability rights movements who support the woman's right to terminate a pregnancy for 'social' but not for eugenic reasons. But this negative view of genetics is not shared by all disabled people. For example, the UK Genetics Interest Group[28] represents families affected by genetic disability, and is generally supportive of research in this field. Their case is simply put: if medical advances can prevent avoidable suffering and discrimination, the moral case for exploiting them for such purposes is much stronger than for many other uses.

6.4.7 Therapies for genetic diseases

At present, the ability to diagnose genetically linked disease and disability far outstrips the ability to correct these conditions by genetic treatments. This is an important reason for questioning the value of genetic testing: if it is impossible to ameliorate a medical problem, is there anything to be gained by telling someone that their child has got it, other than offering the abortion option? But putting it that way paints a picture that might be far too gloomy. Two options are theoretically feasible.

Somatic gene therapy. It is true that, to date, somatic gene therapy, i.e. the attempt to modify tissue function by inserting a gene into the cells of the target tissue, has had only very limited success. A notable example is the trials conducted on patients suffering from cystic fibrosis (a condition that results in severe breathing difficulties), who inhale an aerosol spray of liposomes containing the gene that patients lack. If they were more promising, such therapies might appear to be ethically unproblematical, because any changes induced would be limited to the patient, and not passed on to their children.[29] But it has been pointed out that *'the route from genotype to phenotype is often long and circuitous, and doctors may intervene at any of a number of stages'*.[30] Dietary change to avoid the problem of PKU (6.3.2) is one example of using genetic information effectively. Other conditions can sometimes be countered by pharmaceutical, environmental, or behavioural strategies, or by use of prosthetic technologies. The prospect of ribonucleic acid (RNA) interference therapy being a practical possibility also offers hope for many sufferers (6.3.2).

Germ-line therapy. This is a more radical, and generally more ethically questionable, approach, in which genes are inserted into germ cells or fertilized eggs to correct a hereditary disease. In this case, the modification introduced does not die with the individual treated, but is handed down to future generations. Because it attacks the genetic basis of disease, this might be seen as holding out the promise of permanently eradicating certain diseases which have for long plagued humanity. However, almost universally, medical opinion has been opposed to the practice. In their report in 1992 the BMA listed objections to germ-line therapy as follows:

- because the technology is imprecise, any errors in the therapy (e.g. involving the insertion of a gene in the middle of another gene, disrupting its expression) would also be inherited
- outcomes are unpredictable, e.g. eradicating the gene for sickle cell anaemia might increase the susceptibility of some carriers of the gene (e.g. in tropical Africa) to malaria, against which the gene confers resistance
- manipulation of the embryo is seen by many as intrinsically ethically objectionable (5.7.2).[31]

More recently, however, the ground may have shifted. In line with the recommendations of the Clothier Report, germ-line therapy currently remains illegal in the UK. But since that report based its decision on the fact that there was *'insufficient knowledge to evaluate the risks to future generations'*, so that *'gene modification of the germ line should not yet be attempted'*[32] (i.e. solely on utilitarian reasoning), the door seems left ajar for official approval at some future date.

In fact, the difference between somatic and germ-line therapies may not be clear-cut. This is because genes inserted into the body in the course of somatic therapy (e.g. by using a virus as a vector) might combine with other viruses and infect the germ cells. Moreover, if somatic therapy led to a greater concentration of a mutant gene in the gene pool (by allowing patients, who would otherwise die, to survive and reproduce), it might have a significant effect on the gene pool, albeit an indirect rather than a direct one.[33]

6.5 Stem cells

Common degenerative diseases such as Parkinson's disease, which affects brain function, and type I diabetes, which is due to inadequate insulin production by pancreatic islet cells, are a result of cell failure. For some years attempts have been made, with only limited success, to treat these conditions by transplanting healthy tissue from other sources, e.g. aborted fetuses. However, since 1998 it has been possible to isolate **embryonic stem cells** (ESC) from human embryos, opening up many new avenues for cell therapy.

Stem cells are unspecialized cells that have the ability to divide indefinitely, so giving rise to the 200 + cell types present in the body. ESC are pluripotent, having the potential to make most cell types, whereas **adult stem cells** (ASC) taken from tissues such as skin, muscle, and bone marrow, appear to have more limited abilities - although it is possible that future research on reprogramming will reveal as yet unrecognized potential. However, despite their apparently greater versatility, a technical problem with using tissue cells derived from ESC is **immune rejection**, which would clearly not be encountered if the patient's own ASC could be used. (In this context, the stem cells in young children are also ASC.)

6.5.1 Therapeutic cloning

The proposed strategy for obtaining immunologically compatible ESC is to employ **somatic cell nuclear transfer** (SCNT; see 5.5.3) before IVF, with removal of ESC from the resulting embryo. Using SCNT for **therapeutic cloning** involves taking the nuclei from a few somatic cells of the patient's body (most conveniently, skin cells) and transferring them into unfertilized eggs from which the cell nuclei have been removed. In a culture dish, the eggs are then exposed to a pulse of electricity, which stimulates the cells to divide, producing blastocysts (balls of cells). The inner cell mass, part of the blastocyst, can be extracted and ESC grown from it. These ESC, which contain the patient's DNA, will clearly match the patient's immune system, so that any tissues subsequently produced from this cell line can be used in treating the patient, e.g. by implanting them in the pancreas of a diabetic patient, without risk of immune rejection. SCNT is also considered likely to be useful in enabling a woman with damaged mitochondria to produce a healthy baby by insertion of her cell nucleus into an enucleated donor egg cell.

In 2004, the HFEA granted the first licence in the UK for therapeutic cloning to scientists at Newcastle University,[34] a decision which met with widespread support in the bioscience community. However, it is important to examine both the limitations and the objections to this approach. One concern focuses on the practicability of its becoming a routine medical procedure. In fact, the HFEA's 'initial one-year licence' was granted for research purposes only, which was intended to form the basis for further development in the treatment of serious diseases: it was not aimed at any specific illnesses.

Those scientists who see the technique as primarily a research tool (for which the term **research cloning** might be more appropriate) argue that the technique is likely to be too elaborate, and hence too expensive for routine use. Currently, the SCNT procedure is very inefficient (e.g. the Korean scientists who created the first stem cell line from cloned human embryos used more than 200 eggs) and the only supply of such eggs in the work performed under the HFEA licence will be those produced surplus to requirements in IVF programmes. According to Richard Gardner, who chairs the Royal Society's working group on stem cells and therapeutic cloning, *'There is a growing gulf between what medicine can do and what the health service can afford.'*[35]

6.5.2 Ethical concerns over therapeutic cloning

Stem cell research revives many of the earlier debates over the use of embryos, but gives them a sharper focus because some quite specific medical objectives replace the earlier, rather indeterminate, 'quest for knowledge'. Accordingly, the potential to treat disease and relieve suffering is believed by many people to justify research across the whole range of possible sources of stem cells, including embryos.

The case *against* such research appeals to several ethical arguments. Thus, it has been argued that:

- the human embryo demands respect as a human being from the moment of its conception, so that the intentional production of fetuses for research, or the use of so-called 'spare embryos' in IVF programmes, is tantamount to murder, and uses a human life simply as a *means* rather than as an *end* (5.7.2)
- the technique for therapeutic cloning is essentially the same as that for reproductive cloning (5.5.3), so there is a high risk that its approval for one purpose will be exploited for this other, widely condemned, technology
- use of ESC depends on ART, which themselves have ethically problematical features (5.6).

It has also been argued that the potential for therapies based on ASC has been underestimated, making the utilitarian justification for the use of embryos less valid. The report of a group chaired by Liam Donaldson, the Chief Medical Officer in the UK, made several recommendations in the light of the belief that currently ESC have the greatest potential for medical application. However, it also suggested that *'In the long-term the scientific view is that it will be possible to re-programme adult stem cells with the full potential of embryonic cells but without the morally contestable need to create an embryo.'*[36]

Some people argue both that the potential application of ASC has greatly increased since publication of the Donaldson report, and that there are some serious technical limitations to the use of ESC.[37] However, in Richard Gardner's (6.5.1) view: *'Since it is likely that ESC will be found more appropriate for some therapeutic uses and adult cells for others, there is a pressing need for a dispassionate appraisal of the merits and drawbacks of both.'*[38]

Several developments may modify the ethical cases advanced both for, and against, stem cell research. For example, attempts have been made to use unfertilized eggs from cows to produce the embryos[39] (Figure 6.4). This would have a clear technical advantage in that donated human eggs would be avoided, but it is likely to be widely condemned because of the mixing of material from different species.

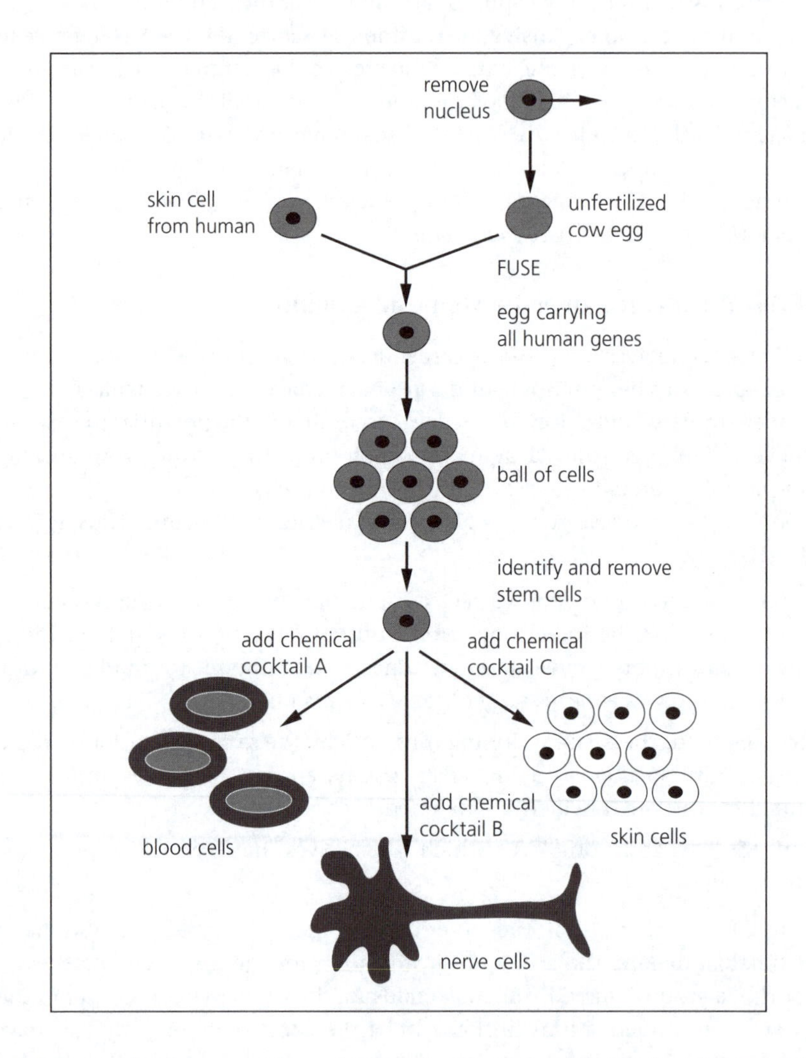

Figure 6.4 The possible future use of somatic cell nuclear transfer and the isolation of stem cells in the production of specialized cells, tissues, and organs. [From: Curran B (2003) A Terrible Beauty is Born. London, Taylor and Francis, p. 138]

6.5.3 Saviour siblings

A closely related issue is the selection of embryos, generally using human leukocyte antigens (HLA) as a means of matching people who are immunologically compatible, and thus able to provide stem cells which can be used to treat a sick patient. Notable cases were the applications from two sets of parents (the Hashmi and Whitaker families) to the HFEA for permission to select embryos (so-called **saviour siblings**) to treat their existing children, who suffered from *beta-thalassaemia* and a rare form of *anaemia*, respectively. The Hashmis were granted permission by the HFEA to create and tissue-type embryos, a process that involved PGD (6.3.3); stem cells were subsequently obtained by extraction of umbilical cord blood from the newborn child. By contrast, the application made by the Whitakers was refused, so the couple went to the USA to receive treatment.

The differences between the two cases are too complex to explore here, but it is important to be aware both of the constraints applied by the HFEA and the concerns expressed over the ethical acceptability of saviour siblings. Chair of the HFEA, Suzi Leather, lists the following *'strict criteria for selecting embryos as matches for other people'*:

- the condition of the affected child must be serious or life-threatening
- the embryos themselves must be at risk from the condition by which the existing child is affected
- all other possibilities of treatment and sources of tissue for the affected child should have been explored
- the technique should not be available where the intended recipient is a parent
- only cord blood should be taken
- appropriate counselling is a requirement
- families should be encouraged to participate in follow-up studies
- embryos may not be genetically modified to provide a tissue match.[40]

Leather identifies the difficulty of separating the selection of undesirable hereditary factors from the optimization of desirable ones, and the 'difficult moral calculus' where the potential risks of PGD fall on one child and the benefits on the other. And she wonders what the impact might be of discovering that you owe your existence to serving another's purpose, however laudable. Perhaps, she notes, instead of saviour siblings one might equally speak of *'spare-part sisters'* and *'bred-to-order brothers'*.

6.5.4 Designer babies

The term **designer babies** is a popular catch-all phrase, applied loosely to choices people make in seeking to determine reproductive outcomes. Saviour siblings (6.5.3) are classic examples of designer babies, but others include:

- use of sex-selection techniques to avoid the birth of a child with an X-linked disease condition (see 6.3.3)

- selection of donors of sperm, eggs, or embryos for particular characteristics
- enhancement or selection of features such as intelligence, physical appearance or athletic prowess.

A notable example of the second case was the selection by a deaf lesbian couple in the USA of a sperm donor who had an inherited form of deafness, so that any child produced by IVF would have a high chance of being deaf. As a result, the couple had two children, who were both deaf. They argued that deafness is not a disability but a *cultural identity*, and they wanted their children to share their own identity. The case caused much controversy on the grounds that the decision: (*a*) intentionally produced children with reduced capabilities; and (*b*) amounted to an instrumental use of another human to suit personal desires.[41] A comparable desire of some parents is to use PGD to select the sex of their child for 'social reasons'. However, following a public consultation exercise in the UK, the HFEA announced in 2003 that it had decided to ban sex-selection for 'family balancing', although permitting it where a child of one sex would risk inheriting a serious genetic disorder.[42]

The third case is illegal in the UK, but there are growing pressures to allow instances of such positive eugenics, as discussed in 6.6.

6.5.5 Ethical evaluation of reproductive choices

Taking a stand on the ethics of reproductive choices is not made any easier by the shifting sand beneath one's feet. In the view of the Chief Medical Officer's expert group, it will be possible, in the long term, to reprogramme ASC for therapeutic purposes, diminishing the force of the arguments for using embryos. But treating serious medical problems is always urgent for those suffering from them, so that taking time to deliberate on what might be thought to be ethical niceties, or waiting for technology to deliver, might seem an unjustifiable, even callous, luxury. This is when we realize, if we hadn't before, that ethics can be hard.

Appealing to ethical frameworks, such as the ethical matrix (5.6.4), may help to clarify the issues, but much depends on the weighing process. We not only need to imaginatively consider ourselves as victims of a debilitating disease, important as it is to do just that, but also to consider the position of those whose interests rarely get into the frame – such as the 'have-nots' (chapter 4), future children, and their children. All our lives are transient, and it may be that in the interests of justice and human dignity there are some practices that ought not to be introduced, because to do so would exacerbate existing injustices or diminish our humanity. Or consideration of the implications of negative and positive rights (5.6.1) might suggest that, when there are competing ethical claims, one demands a higher priority than the other.

6.6 Genetic futures?

Up to this point, we have considered the role of genes in disease and disability, and the ways genetic information might be used to avoid, or cope with, these conditions. But the influence of genes clearly does not stop there. The metaphors often used, such as 'the blueprint of life' and 'the book of man', imply that the genome's role is all-pervasive, albeit modified by environmental factors. Consequently, it is likely that normal behaviour, as well as disease, is genetically determined to a greater or lesser degree. For example, genes may play important roles in intelligence, sexual orientation, and antisocial behaviour.

It might seem to follow that if it is possible to determine individuals' genetic predispositions, people ought to have access to that information, to enable them to modify their behaviour in accordance with their life plans. More contentiously, it might be argued that knowledge of these genetic predispositions could have implications for social policy, e.g. in terms of what sort of educational provisions are made available, or the sentencing of people convicted of certain crimes.

6.6.1 The Nuffield report: genetics and human behaviour

The general tenor the Nuffield report on behavioural genetics was precautionary: it went to great pains to stress that genes play only a contributory role in human behaviour. The report was critical of the common tendency to claim that there were genes 'for' certain types of behaviour, such as *homosexuality* or *aggression*; and this led to a recommendation that researchers should communicate their findings in a responsible manner. Moreover, it was recognized that behavioural genetics has the potential to contribute to the phenomenon of medicalization (5.3 and 6.4.5), which could be damaging if society grew intolerant of those who became labelled as genetically 'abnormal'.

Nuffield shared the concerns of many people that there were dangers to the fabric of society in allowing people to elect for abortion of fetuses following PND (6.3.1), where the decision was based on non-clinical attributes, such as sexual orientation. But decisions based on PGD (6.3.3) were seen in a different light because they did not *'precede the termination of a potential human life'*.[43] On germ-line therapy, the report followed the utilitarian reasoning advanced by the Clothier Committee (6.4.7), stressing that in view of the risks, *'considerable caution should be exercised before contemplating its application to traits that do not have serious implications for health'*.

6.6.2 Reproductive autonomy and positive eugenics

But let us imagine that technical improvements in germ-line therapy are made, so that it carries few risks. Even if this is not imminent, we need to consider the ethical implications in advance. In that case, it can be argued that if individuals wish to *enhance* their prospective children's capabilities by this means, respect for their autonomy should allow them to do so. This is defined as **positive eugenics**, and would entail permanently introducing a genetic trait into the genomes of descendants. A common reaction to this objective is one of outrage, in that it seems to involve *designing* future individuals according to some predetermined plan, rather as the Nazis attempted to do in building a master race. But outrage is not a good basis for ethical deliberation, and we need to examine what is now being advocated to see whether it does amount to the same thing, or to more ethically defensible aims.

Philosopher Philip Kitcher suggests that at least some forms of enhancement are not at all ethically questionable. For example, if we could enhance the immune system, so reducing the vulnerability to infectious and autoimmune diseases, or prevent the loss of hearing which normally accompanies old age, it might seem unreasonable to prohibit such advantages on the grounds that they seek to 'improve on nature'.[44] Kitcher proceeds to argue that eugenic practices do not differ in their aims from those advocated by social reformers who wish to improve society by providing better education, better housing and greater opportunities for living rewarding lives. But conscious of the dangers of a politically imposed programme, as pursued notoriously by the Nazis, he suggests that putting eugenics into practice in ethically acceptable ways (what he calls *utopian eugenics*) requires satisfying four important criteria:

- selection of a group of people whose reproductive abilities will make a difference to future generations
- allowing these people freedom to make their own reproductive decisions
- selecting appropriate characteristics to modify genetically
- employing appropriate scientific information to realize the objectives.

Pointing out that the Nazi programme made bad choices on all four counts, he nevertheless argues that *'Everyone is now to be her (or his) own eugenicist, taking advantage of the available genetic tests to make reproductive decisions she (he) thinks correct'* – the most attractive feature of which is that it attempts to honour reproductive freedom.[45]

6.6.3 Slippery slope or ladder to perfection?

Gregory Stock[46] is another academic in the USA who has few misgivings about germ-line enhancement of human beings. He claims that future opportunities in what he calls **germinal choice technologies** (GCT) *'will far outweigh the risks'*, and that: *'to participate in this unprecedented development is an amazing privilege'*. Stock's strong approval of positive eugenics is based on what he sees as realistic scientific possibilities for good; so in order to examine the ethical justification implicit in this claim we need to explore the assumptions underpinning his case.

His major theme is that GCT is an *inevitable* outcome of the overlapping research programmes currently pursued, in particular those researching the human genome, clinical medicine, animal transgenesis, and human fertility. He argues that to suggest anything else is to underestimate the power of market forces. However, Stock envisages technical difficulties with germ-line therapies, and suggests that GCT will most probably be accomplished by adding auxiliary human chromosomes (numbers 47 and 48), which will carry artificial genes for characteristics promoting attributes like increased intelligence and longevity. To counter the objection that this will fix 'improvements' permanently in the genome, he cites research on the CRE gene, coding for an enzyme that selectively snips out DNA sequences. This could be switched on by taking a drug, so inactivating the auxiliary gene when necessary. According to this scenario, each successive generation could have a genetic update, rather as you currently update your computer! This seems, to Stock, to be the utopian vision, because *'If we could make our baby smarter, more attractive, a better athlete or musician, or keep him from being overweight, why shouldn't we?'*[47]

The parallels with Huxley's *Brave New World* (5.5.3) might seem striking, but supporters of the new positive eugenics argue that eugenic abuses of the past did not concern biotechnology, biology, or human reproduction *'but nationalism, totalitarian regimes...and tyranny'* – abuses that will be avoided when choices are made within the free market.[48] US molecular biologist Lee Silver[49] also speculates that genetic enhancement will enrich people's physical and cognitive attributes, and that in centuries to come changes now unimaginable will become indispensable. But rather more threateningly, he envisages the eventual emergence through genetic enhancement of two distinct species of human, the Naturals, with unadapted genomes, and the genetically enhanced GenRich, who will control *'all aspects of the economy, the media, the entertainment industry, and the knowledge industry'.*

6.6.4 Analysing eugenics ethically

How are we to assess the possibilities of positive eugenics from an ethical perspective? Use of the ethical matrix (5.6.4) provides a framework to guide our deliberations, which focuses attention on the impacts of genetic technologies on respect for well-being, autonomy and fairness. The interest groups to be considered will clearly include the prospective children and *their* prospective children, but it would also be important to consider impacts on society as a whole, both at the time of the interventions and in future generations.

Respect for the child's well-being may seem most evident in potential harms and benefits. Is the technology sufficiently well understood to subject an innocent life to the risks associated with untried technologies? However well techniques are tested in animals, the first human babies will be experimental subjects, and many would argue that a strong version of the Precautionary Principle (13.4) should be applied. Because they are incapable of giving consent, their autonomy is undermined, but possibly no more than for all decisions made on their behalf. Respect for fairness is undoubtedly challenged, but genetic enhancement might compensate for theoretical risks. However, for many people, the perceived instrumental use of prospective children, moulding them to personal desires, would fail to respect human dignity.

The autonomy of the would-be parents might be increased by permitting them to choose wisely (as they see it) for their children, but the uncertainty and risks associated with the procedures might seriously undermine the parents' welfare both through the anxiety caused and as a result of possible adverse discrimination and stigmatization from neighbours and the tabloid press. Moreover, the ability to 'design' their baby can be said to challenge what has been called 'natural humility'. This has been characterized as 'an attitude of unconditional parental love' that is incompatible with what might be seen as the commodification of human life.[50]

The impacts of positive eugenics on society might be profound, even if Silver's 'two human species' outcome is far-fetched. According to the optimistic scenario, a free market in genetic testing and enhancement will ensure a diverse, socially and technically advanced global society. Others, however, might envisage as more likely the possibility of the emergence of a genetic underclass, unemployable, uninsurable, and stigmatized by the *genetic haves*, so amplifying the existing divisions between rich and poor. In fact, we do not have to look far to see recent examples of governments using genetics to improve the levels of intelligence of their populations.[51]

US academic Francis Fukuyama suggests there are two ways by which, in theory, Silver's notion of 'genetic apartheid' might be prevented. It might be declared illegal (but this would entail an enforceable global ban) or the technology could be made available to the least privileged, e.g. by state-sponsored eugenics.[52] An ethical judgement entails assessing, among other things, the likelihood of implementing these alternative scenarios.

THE MAIN POINTS

- Techniques such as PGD coupled with genetic testing allow identification of the risks of children inheriting serious diseases, thus enabling parents to discard rather than implant affected embryos.
- Genetic information in adults (made increasingly possible by the results of the Human Genome Project) can also alert people to desirable lifestyle changes, but there are risks that fatalistic attitudes may offset advantages.
- The roles of genes in determining disability and behaviour (e.g. as affecting intelligence or criminality) raise ethical issues concerning discrimination and privacy.
- The claim that medical research using stem cells derived from cloned human embryos (therapeutic cloning) is ethically acceptable in terms of future benefits is challenged by those who regard it as an excessively instrumental use, which offends human dignity.
- The possible future use of germ-line therapies in eugenics programmes is condemned by some on both utilitarian and deontological grounds, but considered by others to be both an inevitable development and a moral obligation in order to realize human potential more fully.

EXERCISES

These can form the basis of essays or group discussions:

1. Discuss the ethical justification for parents' decision to abort a fetus which they discover, through PND, carries a gene for Huntington's disease (6.3.2).
2. From an ethical perspective, what are the pros and cons of direct marketing of genetic tests to the public? (6.4.3)
3. Does using PGD to select against genes for disability stigmatize disabled people? (6.4.6)

4. Discuss the ethical arguments for and against *saviour siblings* (6.5.2).

5. Is Kitcher's *utopian eugenics* ethically defensible? (6.6.2)

FURTHER READING

- *What Sort of People Should There Be?* by Jonathan Glover (1984). Harmondsworth, Penguin. One of the earliest books on genetic engineering of humans, written by a philosopher.
- *In Our Own Image: eugenics and the genetic modification of people* by David Galton (2001). London, Little, Brown and Co. An engaging account of the new eugenics written for the general reader by a medical scientist (who, despite his name, claims no relationship to the father of eugenics, Francis Galton).
- *The DNA Mystique: the gene as a cultural icon* by Dorothy Nelkin and M Susan Lindee (1995). New York, W H Freeman. A sociological analysis of geneticization.

USEFUL WEBSITES

- **http://www.hgc.gov.org/** *Human Genetics Commission*: the UK Government's advisory body. It has produced several useful reports.
- **http://www.nuffieldbioethics.org/** *Nuffield Council on Bioethics*: a leading UK independent bioethics committee. It has published a number of influential reports over the past decade.
- **http://www.ukbiobank.ac.org** *The UK Biobank*: the newly established database which aims to link medical records to biochemical and genetic analyses.

NOTES

1. Yoxen E (1986) Unnatural Selection? London, Heinemann, p. 111
2. Nuffield Council on Bioethics (1993) Genetic screening: ethical issues. London, Nuffield Council on Bioethics
3. Human Genetics Commission (2004) Choosing the Future: genetics and reproductive decision making. London, Department of Health, p. 11
4. Steinbock B (2002) Preimplantation genetic diagnosis. In: A Companion to Genethics. (Eds) Burley J and Harris J. Oxford, Blackwell, p. 178
5. Xia H, Paulson H, and Davidson B L (2002) siRNA-Mediated gene silencing *in vitro* and *in vivo*. Nat Biotechnol *20*, 1006–1010
6. Chadwick R and Hedgecoe A (2001) In: A Companion to Genethics. (Eds) Burley J and Harris J. Oxford, Blackwell, pp. 334–344
7. Steinbock B (2002) Preimplantation genetic diagnosis. In: A Companion to Genethics. (Eds) Burley J and Harris J. Oxford, Blackwell, pp. 175–190

8. Meikle J (2003) Ten babies born via new IVF technique. The Guardian, 6 August 2003
9. Chadwick R (2001) Genetic screening. In: The Concise Encyclopedia of New Technologies. (Ed.) Chadwick R. San Diego, Academic Press, pp. 193–198
10. Nuffield Council on Bioethics (1993) Genetic Screening: ethical issues. London, Nuffield Council on Bioethics. Paras 10.14 and 10.15
11. United Nations Educational, Scientific and Cultural Organization (1997). Geneva, UNESCO
12. Human Genetics Commission (2002) Inside information. London, HGC
13. The UK Biobank (2004) http://www.ukbiobank.ac.org
14. Human Genetics Commission (2003) Genes Direct. London, HGC
15. Department of Health (2003) Our inheritance, our future – realizing the potential of genetics in the NHS. http://www.doh.gov.uk/genetics
16. Sample I (2005) Infant DNA testing too costly, says watchdog. The Guardian 5.3.05 http://www.guardian.co.uk/print/0
17. Haga S B, Khoury M J, and Burke W (2003) Genomic profiling to promote a healthy lifestyle: not ready for prime time. Nature *34*, 347–350
18. Tilstone C (2003) DNA microarrays: vital statistics. Nature *424*, 610–612
19. Clarke A J (2001) In: The Concise Encyclopedia of New Technologies. (Ed.) Chadwick R. San Diego, Academic Press, pp. 131–146
20. Hoedemaekers R (2001) Geneticisation. In: The Concise Encyclopedia of New Technologies. (Ed.) Chadwick R. San Diego, Academic Press, pp. 157–164
21. Nelkin D and Lindee M S (1995) The DNA Mystique. New York, W H Freeman
22. Hubbard R and Wald E (1997) Exploding the Gene Myth. Boston, Beacon Press
23. Strohman R (2000) Upcoming revolution in biology. In: Made not Born: the troubling world of biotechnology. (Ed.) Walker C. San Francisco, Sierra Club Books, pp. 108–118
24. Shakespeare T (2003) Disability, human rights and contemporary genetics. In: Encyclopedia of the Human Genome. Macmillan Publishers, Nature Publishing. www.ehgonline.net
25. The Paralympic Games (2004) http://www.paralympic.org
26. The International Center for Bioethics, Culture and Disability (2002) http://www. bioethicsanddisability.org/
27. Society for the Protection of Unborn Children (2004) Assisted reproduction technologies. http://www.spuc.org.uk
28. Genetic Interest Group (2004) http://www.gig.org.uk/
29. Hedgecoe A (2001) Gene therapy. In: The Concise Encyclopedia of New Technologies. (Ed.) Chadwick R. San Diego, Academic Press, pp. 123–129
30. Kitcher P (1996) The Lives to Come: the genetic revolution and human possibilities. London, Penguin, p. 108
31. British Medical Association (1992) Our Genetic Future. Oxford, Oxford University Press, pp. 185–188
32. Committee on the Ethics of Gene Therapy (1992) London, HMSO (Cm. 1788)
33. Chadwick R (1998) Gene therapy. In: A Companion to Bioethics. (Ed.) Kuhse H and Singer P. Oxford, Blackwell, pp. 189–197
34. Human Fertilisation and Embryology Authority (2004) HFEA grants first therapeutic cloning licence for research. http://www.hfea.gov.uk/PressOffice/Archive10922338888
35. Sample I (2004) Is there hope behind the stem cell hype? Guardian, 19 August 2004
36. Department of Health (2000) Stem Cell Research: medical progress with responsibility. Chief Medical Officer's Expert Group on Therapeutic Cloning-Stem Cell Research. http://www.doh.gov.uk/cegc

37. Lillge W (2002) The case for adult stem cell research. http://www.21stsciencetech.com/articles/winter01/stem_cell.html

38. Gardner R (2004) Cardiff Centre for Ethics, Law and Science. http://www.ccels.cardiff.ac.uk/launch/gardnerpaper.html

39. Doerflinger RM (1998) Stem cell experiments: renewing the attack on human embryos. http://www.nrlc.org/news

40. Leather S (2004) Saviour siblings transcripts. http://progress.org.uk/Events/Leather

41. Human Genetics Commission (2004) Choosing the future: genetics and reproductive decision making. London, Department of Health, p. 24

42. Human Fertilisation and Embryology Authority (2003) http://www.hfea.gov.uk/Press

43. Nuffield Council on Bioethics (2002) Genetics and Human Behaviour. London, Nuffield Council on Bioethics, p. xxx

44. Kitcher P (1996) The Lives to Come: the genetic revolution and human possibilities. London, Penguin, p. 124

45. Ibid., pp. 193–196

46. Stock G (2002) Redesigning Humans. London, Profile

47. Ibid., pp. 188

48. Ibid., pp. 199

49. Silver L (1998) Remaking Eden: cloning, genetic engineering, and the future of mankind? London, Phoenix

50. Nuffield Council on Bioethics (2002) Genetics and Human Behaviour. London, Nuffield Council on Bioethics, p. xxx

51. Chan CK (1992) Eugenics on the rise: a report from Singapore. In: Ethics, Reproduction and Genetic Control. (Ed.) Chadwick R. London. Routledge, pp. 164–171

52. Fukuyama F (2002) Our Posthuman Future: consequences of the biotechnology revolution. London, Profile, p. 158

PART THREE
Bioethics and animals

Nevertheless, the difference in mind between man and the higher animals, great as it is, certainly is one of degree and not of kind

Charles Darwin (1883)
The Descent of Man

7

Human uses of animals

OBJECTIVES

When you have read and discussed this chapter you should:

- appreciate the animal welfare and animal rights arguments advanced in respect of human uses of animals
- understand the ethical justifications made for meat-eating, vegetarianism and veganism, and challenges made to each of these positions
- be aware of welfare problems experienced by farm animals in the UK, and the legislation which has been designed to protect them
- be able to reach a reasoned personal decision on the ethical acceptability of different forms of animal farming
- appreciate the arguments advanced in support of, and in opposition to, use of animals for various non-food purposes (e.g. as pets, in entertainment and in blood sports)

7.1 Introduction

This chapter is concerned with ethical implications of the ways in which we relate to animals. It chiefly addresses the use of animals for food, and, more briefly, in sport and entertainment. But the ideas discussed also have important implications for other uses of animals, e.g. in biomedical research (chapter 8) and in biotechnological applications (chapter 9).

7.1.1 Attitudes to animals

Thirty years ago philosopher Peter Singer wrote *Animal Liberation*,[1] in which he made a case for the ethical basis of vegetarianism at a time when this was still generally regarded as eccentrically sentimental. Partly a philosophical treatise, partly a graphic account of *factory farming*, and partly a cookbook, it had a major influence on what became known as the **animal rights movement**. (This is somewhat ironic because, as a utilitarian, Singer does not accept the validity of the rights argument.) Now, most people in Western society do acknowledge the ethical standing (3.4) of animals, although there is still much disagreement about how much respect we owe them. But it is worth considering how recently this change in attitude occurred and how it came about.

For the seventeenth-century French philosopher René Descartes, mind and matter were totally separate. Believing that only humans possessed souls and conscious minds, he thought of animals as mindless automata that existed solely for human use. The idea that animals were created *for* humans seems to stem from the dominant Western traditions that had their roots in Ancient Greek thought and Judaeo-Christian religion. Thus, the Book of Genesis in the Bible exhorts mankind to *'have dominion over the fish of the sea, and over the fowl of the air, and over every living thing that moveth upon the earth'*. In the eighteenth century, it was widely believed that animals had been designed for specific human purposes. Thus, it was claimed that *'Apes and parrots have been ordained for man's mirth'*, and singing birds *'on purpose to entertain and delight mankind'*, while cattle and sheep were given life *'so as to keep their meat fresh till we have need to eat them'*.[2] It is difficult to avoid a wry smile at the presumption that God had provided humanity with the equivalent of well-stocked walking fridges!

And yet, for people in general, such attitudes were often combined with an emotional attachment to the animals with whom they lived in close proximity, and to whom they attributed qualities such as affection, obedience, loyalty and courage. Indeed, zoologist James Serpell suggests that hunters saw their quarry as *'equals, or even superiors, capable of conscious thoughts and feelings analogous, in every respect, to those of humans'*.[3] Following the Industrial Revolution, when most people had moved to factory jobs in cities, losing contact with the agricultural roots of their parents' generation, animals came to be regarded sentimentally, as pets who shared the family home and were given human names, or as idealized images of a rural idyll vaguely remembered in folk-history. More recently, many children have become more familiar with cartoon animals than real animals; with Mickey Mouse, Donald Duck, and Teenage Mutant Ninja Turtles, rather than the cattle, sheep, and pigs whose flesh is disguised in plastic-wrapping on supermarket shelves.

7.1.2 The impact of Darwinism

So it is perhaps not so surprising that, even after Darwin first published *The Origin of Species* in 1859, appreciation of the theory of evolution did not rapidly alter the perception of a gulf between humans and non-humans. Only in the last thirty years or so has society at large begun to realize the full significance of Darwin's theory; for now, modern molecular genetics makes the fact of the genetic continuity of humans and animals virtually inescapable. Not only is it increasingly recognized that humans share 98% of their DNA with other higher primates, but doctors are now contemplating using organs from pigs (with but minor genetic modification) in human transplant surgery (9.5.4). The continuity between humans and non-humans is dramatically illustrated by evolutionary biologist Richard Dawkins when he writes: *'But for the accidental extinction of the intermediates linking us to, for example, chimpanzees, we would be united to them by an interlocking chain of interbreeding: a daisy chain of the "I've danced with a man, whose danced with a girl, whose danced with the Prince of Wales" variety.'*[4] The moral implications of Darwinism are certain to be profound.[5]

Despite an almost universal acceptance of Darwinism in Western society, there has been some persistent opposition in the form of **creationism**, especially within the fundamentalist Christian tradition in the USA. Basing their ideas on a literal reading of the Bible, creationists in the so-called 'Bible Belt' southern states reject modern views on the origin of the universe, and especially the notion that humans share common evolutionary origins with apes. Their powerful lobbying has been able to influence textbook committees and even school curricula, resulting, for example, in the insertion of disclaimers in science textbooks, which describe evolution as 'controversial' and stress that it should be regarded as just 'theory rather than fact'.

7.1.3 What is an animal?

Notwithstanding these dissenting opinions, the overwhelming biological consensus on evolution raises two crucial questions: 'What is an animal?' and 'How are we required ethically to treat animals?' A little reflection suggests that the first question begs to be stood on its head. Since we are so much like animals (so that bioethicists now often refer to them as 'non-human animals'), the task is more realistically that of defining what a 'person' is. Because people and animals share so much (genetically and hence physiologically), the answers given often emphasize characteristics that are thought to be unique to humans, such as the abilities to use language, employ reasoning and use tools. But the difficulty with that line of argument is that some *people* (such as all infants, some very ill, some very old and some mentally disabled people), *do not* have these abilities either, whereas certain animals *do* demonstrate 'language' ability (or, at least, marked communication skills), reasoning and simple tool-making abilities. So if our treatment of animals as a source of food or in experiments was considered ethically acceptable simply because they lacked these human attributes, we would be logically required to accept that certain adults and all babies could be treated, ethically, in the same way.

Anthropological studies suggest that the main differences between humans and other primates are a consequence of certain environmental influences. Moreover, while humans of essentially modern anatomical form appeared at least 130,000 years ago, it was only 40,000 years ago that the signs of modern *behavioural* capacities appeared.[6] The distinctively human abilities to speak and read are necessarily made possible by our genotype, but they only develop when the appropriate environmental conditions are present. So the emergence of humanity was probably primarily *culturally* rather than biologically determined. But the other side of the coin is that the human perception of *animality* is also culturally constructed, as we unthinkingly adopt a particular attitude to non-humans, which are marked off from humans by our assumptions, metaphors, customs and language. Treating other living beings differently merely because they are not human is called **speciesism**,[7] a word coined to parallel the concepts of racism and sexism.

7.2 Animal welfare and animal rights

The question of how we should treat animals has traditionally been answered by appealing to two distinct but overlapping types of obligation, corresponding to utilitarian (2.5) and deontological (2.6) theory respectively, namely, **animal welfare** and **animal rights**. Of these, the concept of animal welfare is the least contentious, because few people deny that we ought not to be cruel to animals. Many people have close, affective relationships with animals in domestic and other contexts, but they acknowledge personal contradictions in their attitudes, in that *how much* animal suffering is considered ethically acceptable often depends on the potential benefits that might result from the ways they are used.[8] Sometimes the contradictions are stark – as when the same people treat rabbits as pets, or as food, or as experimental subjects.

7.2.1 Utilitarianism as a constraint on animal use

The benefits derived from animal products (7.4) and from animal experiments (8.3) are usually justified by appealing to utilitarian reasoning (2.5), which weighs the advantages accruing to humans as greater than the costs born by the animals. But utilitarian reasoning can cut both ways (2.5.1), and Singer argues that the benefits derived from eating meat are usually utterly trivial in comparison to the harm inflicted on the animals: *'Pain is pain, whatever the species of the being experiencing it.'*[9] Singer argues that while it might be possible that the pleasure of meat-eating (if great) might exceed the suffering of the animal (if slight, for example, because the animal was reared with great care and killed painlessly, and with maximal humanity), few people buying meat from a supermarket can guarantee that such conditions existed. So according to this line of reasoning we are morally obliged to adopt vegetarianism. The concept of animal welfare is readily understandable, but there are wide differences in how it should be interpreted.

7.2.2 Animal rights

However, many people find the *concept* of animal rights more problematical. A right may be defined as a *'justified claim or entitlement, validated by moral principles and rules'.*[10] Some philosophers argue that despite the physiological similarities with humans, animals cannot have rights because rights can only belong to persons who can understand the concept. The problem with this view is that being consistent would mean that those humans who lack this ability (such as babies and senile people) would also fail to qualify to possess rights. But an alternative interpretation is that rights and duties are *correlative* (see 5.6), so that if we have a duty to treat animals well, and not to harm them, it follows that they have a right to be so treated. According to this view, animals like certain humans, are **moral patients**, owning rights but incapable of exercising responsibilities as fully competent moral agents.[11]

So what are the arguments suggesting that we have a duty to treat animals well? According to philosopher Tom Regan, basic moral rights should be assigned to all beings who are 'subjects of a life', for he claims: *'These animals have a life of their own, of importance to them apart from their utility to us. They have a biography, not just a biology. They are not only in the world, they have experience of it. They are somebody, not something.'* [12] (Basic moral rights exclude those clearly only applicable to humans, such as rights to vote and to a minimum wage.) Regan criticizes Singer's utilitarian argument because in the case of vegetarianism, it ultimately rests on a calculation that the suffering of animals is greater than the pleasure of meat-eating. But consider a different outcome. If it was established that millions of peoples' lives (farmers, slaughtermen, butchers, restaurateurs, waiters/waitresses, beefburger-eaters, etc.) would be seriously upset by a ban on meat production, does that mean killing animals for meat would suddenly become ethical?

At first reading, Regan's argument might seem persuasive, particularly when considering animals such as non-human primates. But there are some real problems over where to draw lines. For example, do insects have ethical standing, or oysters? It seems clear that to be 'a subject of a life' entails a degree of sentience or self-consciousness which is difficult for us to assess in many cases. So accepting Regan's criterion would seem to challenge Moore's 'naturalistic fallacy' (2.3), because it appears to involve deriving an 'ought' from an 'is'. That is to say, we might need scientific evidence on the degree of development of the nervous system of a particular species in order to decide whether it possesses basic moral rights.

Many people would draw the dividing line at sentience, and accept the assessment that *'Only sentient creatures have the characteristic ability to experience value which seems to be so crucial in deciding what entities have moral significance'.*[13] But that still doesn't make the decision straightforward, because we cannot really know what mental experiences other *people* have, let alone individuals of different species. That said, there is now much physiological evidence suggesting that mammals can experience pleasure, pain, fear, anxiety and frustration.[14] Often the line is drawn under the vertebrates, because invertebrates generally seem to demonstrate much less capability for 'intelligent', learned behaviour. But even here there are exceptions, e.g. cephalopods, such as the octopus, exhibit learning skills within the range shown by small mammals such as rats and mice. So a reasonable decision might be to err on the side of caution (13.4), presuming sentience when there is no convincing evidence against it.

However, despite the common use of the term, the concept of animal rights has come in for some serious philosophical criticism. Philosopher Mary Midgley suggests that appeal to rights *'expresses a deep and imperfectly understood connection between law and morality'.* For her, compassion for fellow sentient beings provides justification enough to guide our actions. She is also critical of the notion of speciesism (7.1.3), which does not correspond to its claimed racist model, because *'Race in humans is not a significant grouping at all, but species in animals certainly is'.*[15] Because the needs of different species can be very different, 'well-meaning', as well as intentionally cruel, actions can often cause harm.

7.2.3 **Intrinsic value**

Of course, ascribing rights to animals is not the only factor justifying ethical attitudes towards them. Although dismissive of 'natural rights', Bentham considered that animals' capacity to suffer placed us under certain obligations in the ways we treated them (2.5.1): this is the welfarist argument developed by Singer. Moreover, animals might be considered to have **intrinsic value** (3.6.2), also sometimes called **inherent value**,[16] which is value possessed by the animal, in and of itself, irrespective of any usefulness it might have to others. According to an article of the Swiss Federal Constitution, introduced in 1992, it is important to show respect for the 'dignity of creation' (*Würde der Kreature*) of all living beings, quite apart from the need to protect them from suffering;[17] and the concepts of intrinsic value[18] and the 'dignity of creation'[19] are the subject of much philosophical discussion. In thinking about these issues it might be useful to consider the distinction that can be drawn between *duties to* certain beings, such as sentient animals, and *duties about*, which might be applied to inanimate objects, such as ancient oak trees.[20] These issues are discussed further in chapter 12.

7.2.4 Practical implications

Despite hotly debated differences in the ethical justifications based on utilitarian and deontological theories, there is much common ground when it comes to practicalities. Whether or not ill-treatment of animals is considered wrong primarily because it infringes their rights or because it causes them suffering, we need to assess the extent of such ill-treatment if it is to be avoided. Because we cannot ask animals what they feel, it is necessary to establish ways of examining the nature and scale of the problems.

Animals respond to adverse physical and mental conditions in a variety of ways to help them cope. The brain and adrenal system are involved in providing additional energy when required, while behavioural responses are important in 'fight or flight' reflexes. The endorphin/encephalin system involves production of analgesic peptides, so that by self-narcosis animals may be enabled to cope with pain and stress. In some cases, the difficulties are so pronounced that the coping processes break down and very poor welfare results. When stress is extreme it leads to ill health, and ultimately death.[21] But, as pointed out by animal welfarist Colin Spedding, there are many grey areas where it is difficult to be certain whether animals are suffering or not. Some animals deliberately disguise injury, whereas others feign it.[22]

Because it has only recently become widely appreciated how similar people and certain non-human animals are, we have to be prepared to reconsider the whole basis of the way we have traditionally treated animals.[23] The claims that, on the one hand, genes are critical factors in explaining living organisms and, on the other, that the great apes share 98% of human DNA, can hardly be irrelevant to the ethical acceptability of using them as experimental subjects. To kill and eat non-human animals, to whom we now know we are so closely genetically related, requires, at the very least, ethical justification. And it would not be surprising if, for some people at least, serious ethical analysis also entailed fundamental dietary changes.

7.3 How animals are used

The majority of us *use* animals, and their instrumental value is perhaps the main reason that most people connect with animals at all. Even if we respect or admire their intrinsic value as we keep them as pets, bet on them at the dog-track or

Table 7.1 Human use of animals in the UK (principal species)

Animals	Uses
FARM ANIMALS (millions)	
Cattle (11)	Meat, milk
Sheep (44)	Meat, wool
Pigs (7)	Meat
Poultry (126)	Meat, eggs
Fish (salmon: 35)	Meat
COMPANION ANIMALS (millions)	
Dogs (7)	
Cats (7)	
Horses and ponies (1)	Riding, jumping
Caged birds (2)	
Rabbits, mice, hamsters	
CAPTIVE ANIMALS	
Large cats, elephants, monkeys, bears, birds, reptiles, amphibians, fish	Zoos and circuses
WORKING ANIMALS	
Horses, donkeys	Traction, cultivation, transport
Dogs	Herding, guides, guards
SPORTS ANIMALS	
Horses, dogs, pigeons	Racing, showing
Hawks	
Pheasants	Shooting
WILD ANIMALS	
Foxes, deer, game birds, fish	Quarry in hunting, shooting and fishing
LABORATORY ANIMALS	
Rats, mice, dogs, cats, monkeys (approximately 3 million used p.a.)	Research, drug- and product-testing, education

Note: Approximate numbers in millions, where available, are given in brackets.[24]

ride them for pleasure, we still use them for our own ends. Some people would say that these are essentially exploitative relationships, in which the animals are 'enslaved'. Others deny this, because of what they see as a valid comparison with the ways we treat other people. Although every person is usually assigned intrinsic value, we do often use other people without feeling we are exploiting them, because, in accordance with an **unwritten contract**, we repay them in kind or with money (3.6.3).

So it might be useful to consider our relationships with animals as also being contractual, and hence requiring some form of repayment. Purists might object that entering into a 'human–animal' contract requires *two* willing partners, but others will conceded that, as with the case of 'animal rights', the concept has much to commend it. When it comes to killing animals, for food or in 'blood sports', the concept of a contract seems far fetched – but it is worth exploring whether a case might be made for its validity even here.

As such a wide range of animals is used and in such a multiplicity of ways, it is useful to devise a system of classification in order to identifying ethically relevant features. Table 7.1 identifies seven broad categories of animal use, and gives estimates, where available, of numbers used in the UK. The list is not exhaustive, e.g. it omits animals used in religious sacrifices.

It is clear that, from the animal's point of view, two factors are critical in assessing the impact of our treatment of them. First, it must be assumed that the more developed their nervous system the more intensely will animals be able to experience both pleasure and suffering: seals and slugs are unlikely to be on a par. And suffering is not simply a physical experience; a cat shut in a box or separated from its kittens might suffer intense distress, even though subjected to no physical harm. Second, the intensity of the effect on the animal depends on the way it is treated, which spans everything from taking the dog for a walk to force-feeding geese to produce pâté de foie gras.[25]

A common utilitarian perspective is that the *effects* of our actions are all that matters, for as veterinary scientist John Webster puts it: *'Animals are not affected by how we feel but what we do'.*[26] Although this may not adequately encompass all ethical issues concerning animals, it forms a useful starting point for our ethical analyses. So we need to first define the claimed benefits of animal use, and then the costs against which a utilitarian analysis would weigh them.

7.4 Farm animals

Farm animals provide several products apart from food (such as wool, leather, and bone), but this discussion is largely confined to their provision of meat, milk and eggs. As dairy and poultry farming are closely linked to the meat industry, the following account can be seen as presenting a rationale for meat eating – which is clearly challenged in one way or another by people who practise vegetarianism. Alleged benefits of meat-eating are listed below under five headings: nutrition, co-evolution, ecology, economics, and cultural considerations.

7.4.1 **Nutrition**

Animal products such as meat, milk and eggs are good sources of essential amino acids, and meat and fish provide several nutrients that are scarce or absent in plant foods, e.g. iodine, taurine, vitamin B_{12}, vitamin D and certain long-chain fatty acids. Meat is an important source of bioavailable iron, and dairy products are good sources of bioavailable calcium and phosphorus.

Even so, meat-eating in excess can have serious adverse health implications. For example, compared with non-vegetarians, Western vegetarians have a 25% lower mortality from ischaemic heart disease, and lower risks of suffering from constipation, gallstones, and appendicitis - although whether such benefits are a consequence of not eating meat, or of something else (such as the influence of smoking habits), is uncertain.[27] However, such comparisons are complicated by the problem of definition: many people who call themselves 'vegetarians' do consume eggs, dairy products, and fish, and such foods are conventionally classed as such on restaurant menus. Some 'vegetarians' even eat chicken, rejecting only red meat.

7.4.2 **Co-evolution**

The claim has been made that *'In an evolutionary sense'* modern farm animals have *'chosen us as much as we chose them'*.[28] In pre-historical times, their characteristics of docility, lack of fear, and high reproductive capability encouraged their voluntary association with humans, from whom they derived food (initially from scavenging) and protection from both climatic extremes and predators. Now, they also benefit from veterinary medicine.

Domesticated animals have, in fact, undergone marked evolutionary changes which have made them almost totally dependent on human care, having largely lost their adaptation to the wild. This means that their instincts of dominance and territoriality are greatly diminished and their physical defence mechanisms atrophied. The much-reduced flying ability of domestic fowl is a prime example. So, it has been argued that it would be perverse to attempt to return farm animals to 'the wild' (even assuming such territory still exists) because few of them would be able to survive.

7.4.3 **Ecology**

Traditionally, and still in many organic systems (7.7.2), animals have played key roles in integrated farming systems, maintaining soil fertility through their production of manure and efficiently utilizing crop by-products. Ruminant animals, such as cattle and sheep, can bring into productivity land which is otherwise too poor, erodible or difficult to cultivate. By converting fodder such as grass, inedible for humans, into nutritious food in the form of meat and milk, they sustain human populations on land that could not easily support them from plant foods alone. Although poultry, pigs, and fish are not able, like ruminant animals, to subsist on plant cell walls, they can scavenge feed material inaccessible to humans and convert it into human food.

However, it can be argued the particular ways in which many animals are currently kept in intensive systems in DC are *not* sound from an ecological perspective. Excessive use of fossil fuel inputs in producing, transporting and delivering feed to housed animals, together with major environmental problems associated with the disposal of their wastes, often totally undermine the advantages described above.[29]

7.4.4 Economics

A major reason why farm animals are kept is that they provide desirable products. And whether or not, in DC societies, they provide irreplaceable components of a healthy diet (which the existence of healthy vegans appears to question), for most people animal products provide the basis of appetizing meals. In LDC the roles of animals are often even more fundamental. Animals provide traction power (for pulling the plough and general transportation), fuel (as dung), and fibre (as wool, bone, and leather). World-wide, the activities of many societies revolve around the use of animals, which thus have enormous economic importance. Millions of people's lives and livelihoods currently depend on the farming of animals for food.

7.4.5 Cultural considerations

Finally, it has been proposed that human relationships with animals are a vital ingredient of human culture, which has deeply influenced the way in which people see life – giving society its world view. According to animal scientist John Hodges: *'We are enough like animals to be kept humble; we are different enough from animals to be aware of our unique responsibility as "husbandmen" of the natural world.'*[30] This is an unusual claim, but one which, by putting animal use in a broad cultural perspective, may articulate many people's deeper convictions about the ethical use of animals, and suggest that current intensive systems undermine the 'contract' which should apply (7.3).

7.5 Taking life

For some people of formidable character, such as the French theologian and doctor Albert Schweitzer[31] and the Hindu nationalist leader Mahatma Gandhi,[32] the doctrine of 'reverence for life' means they lead lives of rigorous non-violence towards both people and animals. But in virtually all forms of livestock farming, whatever the primary purpose, animals are ultimately killed and their flesh consumed as meat. Sometimes, when they serve no 'useful purpose' (such as day-old male chicks), they are dispatched summarily. Or they may live very short lives because meat from young animals is tender. Since depriving animals of their life weighs negatively in the utilitarian balance sheet, the act of killing requires ethical justification.

What factors determine whether it is right to kill another sentient being? To clarify the issues, consider the ways people treat other humans. The list of justifications for killing *other people* goes something like this:

- in self-defence
- in a 'just' war

- (sometimes) out of compassion (*euthanasia*)
- for some people, as a justified punishment for a capital crime (although not in the UK)

At the other end of the zoological scale, many believe it right to kill vermin, such as rats; few have any scruples about killing insects if they are a nuisance; and almost everyone considers we ought to kill bacteria that cause infectious diseases.

Farm animals clearly lie somewhere in between these two extremes, and almost everybody considers that we have lesser ethical obligations to them than we do to other people. For example, it is generally accepted that it is right to 'put down' animals that are seriously injured or terminally sick. When human needs demand it (e.g. to alleviate starvation) most people endorse killing animals, while even less acute problems, such as poor health, that could be remedied by consuming animal products, usually also provide adequate justification. Hence, almost without exception, people advance certain reasons to exempt themselves from a *prima facie* obligation not to kill others, whether humans or animals. (Of course, by appealing to the 'majority opinion' this analysis is reflecting the common morality, which as noted in 3.1.1, can only be regarded as a starting point for genuine ethical analysis.)

For farm animals, meat-eaters who justify their habit ethically usually appeal to one or more of the arguments listed in 7.4. Yet in none of these cases is a reason given for treating animals with wanton cruelty. Thus, the arguments advanced for justifiable killing of animals can be said to depend on three conditions:

- having a good (perhaps vital) reason
- ensuring, as far as possible, that the animal has led as good, or better, a life than it is likely to have led in the wild
- ensuring that the animal is not treated cruelly at any stage.

We need to examine how often these conditions are observed.

7.5.1 Vegetarianism and veganism

But before doing so, it is appropriate to consider the position of **vegetarians**. One difficulty is that vegetarianism is open to many different definitions.[33] Some people calling themselves vegetarians often consume certain animal products, such as milk and eggs (when they are called *ovo-lacto* vegetarians), and hence rely on the arguments advanced above (7.4) concerning the utilitarian justification for animal use. Quite a number eat fish, and some eat chicken. Even if vegetarians were to avoid all *meat*, they would not thereby necessarily escape responsibility for the way the animals that provide some of their food live and die. In a sense, unless they are vegans, they rely on other people to eat meat for them. Of course, in utilitarian terms, where what is important is the weighing of costs and benefits, vegetarian practices may significantly reduce overall animal suffering. But the fact that the same might be true of meat-eaters who eat only small amounts of meat, and that derived only from animals raised and slaughtered under the most humane conditions, blurs the ethical distinction which is often drawn between these two practices.

Vegans are the strictest form of vegetarians, rejecting all food of animal origin, e.g. even honey. But even vegan diets and/or lifestyles are not necessarily immune from criticism in terms of their ethical impacts on animals. Thus, intensive crop-production systems, industrial activities, collisions with road vehicles, and predatory activities of companion animals (7.8.5) can all have adverse effects on wildlife, both as individuals and as species.[34] And, of course, there is the awkward question of how vegans would deal with the problem of caring for the millions of ageing farm animals if their desire that other people no longer killed farm animals were suddenly adopted.

In fact, to be at all consistent vegans are forced to adopt some arbitrary ethical positions. If, for example, the vegan's object is not to cause animal suffering then there seems little reason to limit this concern to farm animals. But if individual wild animals are regarded as having ethical standing (3.4), then there would seem to be an ethical obligation to protect the predated animal from its predator, which would, even in the unlikely event that it could be realistically achieved, inevitably have adverse effects on the predator's well-being and survival. So even the most rigorous observance of veganism seemingly turns a blind eye to the extensive animal suffering which occurs in the wild. In practice, vegans' concern for wild animals generally conforms to the environmental ethic described in 3.6.4.

7.5.2 Psychological and sociological influences

Most people's attitudes to meat-eating are not entirely (or perhaps even primarily) based in ethical reasoning, so that it is important to appreciate the impact of sociological and psychological influences on food habits. Sociologist Julia Twigg showed that there is a hierarchy of status and potency in people's attitudes to meat (Table 7.2).[35] Red meat is placed at the top of the hierarchy because its high blood content confers symbolic power on its consumers: blood is conventionally associated with virility, strength, aggression and sexuality. However, it is also seen as somewhat dangerous for such reasons. Twigg argued that because of its symbolic power, traditional practices have ensured that it has been 'tamed' by transforming it in cooking, curing, smoking etc. – thereby wresting it from the realm of nature and bringing it into the realm of culture. In these terms, the meat of certain animals, such as carnivores or uncastrated animals (bulls and boars) is considered too strong. However, bloodless meats such as poultry and fish are traditionally thought of as less powerful, and most suitable for children and those of weak constitution. And according to this hierarchy, vegetables are traditionally considered too weak to adequately provide the basis of a square meal.

In considering the ethics of meat-eating it is worth assessing the extent to which such assumptions might play a role in your own eating habits – and whether these are legitimate factors.

Table 7.2 The conventional hierarchy of food status and potency

TOO STRONG	Uncastrated animals Carnivorous animals Raw meat
Dominant culture boundary	
STRONG	
Powerful: blood	Red meat
Less powerful: non-blood	Chicken Fish
Vegetarian boundary	
LESS STRONG	Eggs Dairy products
Vegan boundary	
TOO WEAK	Fruits Cereals Root vegetables Leaf vegetables

7.6 The welfare costs for different species

The other side of the cost–benefit equation entails assessing the ethical costs of using animals for food. These are experienced in several ways, some of which have been discussed briefly already, e.g. effects on human health (7.4.1) and environmental impacts (7.4.3). The focus of this section is on animal welfare, although passing reference will also be made to other issues that are pursued more fully in other chapters.

Promoting farm-animal welfare aims to satisfy certain basic needs (for food, water, space, clean air, etc.) and avoid certain harms (such as disease, injury, stress, excessive climatic exposure). Such provisions are measured in terms of respect for the Five Freedoms, defined by the UK Farm Animal Welfare Council (FAWC; Box 7.1), and welfare standards in the UK are often reputed to be higher than in most other countries. Nevertheless, it is clear that further significant improvements could be made. The following list highlights some of the remaining concerns over farm-animal welfare.[36]

7.6.1 Broiler chickens

Excessive growth rates of broiler chickens place such a strain on their legs that they are often crippled and/or suffer bone fractures, while heart failure due to ascites affects 5% of broilers worldwide. Intensive housing means that contact with ammonia-impregnated litter (from the birds' excrement) often causes 'burns' on the body surface.[37]

BOX 7.1 THE FIVE FREEDOMS (FARM ANIMAL WELFARE COUNCIL)

- **Freedom from hunger and thirst**: by ready access to fresh water and a diet to maintain full health and vigour
- **Freedom from discomfort**: by providing an appropriate environment including shelter and a comfortable resting area
- **Freedom from pain, injury, and disease**: by prevention or rapid diagnosis and treatment
- **Freedom to express normal behaviour**: by providing sufficient space, proper facilities, and the company of the animal's own kind
- **Freedom from fear and distress**: by ensuring conditions and treatment which avoid mental suffering

7.6.2 Laying hens

Long-term confinement of layers, when they are producing more than 300 eggs p.a., often causes *osteoporosis*, resulting in fractures of the ribs, sternum and leg bones. In the UK, it is estimated that 25% of the 32 million laying poultry (i.e. 8 million p.a.) suffer from bone fractures – but *'as with all suffering, only a small proportion is ever recorded'*.[38] In the UK more than 80% of laying hens spend their lives in battery cages with three or four others. At the current EU approved standards, each bird is allowed 450 cm^2 floor space (although, typically, in the USA only 350 cm^2), which prevents them performing any sort of normal behaviour, such as dust-bathing, foraging, or spreading their wings. By 2012, in the EU only enriched cages will be allowed, giving each hen 750 cm^2, a nest, and 15 cm of perch space.

7.6.3 Pigs

In intensive systems, pig breeders aim to wean the maximum number of piglets per sow p.a., while rearers and finishers aim to achieve their end-points in the minimum number of days. In the UK, under indoor commercial conditions weaning is normally at 3–5 weeks old, but early weaning (which entails a sudden change of diet, a new environment and mixing with unfamiliar animals) makes piglets vulnerable to infections, indigestion and food allergies.[39] To counteract these problems, medication of starter rations is an almost universal practice. Piglets are kept at high stocking density, on bare concrete, or slatted floors, which prevents natural behaviour, such as rooting, foraging and exploration, and may lead to lameness. Deprived of natural stimuli they often resort to aberrant practices, such as biting each others' tails. By contrast, free-ranging sows wean their pigs gradually between 8 and 20 weeks of age.

7.6.4 Dairy cows

Cows may usually appear contented, but this can disguise the fact that they are sometimes subjected to very high metabolic demands, which have been equated to *'a man jogging for 6–8 hours per day, every day'.*[40] High levels of concentrate feeding, combined with genetic selection for high yield, often result in so-called *production diseases*, such as mastitis (udder inflammation), lameness, digestive and metabolic disorders (such as milk fever and ketosis), and infertility. In the UK dairy herd as a whole, the average incidence of clinical mastitis is 35 to 40 cases per 100 cows p.a.,[41] and 25–30% suffer from a foot or leg disorder.[42] Increased milk yields often exacerbate these problems.

7.6.5 Beef cattle

Suckler herds, in which calves are not separated from their mothers but suckle and graze with the cows for at least one summer, usually enjoy a high level of welfare. They are usually housed over winter and generally slaughtered at 1–2 years of age: their mothers might live for 20 years.

However, most calves raised for beef are products of dairy herds. Such animals, separated from their mothers after about 24 hours, are reared initially on milk-replacers (generally based on skimmed-milk powders) and then weaned onto cereal-based starter rations. They are usually confined in buildings and yards throughout their lives of just over one year.

7.6.6 Sheep and lambs

The animal welfarist Audrey Eyton wrote: *'Through a happy accident of nature, sheep have proved singularly resistant to the so-called progress of post-war agriculture* [i.e. factory farming], *and lamb provides one of the kinder choices in comparison to other farmed meat.'*[43] This common perception that sheep lead natural, stress-free lives may be true for some - but for others outdoor life can be harsh. This is because in some cases they may suffer from cold weather, starvation, and infestation with maggots through exposure and neglect – victims of a system that accords them little economic value as individuals.[44] In the UK, 10–25% of all lambs born die within three days.[45]

7.6.7 Fish

Fish-farming has expanded greatly over the last 30 years in the UK, with salmon, trout, and carp the main species involved. For example, approximtely 35 million farmed salmon are produced p.a. They are grown in hatcheries before transfer to floating cages in the sea or lochs, containing 1 million fish. Overcrowding in cages may lead to infestation with sea lice, physical damage, and bullying, while *anaemia* and eye cataracts are examples of production diseases.[46]

7.6.8 Management procedures

Many animals are subjected to procedures that facilitate the ways they are managed on the farm, transported, and slaughtered. This section focuses on the welfare implications of some of these procedures.

Mutilations. These are procedures that involve removing or damaging part of an animal's body as a routine management procedure. The commonest forms of mutilation are shown in Table 7.3.[47] Often animals are subjected to painful mutilations (such as tail-docking and castration of piglets, ear-notching and branding of cattle) without anaesthesia. Some of these are not only welfare concerns in themselves, but adversely affect subsequent animal behaviour, e.g. beak-trimming of poultry restricts the behaviourally important preening of feathers.

Transport of animals, handling at markets and treatment at abattoirs can all entail considerable stress, and sometimes injury, for many animals.[48] For example, some animals suffer bone fractures and joint dislocations, clinical and behavioural responses to food and water restriction, adverse responses to thermal conditions (too hot or too cold), or responses indicative of fatigue. Although there are EU regulations (Council Directive 91/628/EEC, as amended) relating to feeding and watering, methods of loading, journey times, and lorry design, they are difficult to police effectively.[49]

Slaughter. This comprises two stages:

- *stunning,* which renders the animal unconscious
- *sticking,* which follows stunning, and consists of cutting the major blood vessels – so resulting in death due to rapid blood loss.

The stunning procedure used depends on species, e.g. for cattle a captive bolt is used, whereas for pigs and sheep, electrical stunning is applied to the head to induce concussion. After stunning, sticking is performed quickly to deflect blood from the brain before consciousness is regained. That there is much scope for improving the welfare of such animals is evident from a FAWC report, which made no fewer than 277 recommendations for changes in current practices.[50]

For poultry, it is common for the heads of the birds, suspended upside down from shackles, to be passed through an electrified water-bath before being bled by a neck cut.

Table 7.3 Main forms of mutilation in UK farm animals (from Spedding, 2000)

Mutilation	Species
Beak-trimming	Poultry
Declawing	Turkeys
Dubbing (removal of the male comb)	Turkeys, cockerels
Tail-docking	Pigs
Nose-ringing	Pigs (outdoors), bulls, boars
Dehorning	Cattle
Castration and spaying	Cattle, sheep, pigs
Branding	Cattle
Ear-punching	Cattle, sheep

With skilled operators, these operations can be performed efficiently and humanely, but problems arise in some circumstances, for example if:

- the operator is incompetent or inconsiderate
- equipment is faulty (e.g. providing an inadequate voltage for stunning)
- a bird is too small (so that it is insufficiently immersed in the stunning water-bath)
- the throughput is too great, so that operators do not have adequate time to attend to individual cases (in poultry units, 6000 birds are routinely slaughtered per hour).[51]

Although many animals are killed because their carcass is used for meat, in the UK about 40 million chicks are killed in hatcheries because they are males unwanted for egg production or because they are too sickly when hatched. A report from the Humane Slaughter Association revealed significant welfare problems, and in particular it was critical of the use of carbon dioxide for stunning. It recommended that this should be replaced by 90% argon gas or use of *instantaneous mechanical destruction*, which employs mincing devices.[52] All these systems clearly have significant drawbacks. Even if it were certain that the chicks feel no pain, it can hardly be denied that the termination of their lives in a mincing machine shows absolutely no respect for their intrinsic value as sentient beings.

7.6.9 The animal-products food chain

The process in which animals are used for food (from 'farm to fork') in DC such as the UK consists of a chain of interlocking stages, each of which raises its own type of concerns (Figure 7.1).[53] The entire process is called the **animal-products food chain**. Its operation can be said to be governed by the outcome of a tension between commercial objectives, consumer demands and ethical concerns.

Because market forces have a tendency to marginalize factors which reduce profitability, it has long been considered necessary for governments, particularly in DC, to protect consumers from practices which threaten:

- food safety
- farmers' incomes from excessive fluctuations in market prices
- farm animals from procedures which reduce their welfare.[54]

Government regulation is thus often the means by which (some) ethical concerns are recognized and ameliorated. Many of the concerns identified in Figure 7.1 go beyond the specific remit of this chapter: for example, they impact on issues discussed in chapter 11 (concerning diets) and chapter 12 (concerning environmental sustainability). This emphasizes the point that bioethics cannot easily be pigeon-holed. Ethical decision-making on how we should treat farm animals needs to be related to the broader contexts of how agriculture affects society and the environment.

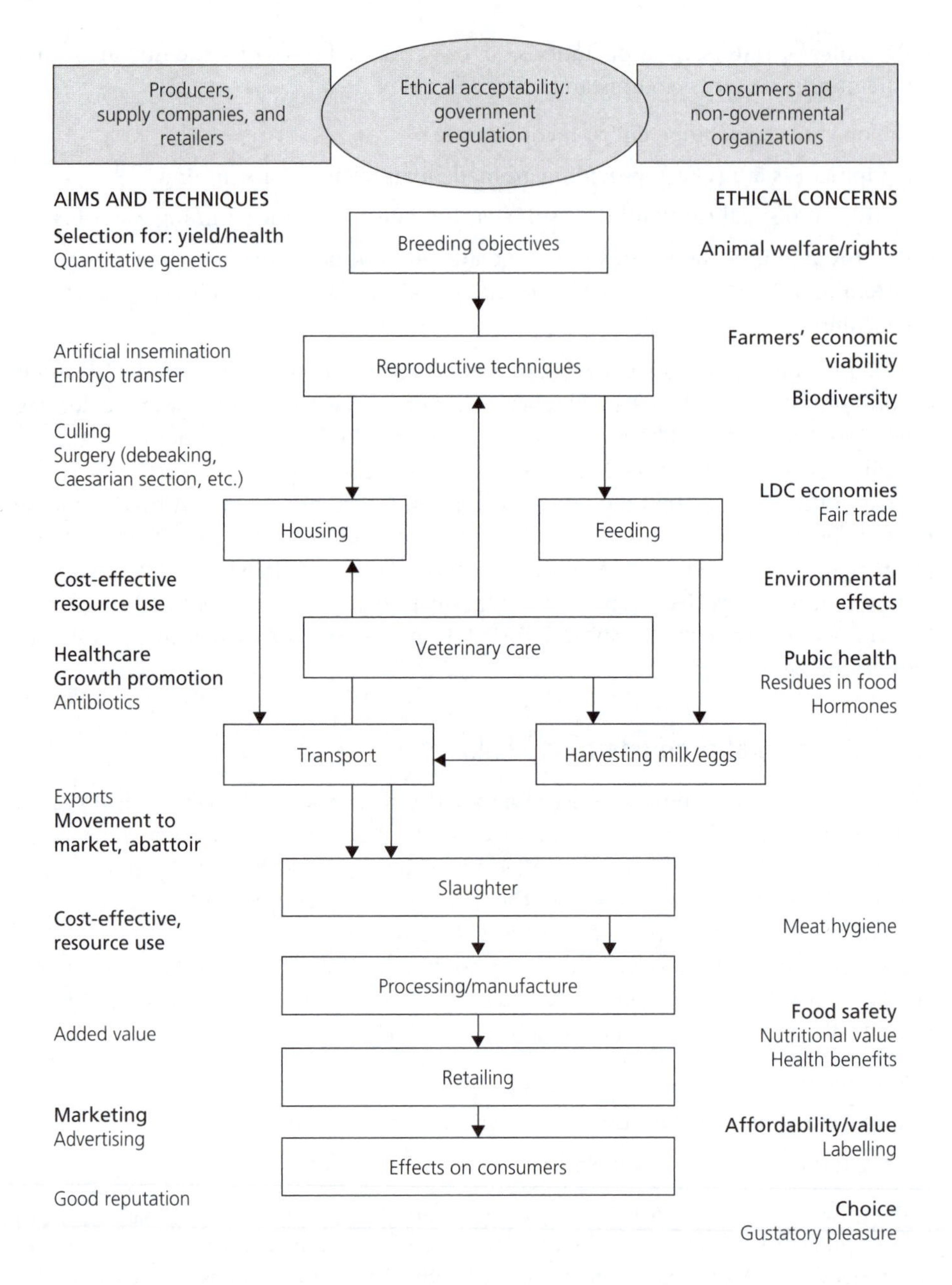

Figure 7.1 Features of the standard *animal-product food chain* in the UK, illustrating (*left*) the principal aims (in bold) and techniques employed at *different stages* (*centre panel*) and (*right*) *common ethical concerns* (major concerns in bold). The principal determinants of how the chain operates, including commercial regulatory and consumer pressures, together with standards of ethical acceptability, are shown at the *top*.

7.7 Ethical perspectives

According to some utilitarian analyses, the claimed benefits of conventional animal production systems are negated by substantial suffering experienced by the animals. In the opinion of philosopher Peter Singer, the costs to animals far outweigh any benefits to humans, and suggest that ethical integrity demands the adoption of vegetarianism. The balance might be said to be even more weighted against meat-eating by including public health consequences such as: heart disease (11.2); the risk of diseases such as Creutzfeldt-Jakob disease (vCJD) which is believed to have been caused by ingestion of beef infected with bovine spongiform encephalopathy (BSE; see Box 13.2); and the adverse environmental impacts of intensive farming practices (12.5.3).

Those arguing rigorously from the perspective of animal rights usually rule out all forms of animal exploitation and adopt veganism. But apart from the problems identified with this stance (7.5.1), they also face the practical difficulty of how to live sociably in a society in which most people do not share beliefs that might well be perceived as 'extreme'. If you can never sit down for a convivial meal because your hosts or guests are liable to question your every choice (of food, clothing and children's pets), the principled position adopted may seem to undermine the charitable intentions you started out with.

7.7.1 Defending the current systems

Those who seek to justify the current intensive systems of animal production argue that, despite the welfare problems discussed in 7.6, the benefits to people - in the form of cheap nutritious food - outweigh all these costs. Many people holding this view deny that animals have significant rights, or that the welfare problems are serious. For example, it might be claimed that animals which are permanently housed are protected from climatic extremes and are well fed - although admittedly primarily in order to optimize yields.

Moreover, it is often argued that animal welfare standards (which are considered higher in the UK than elsewhere) exact a financial cost, and that further improvements in welfare will disadvantage UK farmers in competition with imports from abroad which, because of lower wages and welfare standards, can be sold more cheaply. This pragmatic argument suggests that, accepting that meat-eating is ethically acceptable in principle, improvements in animal welfare can only be introduced to the extent that consumers are prepared, or can afford, to pay for them.

7.7.2 Organic livestock farming

An intermediate claim to those identified in 7.7 and 7.7.1 is that *only* more humane forms of livestock farming, such as organic farming, together with the arguments for benefits presented in 7.4, can provide sufficient ethical grounds for using animals for food. Consequently, organic livestock husbandry places a high priority on animal health and welfare. Disease prevention is based on four principles:

- selection of appropriate breeds or strains
- application of appropriate husbandry practices, encouraging strong resistance to disease, and prevention of infections
- use of high-quality feed which, together with regular exercise and access to pasturage, encourages natural immunological defence mechanisms
- avoiding overstocking.

Use of chemically synthesized allopathic veterinary medicinal products or antibiotics for preventive treatments is prohibited, as is the use of battery-caging of poultry. In consequence, it is likely that animal welfare is significantly better on organic farms (12.5.3) than on many conventional farms.

In that these higher standards are guaranteed by an inspection regime, and paid for by higher food prices, some people might consider organic farming to be an acceptable compromise between a desire to respect animals' rights and welfare, on the one hand, and a legitimate exploitation of them on the other. Implicitly, this appeals to the notion of the *animal–human contract* (7.3). Purchasing organic products, and other products such as 'free-range' eggs or outdoor-raised pork (providing the consumer appreciates the limitations of these definitions), are examples of **ethical consumerism** (14.8), which can influence agricultural practice through market forces.

7.8 Animal use in sport, entertainment, and work

This short section covers a broad range of animal uses, so that the discussion is necessarily brief. However, the arguments from the perspectives of animal welfare and animal rights already discussed (7.2) apply equally to these different practices and motives.

People who engage in so-called **blood sports** may justify their activities (hunting, shooting, and fishing) by arguing that the wild animals they kill experience none of the welfare problems experienced by animals kept in captivity. These animals lead natural lives, are not branded, mutilated, penned, separated from their young, transported, or subjected to the ordeals of the slaughterhouse. Reasons advanced for killing are to obtain food (e.g. fish, birds, and rabbits) or to control pests or the numbers of animals (foxes, pigeons). But there is a common perception that the *pleasure* derived is the prime motive, either from the 'thrill of the chase', or a sense of satisfaction in demonstrating shooting or fishing skills, or outwitting the quarry. Animal suffering may differ substantially between the species, but animal rightists would advocate a ban on all animal exploitation. A few examples will illustrate the ethical issues involved.

7.8.1 Deer

It is believed that deer suffer greatly when being pursued. The Bateson Report (1996) suggested that, after a chase of 15 km, when all the glycogen in their energy store has been expended, muscle enzymes leak into the blood and the deer are completely exhausted. It was assumed that half of the deer escaping the hunt might be similarly affected. However, these results have been challenged by other studies, which suggest the welfare problems had been exaggerated.[55]

7.8.2 Foxes

In the UK there are about 240,000 adult foxes, which produce about 400,000 cubs annually. It is estimated that gamekeepers kill about 75,000 and fox-hunts about 15,000, but the majority die in road accidents. No comparable physiological data to those cited for deer (7.8.1) are available for hunted foxes.

At the time of writing, the Speaker of the House of Commons had just invoked the Parliament Act which led to the banning of hunting of foxes with dogs in England and Wales (and also deer hunting and hare coursing) with effect from February 2005, thus ending years of political wrangling between hunt supporters and hunt opponents.[56]

Although, on a free vote in the House of Commons there had been consistent support for a ban (with substantial majorities in favour on each of the 10 times in 10 years when the issue was debated), opposition in the House of Lords had seriously delayed introduction of the ban. The Government claims that 70% of the general public oppose hunting.

A utilitarian argument for resisting a ban on fox-hunting points to the consequences for horses, hounds and people. Thus, it is estimated that in the UK there are 37,000 horses kept solely for hunting, of which up to 20,000 might be slaughtered if a ban were introduced, a fate which might also befall 12,000 hounds.[57] The Burns inquiry, set up by the Government to investigate the impact of a ban on fox-hunting, suggested that 6000–8000 jobs depended on it, and that it is a *'cohesive social force'*.

On the other hand, Burns reported that hunting was *'disruptive and intrusive'* and *'seriously compromised the welfare of the foxes'*.[58] From a purely utilitarian perspective, smaller numbers of people and of animals are involved than is the case with farm animals, but the larger number of people who object to hunting also needs to be factored into the equation. Many people would claim that the militant activities of some hunt supporters show little respect for the democratically expressed 'will of Parliament'.

7.8.3 Fish

Many people who engage in fishing do not eat, or even keep, the catch - which is returned to the water. Despite this, the fish might suffer appreciably, since in the process they are hooked, played, handled and confined to holding nets. It has been commonly believed that fish experience little or no pain, but recent studies provide significant neurophysiological evidence of nociception in teleost fish, which together with changes in behaviour and physiology suggest that that fish experience discomfort when exposed to noxious substances.[59] Fish tend to be ignored in discussions of animal interests, but this attitude may be ethically untenable, especially as they comprise more than 50% of all living vertebrates.[60]

7.8.4 Other animal uses

Animals perform many other roles, which are often perceived as mutually beneficial to the animals and their owners. Examples are:

- companion animals (encompassing a wide range – from goldfish to primates, and including cats, dogs, ornamental birds and caged birds)
- guide dogs
- working animals (sheep dogs, gun dogs, pony rides).

Others serve roles in education (e.g. zoo animals) and entertainment (circus animals), but little benefit can be thought to accrue to the animals, and sometimes the impact on them seems entirely negative by comparison with their life in the wild.

7.8.5 Motives and consequences

If, with Kant (2.6), we place emphasis in ethics on *intentions,* it is useful to categorize the uses of these animals, based on whether harms are inflicted, and if so whether these are accidental or intentional.

Where animals are not intentionally harmed. This category includes horses for riding and racing, companion animals, working dogs (guide and sheep dogs) and racing pigeons. Animals used are generally well cared for, not least because high performance is the desired objective. However, the categorization might be questioned because some animals are harmed, e.g. through injuries incurred from falls in steeplechases, when they may need to be put down, or through excessive use of the whip in race horses.

Where harm is not intrinsic to animal use, but is almost inevitable. Fish returned to water after the catch (7.8.3) exemplify this category. Other examples are animals kept in zoos and circus animals, which by being confined or trained clearly suffer an infringement of their natural liberty.

A particular, often unrecognized problem, results from keeping animals which, although domesticated, retain many of their wild instincts. The approximately 7 million cats kept as pets in the UK are believed to be responsible for more than 80% of all wildlife killed, figures put at 88 million birds and 164 million small mammals p.a.[61]

Where harm is intended. This category includes most animals that are the objects of hunting, shooting and fishing. Important ethical distinctions might lie in the justification for the intended harm, and the manner in which death is inflicted. However, *'the belief that people who enjoy hunting or fishing do so because they enjoy dominating and inflicting pain is often used to support an argument that field sports are morally repugnant and should be banned'.*[62]

7.9 Ethical analyses

This chapter has raised ethical issues associated with the human use of a wide range of animals. It has suggested that appeal to ethical theory can assist the process of ethical deliberation, but that the marked differences between animals in anatomical, physiological and behavioural features means that scientific understanding is also an important ingredient in any meaningful assessment of how animals should be treated ethically. Thus the ethical implications of our dealings with animals may be thought to refute Moore's concept of the 'naturalistic fallacy' (2.3).

Use of the ethical matrix, with but minor adaptations to that illustrated in Table 3.1, may facilitate ethical analyses of the various animal-production systems; and the case described in 3.6 suggests several issues which are more widely relevant to use of farm animals for food. But in using the matrix, careful attention needs to be given to identifying all the interest groups and the distinctive ways in which respect for the ethical principles is reflected when dealing with non-humans.

THE MAIN POINTS

- In recent years, attitudes to animals have changed from a predominant speciesism to a view more respectful of their welfare and/or rights as sentient beings, sharing many human attributes. Using animals thus requires ethical justification.
- Animal rightists and some utilitarians deny there is any justification for killing animals for food, but others claim that where their intrinsic value and welfare are adequately respected (e.g. in a notional *contract*) some forms of animal use for food are ethically defensible.
- In current animal production practices, despite promotion of welfare through the Five Freedoms, many animals are subjected to close confinement, painful mutilations, and stresses due to high metabolic rates, harsh environmental conditions, and transportation and slaughtering procedures.
- Some animal production systems, such as organic farming, place a high priority on animal welfare – e.g. banning most mutilations, battery housing, routine antibiotic use and high stocking densities.
- Animals used in sport (hunting, shooting and fishing) may experience more natural lives than farm animals, but ethical objections are often raised to the fact that the motive for killing other sentient beings is human pleasure.

EXERCISES

These can form the basis of essays or group discussions:

1. Do farm animals have ethical standing? What are the reasons for your belief? (7.1–7.3)
2. What are the strongest points in *favour* of your decision, and the strongest points *against* it, if you are: (*a*) a meat eater, (*b*) a vegetarian (state which type), (*c*) a vegan (7.4–7.7).
3. Are you persuaded by Hodges' claim that human relationships with animals are a vital ingredient of human culture? (7.4.5) How would you answer the objection that this argument could have been made in support of slavery?
4. In what ways, if any, is organic farming an ethically acceptable way of raising animals for food?
5. Write an essay on whether there is any ethical justification for fox-hunting (7.8).

FURTHER READING

- *Animal Welfare* by Colin Spedding (2000). London, Earthscan. By a former chair of the UK Farm Animal Welfare Council.
- *The Human Use of Animals: case studies in ethical choice* by F Barbara Orlans, Tom L Beauchamp, Rebecca Dresser, David B Morton, and John P Gluck (1998). Oxford, Oxford University Press. A useful compendium of essays on philosophical and practical issues.
- *Biology, Ethics and Animals* by Rosemary Rodd (1990). Oxford, Oxford University Press. A readable philosophical perspective.

USEFUL WEBSITES

- **http://www.defra.gov.uk/animalh/animindx.htm** *Department for the Environment, Food and Rural Affairs* (Defra): its 'Introduction to animal health and welfare' provides links to many other useful sites.
- **http://www.fawc.org.uk/contact.htm** *Farm Animal Welfare Council*: the Government's advisors on farm animal welfare.
- **http://www.ciwf.org.uk/** *Compassion in World Farming*: a leading NGO.

NOTES

1. Singer P (1975) Animal Liberation. London, Jonathan Cape
2. Thomas K (1983) Man and the Natural World: changing attitudes in England, 1500–1800. Harmondsworth, Penguin, pp. 19–20
3. Serpell J (1986) In the Company of Animals. Oxford, Blackwell, p. 142
4. Dawkins R (1999) In: The Genetic Revolution and Human Rights. (Ed.) Burley J. Oxford, Oxford University Press, pp. xv–xvi
5. Rachels J (1991) Created from Animals: the moral implications of Darwinism. Oxford, Oxford University Press
6. Ingold T (1994) (Ed.) What is an Animal? London, Routledge
7. Ryder R D (1989) Animal Revolution: changing attitudes towards speciesism. Oxford, Blackwell
8. Agriculture and Environment Biotechnology Commission (2002) Animals and Biotechnology. London, DTI, pp. 65–66
9. Singer P (1990) The ethics of animal liberation: a summary statement. In: Animal–Human Relationships: some philosophers' views. Horsham, RSPCA, p. 6
10. Orlans F B, Beauchamp T L, Dresser R, Morton D B, and Gluck J P (1998) The Human Use of Animals. New York, Oxford University Press, p. 28
11. Benton T (1996) In: Animal Rights. (Ed.) Garner, R. London, Macmillan, pp. 19–41

12. Regan T (1990) The rights view. In: Animal–Human Relationships: some philosophers' views. Horsham, RSPCA, p. 8
13. Rodd R (1990) Biology, Ethics and Animals. Oxford, Clarendon, pp. 20–21
14. Fraser A F and Broom D N (1997) Farm Animal Behaviour and Welfare. Wallingford, CAB International
15. Midgley M (1983) Animals and Why They Matter. Athens, University of Georgia Press, p. 98
16. Regan T (1985) The case for animal rights In: In Defence of Animals. (Ed.) Singer P. Oxford, Blackwell, p. 22
17. Swiss Ethics Committee on Non-Human Gene Technology (2001) Die Würde des Tiers. Bern, ECNH
18. Dol M, Fentener van Vlissengen M, Kasamoentalib S, Vissert T, and Zwart H (1999) (Eds) Recognising the Intrinsic Value of Animals. Assen, Van Gorcum
19. Schaber P (2000) (Ed.) The dignity of non-human organisms. Special issue of J Agr Environ Ethic *13*, 1–78
20. Rodd R (1990) Biology, Ethics and Animals. Oxford, Clarendon, p. 4
21. Broom D M (1989) Ethical dilemmas in animal use. In: The Status of Animals: ethics, education and welfare. (Eds) Paterson D and Palmer M. Wallingford, CAB International, pp. 80–86
22. Spedding C (2000) Animal Welfare. London, Earthscan, p. 32
23. Zeuner F E (1963) A History of Domesticated Animals. New York, Harper and Row
24. Spedding C (2000) Animal Welfare. London, Earthscan, p. 36
25. Orlans F B, Beauchamp T L, Dresser R, Morton D B, and Gluck J P (1998) Force feeding of geese. In: The Human Use of Animals. New York, Oxford University Press, pp. 227–240
26. Webster J (1994) Animal Welfare: a cool eye towards Eden. Oxford, Blackwell
27. Sanders T A B and Reddy S (1994) Am J Clin Nutr *59* (suppl), 11765–11815
28. Budianski S (1992) The Covenant of the Wild. London, Weidenfeld and Nicolson
29. Turner J (1999) Factory Farming and the Environment. Petersfield, CIWF
30. Hodges J (2000) In: Livestock, Ethics and the Quality of Life. (Eds) Hodges J and Han I K. Wallingford, CAB International
31. Schweitzer A (1933) My Life and Thought. London, Allen and Unwin
32. Gandhi M K (1940) An Autobiography. Ahmedabad, Navajivan Publishing
33. Beardsorth A and Keil T (1997) Sociology on the Menu. London, Routledge, pp. 218–241
34. Sainsbury A W, Bennett P M, and Kirkwood J K (1995) The welfare of free-living wild animals in Europe: harms caused by human activities Anim Welfare *4*, 183–206
35. Beardsorth A and Keil T (1997) Sociology on the Menu. London, Routledge, p. 211
36. Food Ethics Council (2001) Farming Animals for Food: towards a moral menu. Southwell, FEC
37. Compassion in World Farming Trust (2003) The Welfare of Broiler Chickens in the EU. Petersfield, CIWF
38. Spedding C (2000) Animal Welfare. London, Earthscan, p. 45
39. Rolfe P (2004) A bioethical analysis of antibiotic use for growth promotion in animal production. Unpublished MPhil thesis, University of Nottingham
40. Webster J (1995) Animal Welfare: a cool eye towards Eden. Oxford, Blackwell, p. 170
41. Hillerton J E, Stearn M F H, Staker R T, and Mckinnon C H (1995) Patterns of intra mammary infection and clinical mastitis over a 5 year period in a closely monitored herd applying mastitis control measures. J Dairy Res. *62*, 39–50
42. Blowey R W (1993) Cattle Lameness and Hoofcare. Ipswich, Farming Press

43. Eyton A (1991) The Kind Food Guide. London, Penguin
44. Webster J (1995) Animal Welfare: a cool eye towards Eden. Oxford, Blackwell, p. 192
45. Spedding C (2000) Animal Welfare. London, Earthscan, p. 46
46. Lymbery P (2002) In Too Deep: the welfare of farmed fish. Compassion in World Farming Trust. Petersfield, CIWF
47. Spedding C (2000) Animal Welfare. London, Earthscan, pp. 56–57
48. Royal Society for the Prevention of Cruelty to Animals (1998) Live Animal Transport in the UK: the case for change. Horsham: RSPCA
49. Spedding C (2000) Animal Welfare. London, Earthscan, p. 49
50. Farm Animal Welfare Council (2003) Report on the welfare of farmed animals at slaughter or killing. Part 1: red meat animals. http://www.fawc.org.uk/reports/slaughter
51. Gregory N G (1998) Animal Welfare and Meat Science. Wallingford, CAB International
52. Metheringham J (1999) The Disposal of Day Old Chicks: the way forward. Wheathampstead, Humane Slaughter Association
53. Food Ethics Council (2001) Farming animals for food: towards a moral menu. Southwell, FEC
54. Newby H (1979) Green and Pleasant Land? London, Wildwood House
55. Spedding C (2000) Animal Welfare. London, Earthscan, p. 128
56. DEFRA (2005) Hunting with dogs – a promise fulfilled. http://www.defra.gov.uk/rural/hunting/
57. Spedding C (2000) Animal Welfare. London, Earthscan, p. 128
58. The Government's Hunting Inquiry (2000) http://www.huntinginquiry.gov.uk/
59. Sneddon L U, Braithwaite V A, and Gentle M J (2003) Do fish have nociceptors? Evidence for the evolution of a vertebrate sensory system. Proc R Soc Lond B: Bio *270*, 1115–1121
60. Von der Emde G, Mogdans J, and Kapoor B G (2004) The Senses of Fish. Dordrecht, Kluwer
61. Spedding C (2000) Animal Welfare. London, Earthscan, p. 62
62. Johnson A (1996) Animals as food producers. In: Food Ethics. (Ed.) Mepham B. London, Routledge, pp. 49–63

8

Experiments on animals

OBJECTIVES

When you have read and discussed this chapter you should:

- be aware of the philosophical rationale for animal experimentation and the medical and other benefits which are attributed to it
- understand the ethical justifications advanced both in support of, and in opposition to, experiments on live animals
- be aware of the legislation governing animal experiments in the UK
- understand the 'Three Rs' and the progress made in achieving them over the last twenty years
- be aware of the objectives and strategies of the various political groups seeking to protect animals

8.1 Introduction

One of the most distinctive ways in which bioscientists confront bioethical issues is when they use animals in experiments. The numbers of animals used for other purposes, and the ill-effects of the procedures to which they are exposed (such as in certain types of animal farming or in blood sports – see chapter 7), may often be greater, because the estimated 50 million animals used annually in experiments worldwide[1] only account for about 0.2% of all animal use.[2] But animal experimentation attracts particular attention because to many people it appears unnatural or cruel, in ways that using animal products as food does not. A recent survey showed that for 34% of those questioned, the words 'suffering or cruelty' sprang to mind when the phrase 'animals in medicine and science' was raised.[3]

Few, if any, scientists would deny that animal experimentation requires ethical justification, and it is not difficult to imagine circumstances in which even the strongest supporter of **vivisection** (performing experiments on live animals) would consider that the ethical barriers had been exceeded. So accepting that the *ideal* would be for there to be no animal experimentation at all, we need to examine the justifications which are advanced for failing to meet that objective.

This chapter concentrates on vivisection, although that does not encompass all ethical concerns over the use of animals in science. For example, there may be critical questions about whether animals should be used in anatomical dissections, even when they are killed painlessly; and these questions can become acute when the animal is a member of a social group which suffers from its loss, or when it is a member of a threatened species.

8.2 The aims of animal experimentation

In the UK, the types of animal experiment employed break down into three categories, as follows:

- *Fundamental biomedical research*: explores body function and development, and can include surgically invasive techniques or behavioural studies involving no surgical procedures.
- *Applied research*: includes testing of drugs, vaccines, new medical devices and surgical techniques and use of animal disease *models* in which animals are caused to suffer conditions analogous to the human condition in order to develop methods of diagnosis, treatment, and prevention. A small proportion of these experiments is designed to benefit animals directly.
- *Toxicity testing*: there is a legal requirement for all novel substances to which people, animals, or the environment might be exposed to be tested for safety before their use is permitted. Many such tests involve animals, and under UK law all new medicines must be tested on two different species of animal, one of which is normally a rodent and the other a non-rodent, such as a dog or pig.

The total number of experiments (officially called *procedures*) performed in 2003 was 2.79 million, which represented an increase of 2.2% over the 2002 figure, although up until 2001 the figures had fallen steadily since 1987.[4] However, the word 'procedure' might be deceptive, because in itself it gives no indication of the severity of the impacts on animal welfare, which in many cases might be quite minor[5] (but see 8.7.1). The total number of 'normal' animals used has stabilized, but there has been a steady increase in procedures using GM animals, as discussed in chapter 9.

Animal experimentation in the UK is reputedly governed by some of the most stringent regulations in the world, which are designed to protect animals from excessive and/or unjustifiable suffering. These legal provisions are examined in 8.4, but it is important to begin by considering the ethical rationale for animal experimentation, since any suffering that animals are subjected to, or any infringement of their claimed rights, can only be justified ethically by the value of the resulting knowledge acquired. No one claims we have an absolute duty to conduct animal experiments, although some people claim that if experimentation is *necessary*, then we have a duty not to perform it on humans.

In essence then, the justification advanced for animal experimentation appeals to utilitarian theory, whereby the benefits are deemed to greatly exceed the costs. In 2.5.1 utilitarian theory was illustrated by reference to an imaginary research project designed to investigate a possible treatment for obesity in humans, and it will be useful to refer to that section again to be reminded of the features of the reasoning. As noted, utilitarian theory can 'cut both ways', because it is not only the basis of the standard defence of animal experimentation but, by attaching more significance to the interests of the experimental animals, it can also provide grounds for opposing this practice.

8.3 The scientific rationale for animal experimentation

Those who support animal experimentation would thus seem to be under an obligation to provide coherent ethical reasons for doing so. Sometimes the benefits are apparent in the tangible form of new life-saving medical treatments, but more generally they are claimed to take the form of valuable knowledge. So a *prima facie* case (3.2) for animal experimentation, based on its ability to generate scientific knowledge, might be summarized as follows:

- most medical advances since the late nineteenth century have resulted, directly or indirectly, from biomedical research using animals
- halting such research would have serious adverse consequences for human health and welfare
- there are no viable alternatives to animal experiments, because techniques such as cell and tissue culture and computer simulation are, at best, just useful adjuncts
- animal experiments are justified because of the pervasive biological similarities between the physiology of humans and non-human animals.[6]

This set of assumptions forms the basis of the paradigm (1.5.2) which defines experimental physiology, and because physiology is central to the forms of reasoning adopted in related disciplines, such as biochemistry and experimental psychology, it has a major influence on biomedical science as a whole. However, the epistemological validity (1.4.4) of this paradigm has been challenged, and we need to consider the force of the critical argument to see how it affects the experimentalist's case.

8.3.1 The origins of experimental physiology

Experimental physiology originated in post-revolutionary France, as a development of medical science. Prominent among the pioneer physiologists was Claude Bernard,[7] who in 1865 wrote the hugely influential *An Introduction to the Study of Experimental Medicine*,[8] which is widely regarded as laying down the aims and methodology of all subsequent investigations in animal physiology.

Bernard was motivated by several 'external' factors (1.5.2). First, he was anxious to regain for France the mantle of the world's leading nation in the biosciences, which it appeared to have lost to Germany. His efforts to define the epistemological basis of physiology were thus an essential platform for persuading the French government to invest more resources in medical science.[9] Second, he was impressed by the recent progress in physics and chemistry, which outshone a medical science still largely dependent on clinical observations on patients, and not on controlled experiments.

Third, he was an outstanding scientist, responsible for making several crucial discoveries, such as the glycogenic role of the liver and the existence of vasomotor nerves. He also formulated the fundamental concept of **homeostasis**, although the word itself was only later coined by Walter Cannon. Yet most of Bernard's great discoveries, as one of his astonished students commented, came out of the *'narrow, damp cellar'* that was his laboratory. So it is hardly surprising that he poured his energies into trying to convince others of what he saw as the crucial significance of his discipline.[10]

As a pioneer of the *hypothetico-deductive method* (1.5.1), Bernard believed that physiology could only progress if it was modelled on the physical sciences, and employed rigorously controlled experiments on living organisms. He claimed that, although it would be immoral to experiment on people, *'it is essentially moral to make experiments on an animal, even though painful and dangerous to him, if they may be useful to man'.*[11] To the criticism that animals are not necessarily good models for humans, Bernard argued that *'the vital units, being of like nature in all living beings, are subject to the same organic laws . . . and if we [were to dispute this] we should not only deny science in general but also bring into zoology confusion and darkness that would absolutely block its advance'.* He acknowledged that interspecies differences in physiological response existed but considered that in due course these would be revealed as *'particular cases of a general physiological law, which will thus become the scientific foundation of practical medicine'.*[12]

But evolutionary theory (that was only just emerging in Bernard's time, and which he dismissed as merely 'speculative') points to important differences between even closely related species that challenge the assertion that *any* mammal serves as a good model for humans. While all mammals perform similar physiological functions, such as obtaining nourishment, oxygenating blood, excreting wastes, and reproducing, in adapting to its particular ecological niche each species tends to find distinctive ways of achieving these common functions. And because biological subsystems are tightly interlocked they often give rise to physiological mechanisms that are significantly different. To cite just one example, primates have evolved a form of fetal haemoglobin that permits it to bind oxygen more firmly than in non-primates. The emergence of this unique fetal haemoglobin may be a precondition for the developmental changes that allow the extended fetal life characteristic of simian primates, and consequently the larger brain size, and mental abilities, of humans.[13]

Despite evident limitations, the Bernardian paradigm has proved very persuasive, and in the bioscience community continuing use of animal experiments is widely regarded as essential to the future of the discipline. Our task then is to examine the case made by those who challenge the validity of that claim, by subjecting it to attempted refutation (1.5.1).

8.3.2 Causal analogue models

The strongest ethical case for animal experiments would be one in which the animal acted as an ideal **causal analogue model** (CAM) for a human disease, i.e. the physical cause of the disease and the response to drug treatment were closely analogous in humans and the test animal. But as pointed out by philosophers Hugh LaFollette and Niall Shanks, there must also be *'no causally relevant disanalogies between the model* *and the thing modelled'.*[14] However, disanalogies between species are very common. To give just two examples: rats lack gall bladders, and vitamin C is vital only for a few vertebrate species, notably humans and guinea pigs. So it would be unwise to rely on the animal providing relevant information about a human disease or its effective treatment. At best, CAM tend to be *weak* models, and that being the case, ultimately, the clinical trials on humans are often the critical 'experiments'.

To illustrate the point, consider animal testing of carcinogens (cancer-inducing chemicals). Of the known 26 human carcinogens, most have been shown to cause cancer in one or other mammal, but many successful models involve producing injection site sarcomas (cancerous tumours) or using species other than rats or mice, which are the usual test animals. If attention is confined to long-term feeding studies of rats and mice, only 7 of the 19 non-inhalation carcinogens cause cancer, and including inhalation studies to test all 26 human carcinogens, only 12 cause cancer in mice or rats (i.e. 42%). It follows that *'On the basis of probability theory, we would have been better off to toss a coin'.*[15] Moreover, it has been argued that *'Tossing a coin might be a good idea, since the direct cost of a rodent biosassay is $1 million per chemical tested'.*[16]

Testing for other disease conditions might be even more problematical, not only because of physiological disanalogies, but also because of the different lifespans of laboratory rodents and people, and the fact that it is impossible to test for effects on important human attributes such as memory, hearing and speech. But for most people, the closer the experimental animal is to humans (e.g. chimpanzees), the less acceptable it is to use it in experiments.

Even assuming one has chosen the best model for a human disease, there is a problem in deciding on the appropriate scaling of the response. The **thalidomide** tragedy is an oft-cited example of how things can go wrong. The sedative drug thalidomide, introduced in 1956, was used widely by pregnant women to reduce nausea and vomiting. But by the 1960s it was identified as the cause, in babies born to women who had taken the drug during pregnancy, of *phocomelia* – a devastating birth defect in which the limbs do not form properly and the long bones in the arms or legs are absent. The limitations of using rats as CAM are indicated by the fact that Wistar rats are unaffected by thalidomide doses of up to 4000 mg/kg per d, but in women a dose of 1 mg/kg per d is sufficient to produce the defect.[17]

8.3.3 Hypothetical analogue models

If CAM are of limited value, there may be more to be said in favour of **hypothetical analogue models** (HAM), the aims of which are to prompt the formulation of hypotheses which might be of value to biomedical research in general. Thus it is not claimed that HAM reveal data of *direct* relevance to human medicine, but by increasing understanding of biological phenomena the results may be suggestive of possible human physiological mechanisms. In fact, the intention may be to actively seek out non-mammalian species which present a more convenient system for study based on some known characteristic.

For example, mammals reabsorb sodium from the glomerular filtrate in the distal renal tubule and collecting duct by a complex mechanism involving the hormone aldosterone. But in toads sodium reabsorption occurs in the bladder rather than in the kidney, so that effects of aldosterone on sodium reabsorption can be conveniently studied independently of other renal functions by using a sheet of toad bladder separating two compartments. Results obtained can then guide subsequent experiments in mammals.[18] Often however, animal models are chosen for reasons of low expense and convenience, and not because they are deemed the best model from a scientific perspective.

Despite the alleged advantages of animal experiments, some critics have sought to falsify their claimed value in the following terms: *'attempts to justify the practice of applied or basic animal research ultimately rest on the presentation of a rich bouquet of carefully picked examples. In effect, the public, along with their political representatives, are invited to commit the* ***fallacy of selective perception*** *(counting successes and ignoring failures).'*[19] Thus, this argument appeals to Popper's criticism of induction (1.5.1), and stresses the importance of critical testing of the experimentalist's hypothesis.

8.3.4 Arguments defending the case for animal experimentation

Supporters of the value of animal experimentation suggest that the above criticisms focus on specific failures of the methodology and generalize unjustly to *all* experiments. They claim that animals are used for many different purposes, to provide different types of information, which contributes to general understanding. *'Basing the validity argument solely on the relevance of animal models to humans therefore seems unsustainable.'*[20] They also stress the importance of the reiterative process, by which results of animal experiments allow new human experimentation to be performed, which in turn suggest new animal experiments.

A strong argument for the success of animal experimentation is claimed to be evident in the enormous medical advances which have been made since its introduction. Table 8.1 lists some of the advances over the last 100 years which are attributed to animal experiments. The list may indeed appear to provide impressive evidence for the value of animal experimentation. Acknowledging that animal suffering is involved, justification might be claimed in a form of the *double effect* (3.2), whereby in order to alleviate worse suffering in humans, animal suffering is considered an unavoidable consequence, rather than as the means to achieve this.

But the assumed value of animal experiments might not be quite so clear-cut. It is undoubtedly true that there have been impressive advances in medical treatments, but the degree to which animal experiments have been *responsible* for these is less certain. In some cases the technical experience gained in vivisection was almost certainly valuable. Take, for example, the development of surgical skills involved in transplantation or in anaesthesia. But the limitations of CAM suggest that other applications (e.g. in testing drugs to treat mental depression) might be of less value. The fact that a successful treatment *follows* animal experiments does not imply that it was a *result* of the experiments, especially when the experiments were a legal requirement (8.4). Clinical trials in humans may thus carry more weight than is generally acknowledged. But if the major concern is *fundamental* research, HAM may be claimed to justify animal experiments because they add to the stock of understanding that may subsequently be exploited in human medicine.

Table 8.1 Twenty medical advances attributed to animal experiments

Date	Treatment	Animals used
1900s	• Corneal transplants	rabbits
1910s	• Insulin for diabetics	dogs, rabbits, mice
1920s	• Modern anaesthetics (modern inhaled anaesthetics from the 1950s)	rats, rabbits, dogs, cats, monkeys
1930s	• Kidney dialysis • Anticoagulants (e.g. heparin, warfarin)	guinea pigs, rabbits, dogs, monkeys rabbits, guinea pigs, mice, dogs
1940s	• Hip-replacement surgery • Kidney transplants • Cardiac pacemakers	dogs, sheep, goats dogs dogs
1950s	• German measles and polio vaccines • Coronary bypass operations	mice, monkeys dogs
1960s	• Heart transplants • CAT scanning • Chemotherapy for leukaemia	dogs pigs mice
1970s	• Life-support systems for premature babies • Drugs to control transplant rejection	monkeys mice, rabbits, dogs, monkeys
1980s	• Treatment for river blindness • Meningitis vaccine	rodents, cattle mice
1990s	• Combined drug therapy for HIV • Drugs for breast and prostate cancer • Better drugs for depression (e.g. Prozac)	mice, monkeys mice, rats, dogs rats

Notes: adapted from the Research Defence Society (2003) Timeline (http://rds-online.org.uk).
CAT, computer-assisted tomography; HIV, human immunodeficiency disease.

8.3.5 Are the different views due to selective citation of evidence?

In considering responses to a public consultation exercise, an Animal Procedures Committee (APC) report concluded that those who deny the scientific validity of animal experiments tend to use specific examples to infer a general case that all such experiments are invalid, whereas proponents are guilty of a similarly selective choice of favourable results, while maintaining that every new case has to be considered on its own merits. The APC report acknowledged that evaluation of scientific validity on a case-by-case basis suggested that it was not always sufficiently rigorous, but although this did *'not add up to a general proof that animal experimentation as a whole is flawed'*, it nevertheless *'cannot be construed as an absolute case that every potential use of animals is scientifically valid'*.[21] Perhaps not surprisingly, given that the APC represents a range of opinions on animal experimentation, the conclusions appear to be 'middle of the road'.

Readers are likely to differ in the degree to which they consider criticisms of animal experimentation are telling, but even so it seems that the widespread perception that leads two-thirds of the public to *'accept animal experimentation so long as it is for medical research purposes'*[22] begs more questions than it answers.

8.4 Legislation for animal experimentation

As noted (8.3.6), a majority of the UK public support animal experiments conducted with the aim of advancing medical treatments, but their support is conditional on evidence that these procedures are closely controlled. The UK was the first country to introduce legislation on animal experimentation (as The Cruelty to Animals Act, 1876), which came about in response to alleged cruelty in scientific laboratories.[23] Physiologists, who felt threatened by the public concern over their activities, closed ranks and in the same year formed the Physiological Society - which thus became the world's first such professional body, even though other countries (such as France and Germany) had long been more active in this field.

The 1876 Act remained in force for more than a century, only being replaced in 1986 by the Animals (Scientific Procedures) Act, commonly referred to as ASPA. The Act is enforced by the Home Office, so that it is the Home Secretary (in receipt of expert advice) who makes the decisions on which individual researchers may be licensed to perform experiments, the institutions where they may be performed, and the specific programmes of work permitted. Although in principle a very wide range of procedures is permitted (8.2), in practice a number of constraints limit what is allowed. Philosopher Andrew Brennan suggests that the heavy regulation of animal experiments is a result of the following four factors:

- experimental animals are particularly vulnerable to mistreatment, with no possibility of escape
- because the places where research is typically performed, such as university laboratories, are committed to responding to reasonable criticism, a high priority is placed on defending the practice intellectually
- cases where there is intentional infliction of harm (e.g. causing animals to develop cancer) differ from most other uses of animals, where any harm caused is unintentional (e.g. beak-trimming of chickens aims to reduce harm caused by birds pecking each other)
- when animals are used as models for human diseases, their status might seem to imply they are unfairly treated.[24]

A practical outcome is that animal experiments must be seen to be non-trivial, and capable of producing valuable results. For example, section 5(4) of ASPA states that *'in determining whether and on what basis to grant a project licence the Secretary of State shall weigh the likely adverse effects on the animals concerned against the benefit likely to accrue as a result of the programme to be specified in the licence'*. For such reasons, the testing of cosmetics and toiletries ingredients on animals has been banned in the UK since 1998.

But although cost–benefit analysis might seem to imply a form of mathematical calculus, the APC stresses that it has to be recognized that *'It is in the nature of scientific experiments that outcomes are uncertain, so evaluation of costs to animals and likely benefits of studies can be based only on potential, likely and probable, not certain outcomes. Cost–benefit assessment, moreover, is not morally neutral. For example, values come into play in deciding what should count as a "legitimate" cost or benefit, and the relative weights the different*

BOX 8.1 LEGISLATION ON ANIMAL EXPERIMENTATION

- **EU Directive 86/609/EEC (Article 7) 1986**

¶7 (2) An experiment shall not be performed, if another scientifically satisfactory method of obtaining the result sought, not entailing the use of an animal, is reasonably and practicably available.

¶7 (3) When an experiment has to be performed, the choice of species shall be carefully considered, and where necessary, explained to the authority. In a choice between experiments, those which use the minimum number of animals, involve animals with the lowest degree of neuro-physiological sensitivity, cause the least pain, suffering, distress or lasting harm and which are most likely to provide satisfactory results, shall be selected.

¶7 (4) All experiments shall be designed to avoid distress and unnecessary pain and suffering to the experimental animals.

- **UK Animals (Scientific Procedures) Act 1986**

The Secretary of State shall not grant a project licence unless he is satisfied that the applicant has given adequate consideration to the feasibility of achieving the purpose of the programme to be specified in the licence by means not involving the use of protected animals.

kinds of costs and benefit should be accorded.'[25] It follows that it would be misguided to think that the existence of laws governing animal experiments guarantees that ethical questions are adequately addressed.

8.4.1 The Three Rs

An important feature of ASPA is its emphasis on the so-called Three Rs. This phrase, amusingly introduced in the nineteenth century to refer to the skills of *'reading, (w)riting, and (a)rithmetic'*, was adopted by William Russell and Rex Burch in their seminal book *Principles of Humane Experimental Technique* (1959) as shorthand for *replacement, reduction and refinement*. If the total suffering of animals used in experiments is assessed as the product of the average severity of the procedures and the number of animals, then this total can in theory be diminished by one or more of the following:

- **Replacement**: the substitution for conscious living higher animals of insentient material
- **Reduction**: reduction in the numbers of animals used to obtain information of given amount and precision
- **Refinement**: any decrease in the incidence or severity of inhumane procedures applied to those animals which still have to be used.[26]

However, despite the force of the arguments, little attention was paid to these recommendations for nearly thirty years – until, in 1986, the Three Rs formed the basis of both EU and UK legislation (Box 8.1). It is likely that the progress achieved in the

1960s in other fields, such the civil rights of minorities and women's rights, led to the introduction of the term *animals rights* (7.1.1) in the 1970s, and gave impetus to the pressure for change. But, at the same time, increased concern over consumer rights and welfare means that society has also witnessed increasing demands for safety testing of a growing number of new medicinal and non-medicinal products to protect human health and the environment.

The next three sections explore the scientific rationale for each of the Three Rs, and the progress made, and constraints encountered, in implementing change. But it should be noted both that there is much overlap between them and that each affects the others.

8.5 Replacement

Russell and Burch distinguished between *relative* replacement, in which animals are still required but not exposed to actual distress in the experiment, and *absolute* replacement, in which animals are not required at any stage. But ideas on replacement have evolved considerably in the past forty years and it is usual now to speak of **alternatives** to animal experiments, a change given official recognition by the EU in the establishment of the European Centre for the Validation of Alternative Methods (ECVAM) in Italy. If a replacement alternative is defined as a method which does not involve use of a protected animal in a regulated procedure, a range of methods is available (Box 8.2).[27]

BOX 8.2 REPLACEMENT ALTERNATIVES

- improved storage, exchange, and use of information about animal experiments, avoiding unnecessary repetition
- use of physical and chemical techniques and predictions based on the physical properties of molecules
- use of mathematical and computer models (modelling quantitative structure–function relationships; molecular modelling using computer graphics; modelling of biochemical, physiological, pharmacological, toxicological, and behavioural systems and processes)
- use of lower organisms (e.g. invertebrates, plants and micro-organisms)
- use of early developmental stages, before they are protected by ASPA (e.g. before the mid-point of pregnancy in mammals)
- use of *in vitro* methods (subcellular fractions, tissue slices, cell suspensions, perfused organs and tissue culture proper – including human tissue culture)
- human studies, involving volunteers, post-marketing surveillance and epidemiology

However, implementing replacement alternatives encounters a number of problems. For example, better use of information depends on its being freely available, but many results from animal experiments are protected by **commercial confidentiality**. Moreover, use of physico-chemical techniques, and of mathematical and computer modelling, needs to be *validated*, which usually means checking against animal models.

Use of lower organisms is problematical because, although it is generally considered that invertebrates lack sentience, there are no obvious grounds for assuming a sharp distinction, for example between lower vertebrates, such as fish, and higher invertebrates, such as some cephalopod molluscs. Indeed, on the principle of giving the benefit of doubt, ASPA has been amended to extend protection to *Octopus vulgaris*.

The alternative of using human patients also encounters significant difficulties, which are not easily resolved. For example, if informed consent is required for all treatments designed for human patients, there would seem to be no scope for devising remedies for sick children, or elderly people with senile dementia.

Replacement strategies are usefully considered in relation to two different types of animal use: fundamental research and toxicity testing.

8.5.1 Fundamental research

It is almost impossible to establish in advance the *necessity* of performing fundamental research on animals in terms of the likely benefits, because 'knowledge' is not necessarily a good in itself. Indeed, some (trivial) knowledge would not be worth having at any price. The problem is that ASPA requires the costs born by the animals to be weighed against the prospective benefits of this, uncertain and possibly useless, knowledge (8.4). Although all scientists clearly aim to acquire significant new understanding, they can hardly be immune to the criticism that an important objective along the way is obtaining data that can be published in the scientific literature (15.3.2). If animal experimentation is central to your job, the ability to advance your career by producing publishable results is likely to be a high priority, which might conflict with the aims of replacing animals by alternative methods.

Former head of ECVAM, Michael Balls, claims that *'The current situation is very unsatisfactory. Criticism by outsiders is often dismissed by some members of the scientific community as ill-informed and malicious in its intent. Some defenders of animal experimentation argue that even the indefensible must be defended, lest yielding ground in one case should lead to uncontrolled penetration of criticism into animal-based research in general.'*[28]

The problem is perhaps most acute in relation to fundamental studies which are aimed at treating major diseases such as cancer or Alzheimer's disease. Proposed research on these subjects does not become ethically acceptable *merely* because of the severity of the disease conditions. Over and above this, Home Office inspectors have the task of assessing the validity of the hypothesis to be tested, the appropriateness of the experimental design, the competence of the personnel involved, and the likelihood of valuable results being obtained. Judgement is a crucial factor at all stages.

8.5.2 Toxicity testing

Use of animals in toxicity testing (e.g. of drugs, household products, agrochemicals, industrial chemicals and food additives) has two important features with ethical implications:

- the induction of adverse effects is intentional, in order to assess safety limits for humans
- such testing is often a legal requirement (e.g. for 87% of all cases in 2002).[29]

Toxicity testing is also different from many other animal uses because the maintenance of the *status quo* is backed up by very substantial vested interests, in the form of regulatory authorities, industrial and academic laboratories, and contract testing establishments. Thus, if animal testing were to be significantly replaced by alternative methods this would have significant economic implications. These factors clearly impede attempts at replacement. A notable, and controversial, example is the **Draize test** for eye irritancy, which involves placing liquid, flaked, granulated, or powdered substances in the eyes of rabbits, and noting the eyes' progressive deterioration. Data are recorded at intervals over an average period of 72 hours (though the period may extend in some cases to 18 days) and reactions include swelling of the eyelid, inflammation of the iris, ulceration, haemorrhaging and blindness.[30]

As noted (8.4), animal testing of cosmetics is banned in the UK, but some people consider this is an anomalous decision, because other products which might be claimed to be equally trivial, such as clothing dyes and food colourings, are still tested. Moreover, there are notable loopholes, because cosmetics imported from abroad are currently exempt from the animal-testing ban; while the use of BOTOX® (a type of Botulinum toxin) to treat facial wrinkles is also licensed for therapeutic purposes (e.g. for Parkinsonism and multiple sclerosis) and so escapes the ban.[31]

A critical problem is **validation**, i.e. ensuring that the proposed alternatives are as reliable and sensitive as the animal tests they seek to replace (although, as noted in 8.3.3, animal tests are often far from reliable themselves). But in addition to the legislative barriers which *require* animal testing, there are also several other barriers to winning support for alternative methods, such as psychological barriers (e.g. traditional scientists resisting the 'new') and litigation barriers (manufacturers protecting themselves by resort to animal tests as the 'accepted norm').[32]

8.6 Reduction

The number of animals used in experiments can be reduced by measures to improve the efficiency of knowledge acquisition, e.g. by avoiding unnecessary duplication and by ensuring that poor experimental design does not waste animals. Detailed surveys of published papers have indicated the poor design of many experiments, which if they had been better designed could have used fewer animals to obtain results with greater precision.[33] For example, the *Student's t* test is frequently abused, *p* values are interpreted dogmatically, and non-parametric tests are often used when it would be better to employ a more sensitive experimental design using fewer animals.[34]

In the 1980s it was realized that because toxicity tests for the same product varied substantially between Europe, the USA, and Japan, pharmaceutical companies wishing to market their products worldwide had to test them using different protocols to satisfy legal requirements. Calls for **harmonization** of international requirements have led to substantial improvements in animal welfare, although a primary motive was economic savings. Efforts to achieve harmonization were facilitated by establishment of the International Conference on Harmonisation of Technical Requirements for Human Use (ICH), and it is estimated that as a result it will be possible to reduce the numbers of animals used in testing a 'standard' new chemical by 47%.[35] There is clearly scope for others, e.g. in the chemical, food, and pesticide industries, to reduce animal use by a similar harmonization process. But there is a danger that strict harmonization might reduce flexibility in interpreting test guidelines, suggesting that *rationalization* of guidelines might be a better strategy.[36]

A notable case where reduction (and refinement) has been achieved is in introducing an alternative to the somewhat notorious **LD50 test**, which is used to determine the *lethal dose* of a chemical needed to kill 50% of the treated animals. The LD50 is now mainly used for industrial chemicals and agrochemicals, in order to satisfy legal requirements. However, an alternative validated procedure, the **Fixed Dose Procedure**, which needs only 20 animals per test (instead of the 60 to 80 required by the LD50) and does not rely on death as an end-point, has now received international acceptance in drug testing.[37]

One novel way that animal numbers might be reduced is through the use of stem cells (6.5), e.g. to remove the perceived need to perform liver damage tests on animals.[38] But, as with all *in vitro* systems, they are unlikely to entirely replace the perceived need for tests on whole animals, or human subjects.

8.7 Refinement

Refinement refers to reducing the severity of the procedures used, which is a product of the actual procedures employed and their impact on the individual animal. So there are in principle two approaches to achieving refinement – using milder procedures and using animals that are less sentient.

8.7.1 The severity of the procedures

To regulate animal procedures, four categories are defined:

- **Mild**: includes taking small blood and tissue samples, and observing how animals behave, e.g. in a maze
- **Moderate**: a very broad category, including everything from injecting substances into animals to induce antibody formation to implanting microtransmitters for blood

pressure monitoring. If surgery is involved animals are given appropriate anaesthetics to minimize pain

- **Substantial**: includes major surgery, lethal toxicity testing, and certain experiments where animals are used as disease models
- **Unclassified**: animals in this category are anaesthetized before any experimentation begins and are killed without recovering consciousness.

It is important to appreciate that these categories are Home Office *predictions*; they thus represent the overall severity banding of a particular project, but do not indicate what individual animals *actually* experience.[39] In 2003, most procedures were classed as mild or moderate, with about 41% using *'some form of anaesthesia to alleviate the severity of the interventions'*. Administering anaesthesia can itself cause stress, and *'for many of the remaining procedures, use of anaesthesia would have increased the animal welfare cost'*.[40] But, as pointed out by the chair of the APC, because the appropriate information is lacking, currently *'it is not possible to say whether the amount of suffering by animals this year is greater or less than last'*.[41]

Of course, assessing the extent of pain and distress experienced by animals presents a significant problem. Even with our fellow humans we can have little idea of what they feel, as opposed to what they say (or exclaim), so assessments in animals are fraught with difficulties (particularly when we are dealing with highly subjective concepts, like pain, anxiety, stress and distress). Even so, considerable progress has been made in establishing objective behavioural, physiological, and clinical criteria.[42] For example, physiological indices include measurements of heart rate, blood concentration of adrenocorticotrophic hormone (ACTH), white blood cell numbers and antibody production.[43]

A factor often overlooked is the conditions under which experimental animals are bred and housed. Exposure to experimentation itself might account for only a small fraction of their lives, but animals can clearly also suffer if kept in barren, sterile conditions which provide no opportunities for expression of normal behaviour. Environmental enrichment, which as well as being a welfare requirement may be necessary to provide reliable experimental data, can often prove to be very inexpensive, as in supplying nesting materials, chewing blocks, and play-things.

A more problematical issue is the welfare cost of death. While some people claim that a painless death (as in 'unclassified procedures', see above) presents no welfare cost, others contend that depriving an animal of life is a very serious cost, which is made worse if it is not an act of necessity.

8.7.2 The use of primates

Although EU legislation states that *'In a choice between experiments, those . . . with the lowest degree of neurophysiological sensitivity . . . shall be selected'*, there are continuing concerns over the use of cats, dogs, equidae and primates. The following comments are confined to experiments using primates. Because of the legal requirement for drugs to be tested on two species, use of non-rodents has become standard practice, so that more than

70% of primates used are for toxicity testing. Macaques are used more commonly than marmosets and tamarins, whereas baboons have not been used at all since 1999. Most procedures fall in the moderate severity category, and while there was a suggestion that total numbers of animals used had declined over the last decade, in 2003 the number of procedures recorded increased by 20% over the 2002 figure, because some animals were reused (e.g. to take blood samples).[44]

However, an APC report suggested that the overall aim of entirely eliminating the use of primates is likely to be challenged by the pharmaceutical industry's view that its future lies in developing new neuroactive drugs to combat the increasingly important neurological diseases of old age. So the perceived need for effective animal models, capable of simulating human cognitive functions, neuroanatomical structure, and cellular structure and function, might increase the demand for primate research. For example, anthropoid apes are the only species sharing with humans the advanced frontal lobe functions which are commonly involved in neurodegenerative diseases. This may well mean that *'primates will become the only species, rather than merely the preferred option'* for such research. Indeed, the report identified a further concern - that the desire for more appropriate CAM (8.3.3) might lead to pressure for using great apes - notably the chimpanzee, the use of which is currently banned in the UK and elsewhere.[45]

8.8 Regulation

In the UK the regulation of any animal experiment under ASPA is governed by a cost-benefit analysis, which is the result of inputs from several people and processes. These include the researchers themselves, local ethics committees, funding bodies, editorial boards of journals (who may refuse to publish reports which do not meet their ethical criteria), the APC and Home Office inspectors, who act as the final arbiter of the ethical acceptability of a project proposal. Despite these apparently rigorous controls, concerns have been expressed that the process is not sufficiently effective and transparent. For example:

- the layout of the project licence application means that the *'cost-benefit assessment is lost within the complexity of the form'*[46]
- it is argued that 22 Home Office inspectors are insufficient for the task[47]
- there is a lack of clarity about how the cost-benefit analysis should be performed.[48]

8.8.1 The House of Lords report

In an effort to address some of the above concerns, in 2002 a House of Lords Select Committee (14.3.1) published the most thorough review of legislation since the introduction of ASPA in 1986. Key findings are listed in Box 8.3.

BOX 8.3 HOUSE OF LORDS SELECT COMMITTEE ON ANIMAL RESEARCH, 2002: CONCLUSIONS

- It is morally acceptable to use animals in experiments, but wrong to cause unnecessary or avoidable suffering.
- There is currently a continued need for animal experiments.
- Toxicological testing is essential for medical practice and protection of consumers and the environment.
- There is a need to strive for the best regulation, but not necessarily the tightest regulation.
- Good quality up-to-date information on what animal experiments are done, and why, is necessary if the issue is to be discussed productively.
- There is scope for scientists to give greater priority to development of non-animal methods, and to pursuit of the Three Rs.
- Development of valid non-animal systems is important both to improve animal welfare and provide benefits for human health.

A number of recommendations were made, of which the following were among the most notable:

- The setting up of a **Centre for the Three Rs**, to coordinate research units within exiting centres of academic excellence. It was considered that this visible commitment to actively searching for improvements in, and alternatives to, animal experiments (especially in toxicology) would engender increased public confidence on scientists' role within society.
- The Home Office Inspectorate (8.8) should be subject to periodic review, by an independent body.
- The so-called 'confidentiality clause' of ASPA should be repealed, and specific justification made for each class of information that needs to be kept confidential. These might include, for example, the identity of researchers, commercial secrets and intellectual property.
- Project licence applications should be simplified to avoid unnecessary and time-consuming paperwork.

The Government responded favourably to several of these recommendations, but initially appeared unconvinced by other arguments. However, in 2004 it agreed to establish a national centre for research into the Three Rs, to be housed in the MRC's Centre for Best Practice for Animals in Research,[49] and to double funding for such research to £660,000 p.a. initially (out of a total MRC budget of £400 million).

However, some animal welfare groups are unimpressed. For example, the Fund for the Replacement of Animals in Medical Experiments (FRAME), a leading NGO, is concerned that the Centre will favour reduction and refinement at the expense of replacement, which is viewed as an attempt to defend animal-based research at all costs. Moreover, they claim that the early indications were that the Centre *'will not be a completely independent organisation, able to function without interference from individuals and bodies which purport to support the Three Rs, but in fact may devote most of their energy to maintaining the independence of biomedical research and testing on animal procedures'.*[50] Other animal interest groups have refused to collaborate with the Centre, seeing its remit as biased.

8.9 Political activity

The use of animals in experiments has for long been the focus of much political protest. The term *animal rights* was used in 7.2.2 to describe a philosophical concept concerning the ethical status of animals. But, as more commonly understood, it refers to the 'animal rights movement', a political activity that seeks an end to all animal experimentation and animal farming. Some militant animal rights organizations have even adopted violent practices against people and property to achieve their ends. Among the targets have been scientific laboratories, cosmetic and pharmaceutical companies, slaughterhouses and butchers' shops, food retailers, circuses, zoos and hunters.[51] Supporters appear to employ a form of utilitarian justification for their actions which counts threats to human life and welfare as an acceptable cost in achieving their aims. Unsurprisingly, this has led to the development of a 'bunker attitude' in the scientific community, which frequently prevents open debate about the aims and methods of animal research.

But the prominent philosopher and utilitarian antivivisectionist Peter Singer has claimed that the threats of violence on researchers and others by animal liberation activists undermine the ethical basis of the movement. For he argues: *'It is difficult to find democratic principles that would allow one group to use intimidation and violence, and deny the same methods to the other'.*[52] Instead, he cites Mahatma Gandhi and Martin Luther King as exemplars of civil disobedience, which he considers is an effective means of demonstrating *'sincerity and commitment to a just cause'.*

There is, however, much pressure for more openness on the nature and justification for animal experiments, especially since the introduction of the Freedom of Information Act (2000). Proponents of greater openness (e.g. animal protection organizations) argue that the secrecy surrounding animal experiments fuels concerns over whether animal use is always justified and the level of animal care adequate. Opponents (e.g. researchers and pharmaceutical companies) claim that there is no need for more openness, and that:

- making more information public may compromise researchers' safety (due to the activities of extremist animal rights organizations)
- for commercial organizations to divulge more information would undermine commercial competitiveness.

A proposed compromise solution is for more information to be released, so allowing members of the public to understand, and question, decisions made to permit experiments, but at the same time to keep confidential both personal details of researchers and information that would compromise commercial considerations.[53]

8.9.1 Classifying attitudes to animal use

Everyone who takes any sort of political stand on animal use can be classified according to their beliefs, goals, and the strategies they advocate to achieve them. Table 8.2 (adapted from a scheme proposed by James Jasper and Dorothy Nelkin)[54] categorizes these different views.

Pro-research advocates see no alternative to animal experiments, but they advocate developments which both improve the efficiency of experimental procedures and animal welfare. **Welfarists** accept most current uses of animals (e.g. for research, food and sport), but seek to improve their lot through the established channels of education, support for animal welfare charities, and through purchase decisions. **Pragmatists** adopt a more proactive stance, supporting legal actions, political protest, and negotiation with those who use animals. In the UK, they might typically support organizations such as FRAME, the Royal Society for the Prevention Cruelty to Animals (RSPCA), the Dr Hadwen Trust and Compassion in World Farming (CIWF), adopt vegetarianism and oppose fox-hunting. **Fundamentalists** believe in more radical approaches to end what they see as unacceptable animal exploitation: they are likely to be vegans, opposed to all animal experiments, and in

Table 8.2 Types of animal protectionist (From: Jasper and Nelkin, 1992)

Category (examples of organizations)	Beliefs about animals	Major goals	Primary strategies
Pro-research with animals (Research Defence Society; AMRIC)	• Animal experiments are essential for medical and scientific advances	• Humane use of animals in experiments	• Encourage self-regulation by scientists within legal constraints
Welfarists (RSPCA)	• Objects of compassion, deserving protection • Clear boundaries between species	• Avoidance of cruelty • Limitation of unwanted populations	• Protective legislation • Humane education • Shelters
Pragmatists (FRAME; UFAW; Dr Hadwen Trust)	• Deserving of moral consideration • Balance between human and animal interests	• Eliminate unnecessary suffering • Reduce, replace, and refine uses of animals	• Public protests but pragmatic cooperation, negotiation, and short-term compromises
Fundamentalists (NAVS; BUAV; Animal Aid; Advocates for Animals)	• Have absolute moral rights to live without human interference • Equal rights across species	• Total and immediate elimination of all animal exploitation	• Moralist rhetoric and condemnation • Direct action, civil disobedience • Animal sanctuaries

Notes: AMRIC, Animals in Medicine Research Information Centre; FRAME, Fund for the Replacement of Animals in Medical Experiments; RSPCA, Royal Society for the Prevention of Cruelty to Animals; UFAW, Universities Federation for Animal Welfare; NAVS, National Anti-Vivisection Society; BUAV, British Union for the Abolition of Vivisection.

some cases willing to break the law in pursuit of their objectives. The distinctions are clearly not rigidly defined, e.g. some who have fundamentalist beliefs might support pragmatic strategies.

Organizations classified in Table 8.2 as *pragmatists* are actively involved in seeking practical ways of implementing the Three Rs by supporting research in alternative methods. Two are noteworthy. FRAME (which also publishes the academic journal *ATLA: Alternatives to Laboratory Animals*) supports all research in alternative methods specified in Box 8.2. The Dr Hadwen Trust only funds *non-animal* techniques, such as those exploiting human tissue cells, magnetic resonance imaging, and computer simulation approaches.

8.10 Ethical analysis

The lack of a clear framework for ethical assessment of proposals for animal experiments (8.8) under ASPA has prompted proposals to rectify this deficiency. The following scheme suggests features of a detailed checklist to be completed by applicants for all proposals.[55] The criteria comprise the following:

- *Justification*: specifying the purposes of the work and how the Three Rs had been applied
- *Scientific relevance*: demonstrating the scientific quality of the work and the quality of the workers and their facilities
- *Animal suffering*: establishing the severity of the procedures, and factors such as the impact of housing conditions and transport
- *Wider social, economic, and environmental impacts*: assessing potential impacts on various interest groups (e.g. animals in general, human patients; consumers, society in developed and developing countries, the biota).

In assessing each criterion, decisions would be assigned to one of the following categories:

- *Approval*: information provided is acceptable, within provisions of ASPA
- *Conditional approval*: approved subject to specified additional conditions
- *Provisional rejection*: substantial modifications are necessary before reconsideration
- *Rejection*: information provided is unacceptable
- *Referral*: insufficient information provided to reach a decision.

The scheme outlined may be seen to incorporate many of the principles of the ethical matrix (3.5). Its representation in this form pays more explicit reference to issues of justification and scientific relevance than have traditionally been identified in legislation governing animal experimentation.

THE MAIN POINTS

- Animal experiments are performed to further fundamental knowledge and to test and develop drugs, medical procedures and new chemical products.
- The rationale for animal experiments usually depends on their serving as good *causal analogue models* for humans, but there is disagreement over whether this criterion is often met, or whether more realistically their main role is to serve as *hypothetical analogue models*.
- Supporters of animal experiments claim they are necessary for advances in medicine, whereas opponents argue that their significance is exaggerated and that there are often satisfactory alternatives, such as tissue culture and computer simulation.
- UK legislation requires consideration be given to replacement, reduction and refinement of animal experiments (the Three Rs).
- Progress in animal welfare has been achieved, but concerns are expressed that numbers of GM animals used are increasing, that use of non-human primates is set to increase, and that insufficient attention is being paid to alternative methods.

EXERCISES

These can form the basis of essays or group discussions:

1. Discuss the limitations of animals as causal analogue models (CAM) (8.3.2).
2. Assess the ethical justifications advanced for using animals as hypothetical analogue models (HAM) (8.3.3).
3. Should non-human primates ever be used in medical research? Give your reasons and specify any extra limitations on such work which you think are necessary (8.7.2).
4. In what ways is it ethically acceptable in democratic societies for citizens to try to influence what scientists are allowed to do in research on live animals? (8.9)
5. Imagine you are a Home Office Inspector faced with a request to approve a project involving research on rabbits which aims to increase knowledge about muscle function. What conditions would need to be satisfied for you to approve the application? Or if no conditions would satisfy you, explain why (8.10).

FURTHER READING

- *Brute Science: dilemmas of animal experimentation* by Hugh LaFollette and Niall Shanks (1996). London, Routledge. A powerful critique of animal experimentation.
- *Lives in the Balance: the ethics of using animals in biomedical research* edited by Jane A Smith and Kenneth M Boyd (1991). Oxford, Oxford University Press. A comprehensive account of the costs and benefits of experimental techniques employed.

- *The Animals Issue: moral theory in practice* by Peter Carruthers (1992). Cambridge, Cambridge University Press. Unlike most books on animals, this is by a philosopher who denies that they have rights.

USEFUL WEBSITES

- **http://www.homeoffice.gov.uk/comrace/animals/** *UK Home Office*: This site provides much information on the law concerning animal experimentation.
- **http://frame.org.uk/** *Fund for the Replacement of Animals in Medical Experiments*: an NGO pioneer of the Three Rs.
- **http://www.rds-online.org.uk/** *Research Defence Society*: the leading organization promoting the continued use of animals in medical research.

NOTES

1. Nuffield Council on Bioethics (2003) The Ethics of Research Involving Animals: consultation paper. London, Nuffield Council on Bioethics
2. Brennan A (1997) Ethics, codes and animal research. In: Animal Alternatives, Welfare and Ethics. (Ed.) van Zutphen LFM and Balls M. Amsterdam, Elsevier, pp. 43–54
3. Medical Research Council (1999) Animals in Medicine and Science (MORI research). London, MRC
4. Home Office (2004). Animals in scientific procedures: annual statistics 2003. http://www.homeoffice.gov.uk
5. Banner M (2004) New animal research data should be treated sceptically. Daily Telegraph. http://businessawards.co.uk/connected/main
6. LaFollette H and Shanks N (1996) Brute Science: dilemmas of animal experimentation. London: Routledge, p. 10
7. Lesch JR (1984) Science and Medicine in France: the emergence of experimental physiology 1790–1855. Cambridge, Mass, Harvard University Press
8. Bernard C (1949) [1865] An Introduction to the Study of Experimental Medicine (trans. HC Greene). New York, Abelard Schuman
9. Coleman W (1985) The cognitive basis of the discipline: Claude Bernard on physiology. ISIS *76*, 49–70
10. Olmsted JMD and Olmsted EH (1952) Claude Bernard and the Experimental Method in Medicine. New York, Abelard Schuman, p. 108
11. Bernard C (1949) [1865] An introduction to the study of experimental medicine (trans. HC Greene). New York, Abelard Schuman, p. 102
12. Ibid., pp. 124–126
13. LaFollette H and Shanks N (1996) Brute Science: dilemmas of animal experimentation. London, Routledge, p. 100
14. Ibid., p. 133

15. Salsburg D (1983) The life time feeding study in mice and rats: an examination of its validity as a bioassay in human carcinogenesis. Fund Appl Toxicol *3*, 63–67
16. LaFollette H and Shanks N (1996) Brute Science: dilemmas of animal experimentation. London, Routledge, p. 163
17. Millstone E (1989) In: Animal Experimentation: the consensus changes. (Ed.) Langley G. Basingstoke, Macmillan, pp. 74–75
18. LaFollette H and Shanks N (1996) Brute Science: dilemmas of animal experimentation. London, Routledge, p. 197
19. Ibid., p. 204
20. Animal Procedures Committee (2003) Review of Cost–Benefit Assessment in the Use of Animals in Research. London, Home Office, pp. 87–91
21. Ibid., p. 25
22. Medical Research Council (1999) Animals in Medicine and Science (MORI Research). London, MRC, p. 18
23. Richards S (1990) In: Vivisection in Historical Perspective. (Ed.) Rupke N A. London, Routledge, pp. 125–148
24. Brennan A (1997) Ethics, codes and animal research. In: Animal Alternatives, Welfare and Ethics. (Eds) van Zutphen L F M and Balls M. Amsterdam, Elsevier, pp. 43–54
25. Animal Procedures Committee (2003) Review of Cost–Benefit Assessment in the Use of Animals in Research. London, Home Office, p. 79
26. Russell W M S and Burch R L (1959) The Principles of Humane Experimental Technique. London, Methuen, p. 64
27. Balls M (1994) Replacement of animal procedures: alternatives. Lab Animal *28*, 193–211
28. Ibid., 193–211
29. Home Office (2003). Animals in scientific procedures: annual statistics 2002. http://www.homeoffice.gov.uk
30. Group for the education of animal-related issues (2004) http://www.geari.org/faqdraize.html
31. FRAME (2003) Growing old disgracefully: the cosmetic use of Botox. FRAME News no. 57, 1–3
32. Balls M (1994) Replacement of animal procedures: alternatives. Lab Animal *28*, 193–211
33. Festing M F W (1994) Reduction of animal use: experimental design and quality of experiments. Lab Animal *28*, 212–221
34. Overend P (2001) In: Reduction. FRAME News, June 2001. (Ed.) Festing M. Nottingham, Fund for the Replacement of Animals in Medical Experiments, pp. 13–14
35. Combes R (2001) In: Reduction. FRAME News, June 2001. (Ed.) Festing M. Nottingham, Fund for the Replacement of Animals in Medical Experiments, pp. 2–4
36. Smith J A and Boyd K M (1991) Lives in the Balance: the ethics of using animals in medical research. Oxford, Oxford University Press, p. 209
37. Fund for the Replacement of Animals in Medical Experiments (1993) Developing Alternatives to Animal Experimentation. FRAME/Glaxo Group Research. Cambridge, Hobson Publishing
38. Twine R (2004) Stem cells to reduce animal testing? http://www.shef.ac.uk/bioethics-today/archives/files/Animalexperimentationcomm.htm comm1816
39. Home Office (2002) Use of primates under the Animals (Scientific Procedures Act (1986)). London, Home Office, p. 15
40. Home Office (2003). Animals in scientific procedures: annual statistics 2002. http://www.homeoffice.gov.uk

41. Banner M (2004) New animal research data should be treated sceptically. Daily Telegraph. http://businessawards.co.uk/connected/main
42. Smith J A and Boyd K M (1991) Lives in the Balance: the ethics of using animals in medical research. Oxford, Oxford University Press, pp. 78–90
43. Broom D M and Johnson K G (1993) Stress and Animal Welfare. London, Chapman and Hall
44. Home Office (2002) Use of primates under the Animals (Scientific Procedures Act (1986)). London, Home Office, p. 16–17
45. Ibid., p. 29
46. Animal Procedures Committee (2003) Review of Cost–Benefit Assessment in the Use of Animals in Research. London, Home Office, p. 66
47. Straughan D W (1995) The role of the Home Office inspector. ATLA *23*, 39–49
48. Animal Procedures Committee (2003) Review of Cost–Benefit Assessment in the Use of Animals in Research. London, Home Office, p. 66
49. Home Office (2004) Report of the Inter-Departmental Group for the 3Rs. http://www.homeoffice.gov.uk/docs2/interdept3rs040818.htm
50. Balls M and Combes R D (2004) Pathway to progress or mere fig leaf? FRAME News *58*, 1–3
51. Jasper J M and Nelkin D (1992) The Animal Rights Crusade. New York, Free Press, p. 3
52. Singer P (2004) Humans are sentient too. http://www.guardian.co.uk/print/0,3858,4982236 103677.00.html
53. Parliamentary Office of Science and Technology (2004) Openness and Animal Procedures. London, POST
54. Jasper J M and Nelkin D (1992) The Animal Rights Crusade. New York, Free Press, p. 178
55. Delpire V C, Mepham T B, and Balls M (1999) A proposal for a new ethical scheme for addressing the use of laboratory animals for biomedical purposes. ATLA *27*, 869–881

9

Animals and modern biotechnology

OBJECTIVES

When you have read and discussed this chapter you should:

- be aware of the aims and limitations of reproductive technologies used in farm animals, such as artificial insemination and multiple ovulation/embryo transfer
- appreciate the aims of genetic technologies, such as generation of transgenic (GM) and cloned animals, and the ethical implications of these techniques
- be aware of the range of actual and proposed applications of GM animals, e.g. in biomedical research, as bioreactors, as sources of xenograft organs, in animal agriculture, in species conservation, in pest and disease control, and in relation to companion animals, and appreciate concerns expressed over animal welfare
- understand the deontological and other concerns expressed over genetic technologies, e.g. relating to animals' intrinsic value, public attitudes and patenting law
- be able to reach reasoned personal decisions on the ethical acceptability or otherwise of producing different forms of GM animal

9.1 Introduction

People have been using animals instrumentally (3.6.3) from time immemorial, but even limiting consideration to the *domestication* of animals, there is archaeological evidence for this dating from 10,000 BC.[1] Undoubtedly, humans and the animals we have domesticated have been partners in a process of co-evolution (7.4.2); and arguably, this has been of mutual benefit. People have been provided with: nutritious food; muscle power for draught purposes and transportation; materials in the forms of wool, skin, leather, and horn that are used for clothing, building materials, and tools; and manure, which is used as fuel for heating and cooking. In many LDC societies, animals continue to perform these vital, multiple roles. For their part, the animals receive food, protection from predators, and often shelter; and in many cases (most usually in DC), provisions now extend to housing and veterinary care. This symbiotic relationship has had a great impact on human value systems.[2] It has also markedly affected the character of the animals we have domesticated (7.4.2), making them largely dependent on human care.

But over the last fifty years or so, this relationship has been transformed by scientific and technological developments, many of which are designed to make animal production more economically efficient. Largely through the application of selective breeding and cost-effective feeding and management regimes, animal productivity has been increased dramatically. For example, whereas forty years ago it took broiler chickens twelve weeks to reach a slaughter weight of 2 kg, it now takes about five weeks; while over the same period, average milk yields of dairy cows have doubled.[3] In the last twenty years, entirely new technologies have emerged - such as genetic modification (GM) and cloning - ethical consideration of which forms the focus of this chapter.

A number of the ethical concerns over modern systems of animal production (7.6) focus on the extent to which animal welfare might be diminished by pushing them beyond the limits of their homeostatic control processes and/or subjecting them to psychological stresses. The range of uses to which animals are put has also increased. For example, in addition to use in agriculture and biomedical research, animals are now being used (or being developed) for production of pharmaceutical substances and industrial materials, and to supply organs for transplantation into human patients.

The new animal biotechnologies to be considered here fall under two broad categories - reproductive and genetic technologies - although the latter are almost invariably used together with the former. The main purpose of sections 9.2–9.5 is to describe the methods used, but brief accounts will also be included of the impacts of the techniques used on animal welfare. The latter are important in assessing costs and benefits in the manner required by utilitarian justifications of these technologies. Fuller discussion of these impacts, and of other ethical impacts, is reserved for later sections (9.6–9.7).

9.2 Reproductive technologies

These aim to control and/or accelerate the breeding process. This account focuses on farm animals, although corresponding techniques are applied to other animals, e.g. those used in laboratory experiments.

9.2.1 Artificial insemination

Artificial insemination (AI) avoids the necessity for natural mating. It has been practised commercially for more than fifty years, and is used extensively in the UK dairy industry, in which more than two-thirds of cows are bred exclusively by AI. It is also used for sheep, often for pigs, and in virtually all poultry and turkey breeding. As well as being a tool for genetic improvement, AI permits more effective control of diseases.

In cattle, semen is collected by arranging for a sexually aroused bull to ejaculate into an artificial vagina. Insemination is achieved by passing a catheter containing the semen (previously stored in the frozen state) through the cow's cervix. This requires the development of skills through training, but performed expertly it does not usually entail anaesthesia or sedation. Animal welfare is sometimes compromised by the procedures of semen collection and insemination, but the natural mating process itself is not free of such concerns, especially when there is a large mismatch in the sizes of cow and bull.

More indirect effects on animal welfare can also result from use of AI, for example because inbreeding may lead to populations which are vulnerable to disease. In the UK, about 8% of all heifers born in 2000 were sired by sons or grandsons of a single bull, called Starbuck.[4] It is estimated that, globally, 110 million inseminations are performed annually on cattle.[5]

Semen collection in pigs is similar, the boar mounting a 'dummy' sow before ejaculating into an artificial vagina. However, in sheep, the cervical route of insemination is less successful, and more usually laparoscopy (keyhole surgery) is used to deposit semen in the ewe's uterus. In poultry, semen is obtained by pressing the male's abdomen near the vent to 'milk' the semen. Insemination involves inserting a pipette into the female's cloaca. The almost universal use of AI in turkeys is largely due to the fact that selective breeding has resulted in them reaching such a size that they are incapable of breeding naturally.

9.2.2 Multiple ovulation/embryo transfer

This is a means by which genetic changes induced in the female can be readily propagated. In cattle, multiple ovulation (**superovulation**) involves injection of hormones into cows to increase the ovulation rate (up to twentyfold), followed some days later by a further hormone injection to cause luteolysis. However, responses are variable, both in terms of ovulation rate and the yield of viable embryos. In 1993, an expert review noted: *'the practice of superovulation is still a primitive art form that has not advanced appreciably since it was first performed over sixty years ago'.*[6]

Embryo transfer involves removing the embryos from the 'donor' cow before they become implanted in the uterus and transferring them to a number of recipient, surrogate, animals. To achieve this, after AI, embryos are flushed from the donor cow using a flexible catheter incorporating a plastic balloon. As the procedure is performed one week after oestrus, it is often difficult to penetrate the cervix, resulting in risks of bleeding and uterine rupture. Transfer to recipient cows (usually following hormone injections) is generally non-surgical, although surgical procedures are sometimes employed for valuable embryos. Because the 'gun' has to pass further into the uterus than for AI, great care is needed to avoid traumatizing the uterus and introducing infection, which entails skills that can only be acquired with much practice.[7]

Perhaps the most significant potential welfare problems relate to mismatches between the embryo and the recipient cow. *'Major problems arise when embryos, obtained from large dairy breeds and particularly from large double-muscled beef breeds, are placed in recipients which are unlikely to be able to give birth to them and therefore require surgery to deliver the fetus.'*[8] The use of multiple ovulation/embryo transfer (MOET) in other farm animals is much less common, either because it confers fewer advantages (as in pigs) or because surgery is required to recover embryos (as in sheep).

9.2.3 Welfare consequences of breeding objectives

In addition to possible reduced welfare as a result of these procedures (9.2.1 and 9.2.2), risks may arise as a result of the breeding objectives, because the marked increases in productivity may exceed the animals' physiological capability to cope with them. For example, leg weaknesses such as *tibial dyschondroplasia*, *femoral head necrosis*, and *dislocated hock joints* are very common in intensively reared poultry. Such conditions, which may affect one-fifth of the birds by the time they reach slaughter age, severely impair their ability to walk or run: in the worst cases they cannot even stand. Another consequence of rapid growth rates is cardiac failure, as the heart is unable to meet the demands placed on it.[9]

In dairy cattle, genetic selection for milk yield, together with other practices, such as concentrate feeds, have also led to adverse effects on animal health, reflected, for example, in increased incidences of mastitis (udder inflammation), lameness, digestive disorders, metabolic diseases (such as milk fever and ketosis), and calving difficulties.[10] Other welfare impacts of intensive farming systems are summarized in 7.6.

It is apparent from this brief account that the benefits of AI and MOET, in terms of improved efficiency in achieving production objectives, economic savings, and convenience, are associated in some cases with adverse effects on the animals' welfare. However, an inverse relationship between costs and benefits is not always obvious, and in better-managed farms adverse effects on welfare can often be ameliorated by higher welfare standards.

9.3 Genetic technologies

This term is used here to encompass **genetic modification** (GM, also referred to as **transgenesis**) and **cloning** (especially by the technique of **somatic cell nuclear transfer**). The list below refers only to the principal methods used: it excludes techniques such as retroviral transfection and sperm-mediated transfer.

9.3.1 Producing genetically modified animals by pronuclear microinjection

The technique of pronuclear microinjection (PMI) allows the direct modification of the genome of animals in ways not possible by mating, because genes from entirely different species may be introduced. Although transgenesis has not been used commercially in farm animals to date, several GM animals (GMA) have been employed in experimental projects, and growing numbers of GM laboratory animals are used in biomedical research. Table 9.1 summarizes the main types of GM and cloned animals.

In PMI, a *transgene* (DNA containing the appropriate genetic code, constructed *in vitro*) is incorporated into the genome of the prospective GMA. This entails micro-injection of DNA into one of the two pronuclei of a newly fertilized single-cell embryo, followed by a number of surgical and hormonal interventions (Figure 9.1). Thus, embryos are extracted from pregnant female animals, cultured *in vitro*, and then transferred to a

Table 9.1 Existing and proposed animal biotechnologies (discussed in 9.5)

Application (GM or cloning)	Purpose	Animals (examples)
Biomedical research	• Disease models • Toxicity testing of drugs and chemicals • Fundamental studies in biology	mice, rats mice, rats, dogs mice, rats, primates
Animals farmed for food and non-food products	• More rapid growth • Modification of food quality • Disease resistance • Altering wool properties • Reduced environmental impacts	salmon, pigs pigs poultry sheep pigs
Bioreactors	• Pharmaceutical production • Nutraceuticals • Material production	sheep cattle goats
Xenotransplantation	• Replacement human organs	pigs
Applications to companion animals	• Replacement of pets by cloning • Avoidance of allergies in owners	cats
Species conservation	• Revival of threatened species	gaur, panda
Pest and disease control	• Reduction in crop wastage • Resistance to diseases, e.g. malaria	insects mosquitoes

GM, *genetic modification.*

hormone-treated, pseudopregnant surrogate mother. In pigs, sheep, and goats, this generally involves two rounds of surgery (although not on the same animal), whereas in cattle non-surgical techniques are used.[11]

Incorporation of the transgene into the genome of the prospective GMA can significantly reduce animal welfare because of the effects of the methods used. Microinjection causes breakage of the chromosomes, so enabling the transgene to become incorporated during the course of spontaneous chromosomal repair. But incorrect repair may cause numerous adverse effects (deletions, translocations and inversions), while the transgene itself may become incorporated within an existing gene. The result is that mutations are common. A large proportion of mutations are lethal, leading to the early death of the embryo, and these are unlikely to cause significantly reduced welfare. However, in some cases the fetus survives to term, which often results in the birth of deformed young. Because of the low efficiency of the whole process in farm animals, less than 1% of microinjected embryos survive to become functional GMA.

Other effects (due to, so-called, epigenic and pleiotropic properties of genes: see 6.4.5) mean that the consequences of random transgene insertion are liable to be

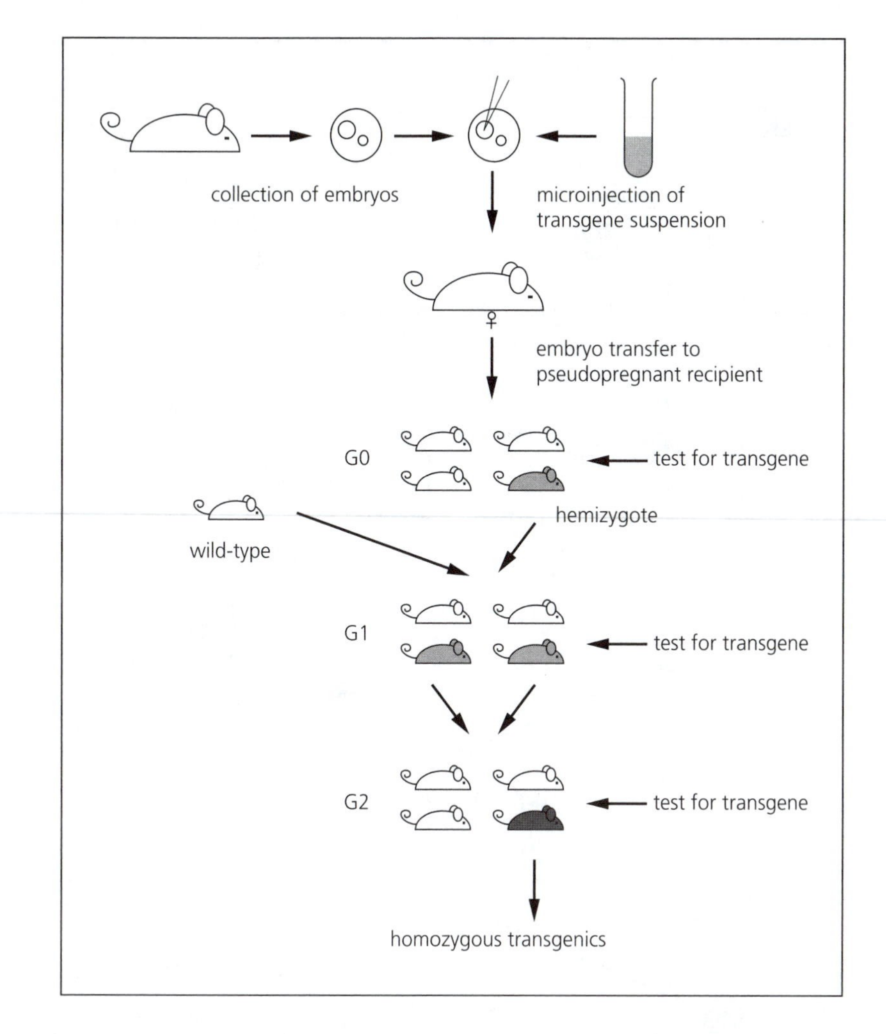

Figure 9.1 A schematic representation of the generation of genetically modified mice using the technique of pronuclear microinjection. G0, generation 0; G1, generation 1; G2, generation 2. [From: Moore C J and Mepham T B (1995) Transgenesis and animal welfare. ATLA *23*, 280–397]

unpredictable and cause widespread physiological changes, some of which may only become apparent after several generations of breeding.

9.3.2 Cloning by somatic cell nuclear transfer

Most famously used in the production of the sheep Dolly (Figure 9.2), the technique can also be used to produce GMA. In sheep, the procedure involves recovering the egg cells by laparotomy (a surgical incision), following the induction of superovulation by hormone injection and insertion of a vaginal tampon. The genetic material is then removed from the egg cells by micro-suction and replaced by DNA from the donor cell, when the

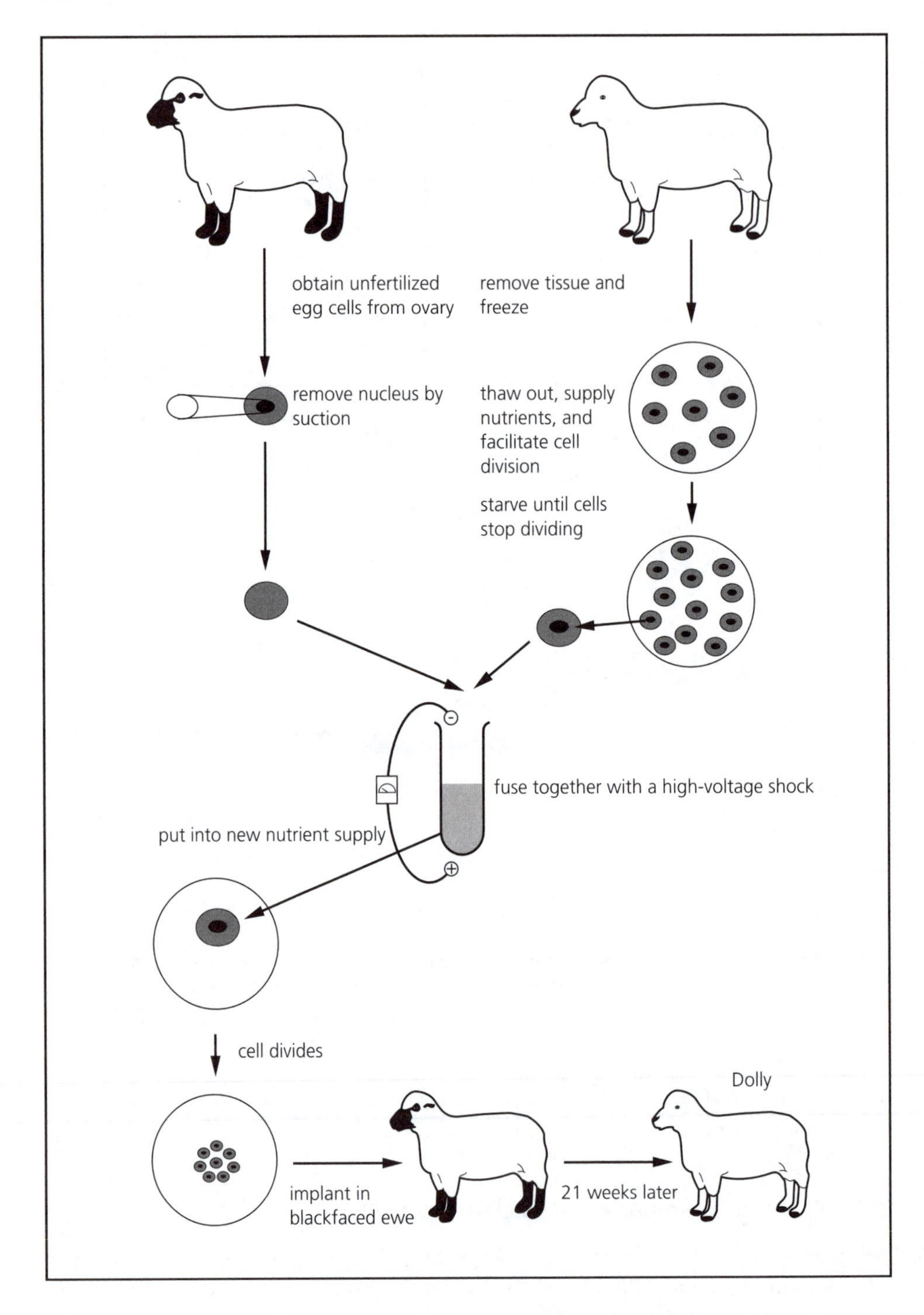

Figure 9.2 The technique of somatic cell nuclear transfer used in the production of the first cloned mammal 'Dolly'. [From: Curran B (2003) A Terrible Beauty is Born. London, Taylor and Francis, p. 11]

process of cell reproduction in the egg cells has been halted by 'starving' the cells in a nutrient-poor medium.[12]

Following application of an electrical current to stimulate cell fusion and reactivate the quiescent cells, the cells are cultured *in vivo* in the ligated oviduct of a live sheep for a period of seven days, one ewe being able to carry large numbers of eggs for this purpose. After this period, the ewe is killed and the developed blastocysts transferred to hormonally synchronized recipient sheep by laparotomy. The blastocysts are then allowed to develop to full term to be delivered. The cloned offspring are not genetically identical with the nuclear donor because they also contain some DNA which was present in the mitochondria of the egg cell cytoplasm. (An engaging personal account of the events leading to the birth of Dolly is provided by the two principal scientists involved, Ian Wilmut and Keith Campbell, and science writer Colin Tudge.[13])

However, the success rates of this procedure are poor. In the series of experiments resulting in the birth of Dolly, although 21 ewes were confirmed as pregnant, only 8 produced live lambs. In an earlier series of experiments resulting in the birth of the lambs Megan and Morag, two lambs died and at autopsy revealed signs of developmental abnormalities in liver and kidney. In its report on cloning, FAWC identified a number of potential welfare problems associated with the technique:

- large offspring syndrome, afflicting some animals born
- wastage of life, due to *'an unacceptably high loss of embryos, fetuses, and mature animals which are killed as part of the procedure'*
- in the experiment in which Dolly was produced, only one successful live birth resulted from 277 attempts using cells in which mature DNA had apparently been transferred
- possible problems of aged DNA, which might result in shorter lifespans, or other abnormalities, in cloned animals
- adverse effects of surgical and other procedures used in the *in vivo* culture procedures.[14]

Even so, instances of the successful use of this cloning technique in cattle, goats, pigs and other animals have now been reported.

9.3.3 Use of embryonic stem cells

GM mice, but not other mammals, have also been produced by using ESC (6.5) subject to *in vitro* manipulation. ESC offers certain advantages over PMI (9.3.1). Thus, because any modification of the cells occurs *in vitro*, before reconstructing an embryo, use of selectable markers permits identification of modified cells, so reducing the wasteful production of non-GMA. Moreover, screening can be used to identify cells that have integrated the transgene at a specific location – whereas in PMI the use of whole embryos prevents extensive *in vitro* manipulation.

9.4 The potential for harm to animals

It is clear that several of the reproductive and genetic technologies described above have the potential to harm the animals, and according to the laws governing animal experimentation these costs must be weighed against the likely benefits (usually to people) that are expected to result from their use. In an attempt to clarify the position over harms, the report of the Government's 'Banner Committee'[15] proposed that three principles should govern the humane use of animals in such technologies:

- Harms of a certain degree and kind ought under no circumstances to be inflicted on an animal.
- Any harm to an animal, even if not absolutely impermissible, nonetheless requires justification and must be outweighed by the good which is realistically sought in so treating it.
- Any harm which is justified by the second principle ought, however, to be minimized as far as is reasonably possible.

The main problem with these principles is that phrases such as 'a certain degree and kind', 'outweighed by the good realistically sought', and 'minimized as far as is reasonably possible' beg almost as many questions as they might appear to address, and according to bioethicists Michael Reiss and Roger Straughan, *'it may well prove impossible in this area to produce any absolute principles that take account of all eventualities'.*[16]

Harms are not confined to impacts on welfare, but it will be convenient to consider these first, since they are the principal focus of 'cost–benefit analysis'. Harms of a general nature have been considered above (9.2 and 9.3), but individual technologies demonstrate distinctive features, which it is the purpose of the section 9.5 to explore briefly.

9.5 Specific applications

This section identifies some important features of the different ways biotechnology is applied to animals, which are listed in Table 9.1.

9.5.1 Biomedical research

Following 1987, the total number of animal procedures (8.2) used in biomedical research in the UK declined year on year, but since 2001 there have been annual increases, due entirely to increased use of GMA. Between 1990 and 2001 the number of procedures involving GMA increased from 50,000 to more than 630,000;[17] and in 2003 use of GMA accounted for 27% of all procedures (of which 98% were performed on rodents).[18] Because this trend appears to run counter to the Three Rs principle (8.4.1), it is necessary to examine the claimed ethical justification.

The rationale is exclusively utilitarian (2.5), being based largely on the increased value of the benefits of the research through the animals' more suitable roles as CAM (8.3.3).

Moreover, it is claimed that in some cases use of GMA might lead to refinement of procedures used and reduction in animal use. Thus, pioneer animal biotechnologist Jon Gordon argues that *'If animals can be engineered to detect mutagens or carcinogens with high efficiency, fewer will be needed for the conduct of such tests'*, while *'animals engineered to develop human diseases* (will serve) *as simpler and less demanding subjects for vaccine testing and pathological studies.'*[19] However, as noted by an expert EU group in 1998: *'There is a substantial risk that the current intense interest in developing novel transgenic strains will, in fact, lead to an increase in overall animal use'*[20] – a prediction that was realized within two years in the UK. Even were the numbers of GMA to decline, in the process of generating a new GM strain many additional animals are required, such as surrogate mothers and vasectomized males. This can substantially increase the total 'welfare burden'.

Despite the fact that GMA are vulnerable to a number of particular welfare risks identified above and to exploitation in several ways not possible by other procedures, current EU legislation (Directive 86/609/EEC) does not explicitly refer to GMA. This leads to much variability in the control of GMA use in different Member States. Consequently, an EU expert group has proposed that there should be an enquiry into current international regulations with a view to their harmonization and rationalization. It also called for more extensive collection of statistics on GMA use, and for several measures designed to respect the Three Rs, such as application of accurate gene targeting techniques, storage of cryopreserved GM embryos, and increased application of *in vitro* methods for characterizing new gene constructs.[21] A subsequent report by several UK animal welfare organizations specifically addressed the welfare of GM mice.[22]

Disease models. These currently represent the most extensive use of GMA. The typical GM disease model is a mouse which has been manipulated so that it will develop a disease state corresponding to that suffered by humans. In so-called *knock-out* versions, an animal gene is disabled or removed in order to study its function. In either case, the aims are to discover factors which trigger the human disease and the way it develops, and to test the efficacy of potential therapeutic measures. It is now possible, almost routinely, to introduce a specific mutation into GM mice to mimic inherited disorders experienced in humans: common examples are those which predispose humans to various forms of cancer, muscular dystrophy, high blood pressure and cystic fibrosis. In the USA in 1988, the **Harvard oncomouse**, which was developed as a model for human cancer, became the first GMA to be patented.

A critical ethical question is powerfully expressed by veterinary ethicist Bernard Rollin: *'The creation of such animals can generate inestimable amounts of pain and suffering... since genetic diseases often involve symptoms of great severity. Given that such animals will surely be developed wherever possible for the full range of human genetic diseases* (estimated at more than 3000), *how can one be sure that vast numbers of animals will not live lives of constant pain and distress?'*[23] Another important question concerns the adequacy of the cost–benefit analyses (8.4) that are used to justify their creation, in terms of their efficacy as disease models and the existence of alternative approaches.

Toxicity testing. GMA have been developed to test the effects that new drugs and chemicals might have on genetic material – so-called *genotoxic* and *mutagenic* effects. Cancer is a multistage process, so that using GMA with a mutation in a single gene of the process can provide a sensitive indicator of the chemicals that might cause a mutation at another stage. This suggests that GM mice would be required in smaller numbers than normal mice in testing for genotoxic chemicals, and that more humane experimental end-points might be achieved. Consequently, they are now used routinely for this purpose.[24]

But ethical concerns focus on whether many of the new drugs and chemicals are *necessary*, and the adequacy of the tests (8.3.3). Critics point out that for GM mice to be adequate models would require them to have a *'human metabolism writ small'*, but *'merely inserting single genes into mice will not produce a human metabolism dressed up differently'*.[25] The issue of the inadequacy of animals as effective CAM (8.3.3) may apply almost equally to GMA.

Fundamental studies in biology. Undoubtedly, use of GMA can provide useful information and insights about fundamental biology, principally through their use as HAM (8.3.4). However, an ethically problematical issue concerns how such knowledge can be assessed according to a cost–benefit analysis, since, by definition, fundamental research attempts to acquire new understanding, the value of which cannot be weighed *in advance* against the likely suffering caused to animals.

9.5.2 Animals farmed for food and non-food products

Although most GMA are rodents, there is increasing interest in genetically modifying farm animals to improve their growth rates and/or the quality of animal products. A number of examples are discussed below.

Quantitative and qualitative changes in food products In the 1980s experiments with GM mice transgenic for growth hormone (GH) showed that the resulting increased blood concentrations of this hormone markedly stimulated their growth. **Supermouse** grew twice as big its non-GM littermates. This discovery led to efforts to produce similar growth stimulation in farm animals, although the aim was to accelerate growth rather than increase final body size.

The resulting GM **Beltsville pigs** (named after the research station in the USA where they were produced) achieved a certain notoriety, because although backfat was reduced, growth was not stimulated by the expression of the human GH gene inserted into the genome. More significantly, the pigs experienced severely reduced welfare, including lameness, arthritis, skin and eye problems, loss of sexual libido and mammary development in males. Of the 19 pigs produced expressing GH, 17 died within the first year, mostly from pneumonia, pericarditis and peptic ulcers.[26] Since then, some progress has been made in reducing adverse effects on welfare, but a recent Royal Society report claimed that *'further research is needed before these developments offer any prospect of commercial application'*.[27]

The similar stimulation of growth in fish is, however, closer to commercial exploitation, especially in North America. GM salmon transgenic for GH can show a threefold increase in weight, and the potential for exploiting colder waters. In a competitive situation, GM salmon appear to have an increased feeding motivation, because when size-matched with non-GM salmon they consume 250% more feed.[28]

In consequence, the AEBC recommended a ban on the commercial production of GM fish in offshore aquatic net pens while significant uncertainty remained about the environmental consequences of fish escapes.[29] But productivity gains also produce some markedly adverse effects on welfare, with GM salmonids expressing GH *'showing morphological disruptions analogous to acromegaly'*, the disease state characterized by an enlarged head and excessive and deleterious deposition of cartilage.[30]

Changes in milk composition have been sought by altering the casein and whey protein content of milk in order to improve cheese production, and removing human allergenic proteins (such as lactoglobulin) from cows' milk or replacing them with less allergenic proteins.[31] Research on GM sheep in New Zealand aims to alter wool composition, so that it takes up dyes more readily and is less prone to shrinkage. This can be achieved by changing the keratin content of the wool or by altering the biochemical pathways of cysteine metabolism to improve wool fibre strength.[32]

Ethical objections focus on what are considered the trivial objectives which necessarily, with the current inefficient means of generating GMA, entail substantial animal suffering.

Disease resistance Disease in farm animals is a significant constraint on both their welfare and their productivity. In some cases, such as BSE, it also has serious public health implications. Susceptibility to animal disease might be partly due to adverse effects of breeding for increased productivity, and partly to management procedures, notably reflected in feeding and housing regimes (7.6). Various GM strategies are being explored to address these problems. For example, resistance to Marek's disease, a herpes-virus-induced cancer of lymphoid tissue, which causes lameness, anaemia, and chronic wasting and costs the UK poultry industry £100 million p.a., seems to be controlled by very few genes.[33] This raises the prospect that GM disease-resistant strains might soon be developed. Other projects are seeking to produce sheep and cattle resistant to prion diseases (such as BSE) by knocking out the PrP gene in these ruminant animals.

Despite the apparently benevolent objectives of such programmes, there are some concerns that relying on such strategies might encourage continued, or increased, intensification of animal production systems in which welfare might be assigned low priority if it comes to be thought that disease risks can easily be technically 'managed'.

Reducing environmental impacts Eutrophication of waterways is a major environmental problem, to which pigs contribute a significant amount of phosphate because they are unable to digest phosphorus in the phytate form in which it occurs in their plant food. So-called *enviro-pigs*, GM for a gene coding for the enzyme phytase, are reported to reduce phosphate levels in manure by 65%. While this could have a significant beneficial environmental impact, critics again suggest that it overlooks the welfare burden of generating GMA, and discourages moves to more sustainable, less intensive farming systems.[34] Moreover, the same result can be achieved by dietary means.

9.5.3 Bioreactors

This term was originally applied to the technologies by which pharmaceuticals were produced in farm animals (mostly in the milk of GM ruminant animals), but the term now has wider potential applicability, and refers to prospective production (most usually in mammary tissue) of foods classed as *functional foods* (11.3), such as nutraceuticals, and of certain non-food products. Projects to produce therapeutic proteins required in human medicine have been pursued in GM farm animals for three main reasons:

- many proteins require post-translational modification (e.g. glycosylation), which cannot be performed efficiently in bacterial, yeast or plant cells
- the amounts of product required can be produced much more economically than in alternative systems, such as mammalian cell culture
- risks associated with extracting material from human tissues, e.g. the AIDS virus, are avoided.

An early attempt at bioreactor technology was the project involving *Herman the bull,* GM for human lactoferrin. This iron-binding protein, which occurs in human breastmilk but is absent from normal cows' milk, offered the prospect of a lucrative market in 'improved' artificial baby foods. It is a rare example of a GMA nutraceutical (Box 11.2). It was disappointing for the Dutch biotechnology company involved that the only animal expressing the transgene was a male, so that a delay was experienced in waiting for him to mature and to sire cows carrying the lactoferrin transgene. However, although the official objective of the project was altered to production of lactoferrin for treatment of gut infections, public resistance to the project in the Netherlands, together with technical limitations, led to its abandonment.

Perhaps the best-known example is production of *human alpha-1 antitrypsin* (AAT) in the milk of GM sheep, one of which, Tracy at the Roslin Institute in Scotland, produced the high concentration of 35 g AAT/litre milk. The aim was to treat patients suffering from the lung diseases cystic fibrosis and congenital emphysema, who require large amounts of AAT, which are currently only available by extraction from human blood. But because of various delays in the clinical trials, regulatory approval for AAT use, originally anticipated by 2001, is now highly uncertain. In 2003 it was reported that 3000 transgenic sheep were to be slaughtered and that the company involved, PPL Therapeutics, was facing bankruptcy.[35]

Cloned bioreactors. The research programme which culminated in the dramatic announcement of the birth of Dolly originated in the desire to multiply copies of highly exceptional, 'successful' GM animals such as Tracy. As a result, the 'hit and miss' procedure used for producing Tracy can now be replaced by a technique with much higher precision. When donor cells are bathed in a solution of the desired transgene, some will incorporate this exogenous genetic material into their chromosomes; and if a selectable marker is used modified cells can be selected and used for nuclear transfer (9.3.2). The resulting animal will contain DNA from the original cell with the additional transgene (Figure 9.3).

The technique was first demonstrated with the birth of the lamb Polly, from the same laboratory that produced Dolly. Polly was cloned from fetal fibroblasts carrying the gene for human factor IX, leading her to secrete human factor IX in her milk. Lack of factor IX in humans causes Christmas disease (haemophilia B) – so that Polly represented

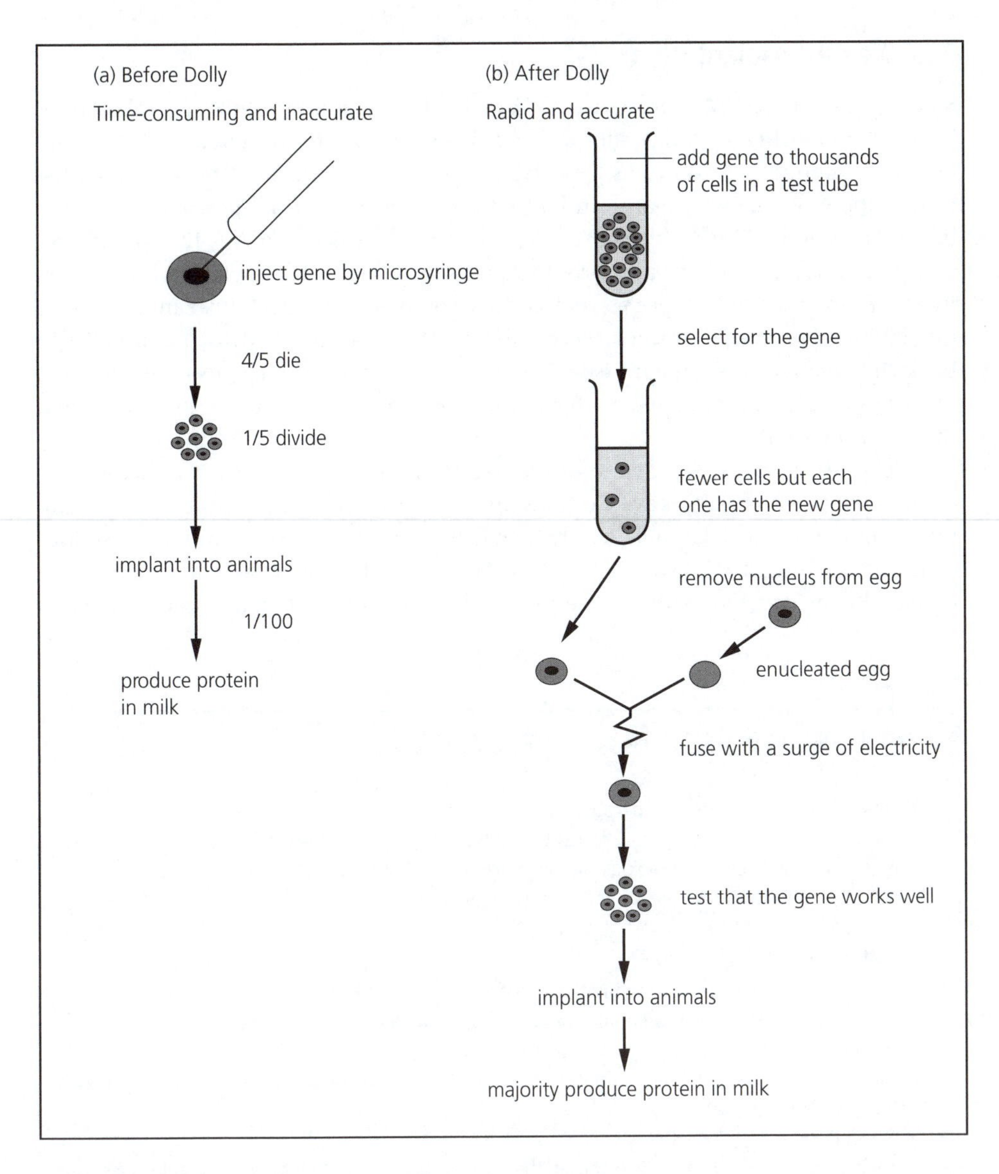

Figure 9.3 Use of somatic cell nuclear transfer to facilitate production of genetically modified bioreactor animal. [From: Curran B (2003) A Terrible Beauty is Born. London, Taylor and Francis, p. 133]

a significant breakthrough in demonstrating the commercial and medical potential of so-called **pharming**.

GM industrial products. A novel use of bioreactor technology is that employing GM goats to secrete spiders' silk in their milk. The protein is the 'dragline' silk used in spiders' webs, which is one of the strongest materials known. Called *Biosteel*®, it is being developed for use in lightweight body armour, and for medical applications.[36]

9.5.4 Xenotransplantation

Since the first successful heart transplant by Christian Barnard in 1967, transplantation into patients of hearts, lungs, kidneys and livers, derived from human cadavers, has become a routine procedure. As expectations have grown, demand has exceeded the ready supply of suitable organs, and attention has turned to the possibility of using organs from animals as viable substitutes. Pigs, because, like humans, they are monogastrics and are of similar body size, have been explored for this purpose. But, as an *immunologically discordant* species, their organs would be rejected if transplanted into the human body, as a result of the **hyperacute rejection reaction**, in which antibodies in the recipient attack the foreign tissue. However, this problem can be prevented by using organs from GM pigs modified so that the *complement control* proteins no longer cause the rejection reaction.

Ethical issues raised by xenotransplantation were addressed by two major UK enquiries,[37,38] which both gave cautious approval. However, the prospect of an additional supply of transplantable organs needs to take account of a number of serious concerns, such as those listed in Box 9.1. Some scientists suggest that whatever the potential benefits to individuals, the risks of causing future human disease epidemics

BOX 9.1 SOME CONCERNS OVER XENOTRANSPLANTATION

- Animal welfare and rights, e.g. because of:
  - the need to exclude all sources of possible infection, which may lead to housing animals under excessively sanitized, microbiologically sterile (*gnotobiotic*) conditions
  - the likely pressure to use higher primates as organ 'donors'
  - a process of 'humanization' of the donor animals.
- Human health, e.g. because of:
  - the possible existence of porcine endogenous retroviruses (PERV) which, though dormant in the animals, may become activated after organ transplantation, with serious public health consequences (possibly causing epidemics)
  - chronic organ rejection (over periods of months or years) even if the acute rejection reaction is successfully prevented
  - transmission to patients of zoonotic diseases
  - doubts that animal organs, with their different physiological and metabolic characteristics, can effectively sustain a human life.
- Equitable allocation of medical resources, e.g. because:
  - initial xenotransplants (at least) will necessarily be experimental
  - of difficulties in deciding which patients should receive (presumably preferred) human organs
  - of the prospect of some (wealthy) people prolonging their lives by having serial transplants, to the detriment of a fair allocation of medical resources.
- Public acceptability, because Eurobarometer (2002) shows that:
  - xenotransplantation is only a weakly approved form of biotechnology, with many respondents judging it 'risky' and only marginally 'morally acceptable' – although a trend to greater acceptance was evident between 1996 and 2002.

should prohibit use of this technology.[39] It is also little appreciated that, based on evidence of the transplantation of a baboon's liver into a human, xenotransplantation may produce *human-animal chimeras*, in that cells from the transplanted organ become dispersed throughout the body of the recipient, and, on the evidence of human organ transplants, may persist in the body for many years.[40]

In the UK, proposals for use of xenotransplants in human surgery are considered by the UK Xenotransplantation Interim Regulatory Authority (UKXIRA) established in 1997. But, to date, no application has been received and no research in this field is currently conducted in the UK. As noted, ethical analysis of biotechnologies always needs to consider alternative approaches, and in the case of xenotransplantation several of these have been proposed.[41] They include:

- improved biomechanical devices (e.g. mechanical hearts, dialysis machines)
- improved transplantation services
- improved recruitment of donors (e.g. by introducing measures such as **presumed consent**, **mandated choice** and **altruistic donation**)
- use of stem cell technologies (6.5)
- improved public education to encourage healthier lifestyles (e.g. exercise and less smoking and drinking).

9.5.5 Applications to companion animals

A company in the USA has established a programme to clone cats and dogs, for an anticipated fee of US$25,000. The first cloned cat was reported in 2002, the only live birth from 87 transferred embryos. Another company plans to produce GM cats, which it proposes to sell for US$1000, in which the gene responsible for a high proportion of people's allergic reactions to cats has been 'knocked out'. The AEBC report noted that given the inefficiency of both GM and cloning procedures, it must assumed that in each case there will be a substantial welfare burden.[42]

9.5.6 Reviving threatened species

Several attempts have been made to clone animals belonging to extinct and endangered species, including the Asian gaur (a wild ox), mouflon lamb, woolly mammoth and panda.[43] Few of these attempts have resulted in the birth of live young, and very few in young that survive more than a few days.

It has been argued that *'modern biotechnologies, including cloning, have a great deal to offer the conservationist'*,[44] but ethical concerns focus on the welfare burden associated with highly inefficient cloning procedures, and on the possible adverse impacts of such efforts on more sustainable approaches to reducing the environmental causes of species loss. On the other hand, if species do become extinct cloning may prove an inadequate remedy, e.g. if ecological considerations make the species nonviable in the wild.

9.5.7 Insects and pests

A strategy for controlling malaria has been proposed in which mosquitoes, which carry the disease, are replaced by a refractory GM strain. It is suggested that reducing or eliminating the natural variety could spare millions of human lives each year. Strategies have been proposed in which GM forms of insects that are agricultural pests might be introduced to control the numbers of the pests. One approach involves the addition of beneficial characteristics to predators or parasites of the pest, while another involves the addition of new, deleterious, characteristics to the pest species itself.[45] There is a clear need for careful assessment of the ecological implications of such programmes.

9.6 Wider ethical issues

The ethical issues raised above have mostly concerned impacts on animal welfare. They have conformed to the category of 'costs' taken into account when Home Office inspectors perform the cost–benefit analysis which UK law requires before projects involving animals may be licensed. To a large degree, scientific measurements can now establish the extent to which animals experience poor welfare (8.7.1).

But for many people ethical analysis extends beyond these utilitarian concerns to encompass those of a deontological nature (2.6). A number of these concerns relate to animal rights, and impacts on the animals' intrinsic value (7.2.3), neither of which is capable of objective measurement. But to ignore them for that reason alone would undermine a substantial part of ethics. Two constructive approaches are to:

- consider the validity of the arguments from our personal perspectives
- examine the results of sociological research on public attitudes, which are certain to have important political and legal implications.

9.6.1 Telos, intrinsic value and intrinsic objections

Assigning intrinsic (or inherent) value to animals usually implies that it is wrong to seek to alter their nature by GM, or to treat them in the highly instrumental way entailed in cloning, for purely human ends. According to this view, e.g. in the terms described by Tom Regan (7.2.2), the categorical imperative that forbids such actions on humans extends incontestably to animals, or at least to those who are 'subjects of a life'.

An alternative concept, which is often interpreted in rather different ways, is **telos**, which, literally, refers to every creature's 'end' or 'purpose'. For Aristotle, in ancient Greece, it referred to the full, flourishing stage of an animal's existence (*'for the sake of which it came to be'*). But more recently it has been interpreted as *'the interests and nature of an animal which it has as a member of a species'*.[46] Some rather subtle philosophical arguments result, depending on whether telos is thought to be located in the individual animal or in the species to which it belongs, and whether, since in domestication mankind has, perhaps unwittingly, altered animals' telos already, it raises any new ethical problems to do so intentionally by use of GM.[47]

It is often useful in such cases to clarify the arguments by appealing to a 'thought experiment'. For example, consider the, as yet hypothetical, case where animals were genetically modified to reduce their sentience. From a utilitarian perspective that sees no problem in changing the animals' telos, the benefits might be great (in that they could be kept under highly efficient factory farming conditions to maximize their output) and the welfare costs low (because their new telos, associated with a reduced sentience, would make them oblivious to the harsh conditions under which they were farmed). But for those who hold that an animal's telos demands our respect due to its intrinsic value, such a technology would be ethically impermissible because the animal's whole nature would have been subordinated to this production objective.

Indeed, the dangers of adversely undermining animal telos are conceded, without apparent irony, by animal biotechnologist George Seidel Jr, who wrote: *'Transgenic procedures often produce extreme phenotypes, and nature tends to select against such extremes... [but] although animals with such phenotypes would not survive in nature, the farmers who use them in production agriculture may survive well economically'.*[48]

The FAWC report on cloning was explicit in its criticisms, stating that cloning could be considered intrinsically objectionable if it involves an unacceptable violation of the integrity of a living being. It added that *'an attitude may be developing which condones the moulding of animals to humankind's uses, irrespective of their own nature and welfare'.*[49] Indeed, for many people, it is unnecessary to appeal to prospective or hypothetical developments. In discussing the ethical matrix (3.5) it was suggested that the notion of fairness could be extended to animals in terms of a need to respect their intrinsic value. Thus, it might be considered *unfair* to animals to treat them in the highly instrumental fashion entailed by several biotechnologies (9.2–9.3).

9.6.2 **Patenting**

A patent is a monopoly granted by the State to an inventor in return for disclosure of the invention: it acts as an exclusive right to stop others making commercial use the invention. In a sense, the patent can be regarded as a kind of 'deal' between the State and the inventor; because the inventor gets a twenty-year monopoly, but has to disclose to the public how to work the invention. In Europe, patent law is based on the European Patent Convention (EPC) which sets out the criteria for patentability (see Box 4.2). These have to be satisfied if an inventor wishes to secure a patent with the European Patent Office (EPO). The current legal assumption is that the EPC can accommodate all the special problems raised by biotechnology, without any need for amendment. This follows from the claim that the patenting criteria in Box 4.2 apply to all inventions, whether or not they consist of living matter.[50]

However, as patents are only granted for 'inventions', some people consider that patenting animals offends the respect due to the animal's intrinsic value, while those holding particular religious views may regard patents as tantamount to blasphemy. In fact, as noted (Box 4.2), according to EU law,[51] inventions shall be considered unpatentable where their commercial exploitation would be considered contrary to **public order** and **morality**, or where the processes involved *'are likely to cause* (the animals) *suffering without any substantial medical benefit to man or animal, and also the* *animals resulting from such processes'*. Moreover, *'processes the use of which offends against human dignity, such as processes to produce chimeras from germ cells of humans and animals, are obviously excluded from patentability'*.

But it must be admitted that such constraints may not necessarily be very restrictive. Because patent law primarily exists to stimulate biotechnological innovation, it rarely appeals to the Precautionary Principle (13.4) in cases where adverse effects are not clearly established.[52] Opponents of the patenting of GMA also stress that *'the wider availability of patents will give a massive commercial boost to genetic engineering, a process that too often poses very considerable threats to the health and welfare of the animals'*.[53]

As the EU Directive acknowledges, to some degree, concern for the intrinsic value of animals may be an expression of the way such procedures impact on the concept of *human* dignity. Respecting the intrinsic value of animals may thus correspond to 'conferring' a quality on them by responding to a perceived human duty so to act. According to this view, some animal biotechnologies, by treating animals with insufficient respect, may be thought to violate the people who use the animals as much as they do the animals themselves.[54]

9.6.3 Social research on public attitudes to animal biotechnologies

In its enquiry into animal biotechnologies, the AEBC commissioned both quantitative and qualitative research on public attitudes.[55] The qualitative research involved discussion group methods with a spread of social groups in the north and south of England in 2001, and included people with a range of experiences with animals (e.g. pet-owners and non-pet-owners; farmers; supporters of field sports; etc.). The main conclusions are summarized in Box 9.2.

BOX 9.2 PUBLIC ATTITUDES TO ANIMALS AND ANIMAL BIOTECHNOLOGIES: KEY POINTS

- Many people have close, affective relations with animals, in domestic and other contexts.
- People often recognize personal contradictions in their attitudes (e.g. keeping pets but eating meat).
- Attitudes to animal biotechnologies are greatly influenced by the purposes, e.g. medical research is much more acceptable than cosmetics testing.
- Most people regard GMA as unnatural. Though few reject this research out of hand, there is much concern about the pace of developments, the degree and precision of planned interventions, and the likelihood of unanticipated mistakes.
- These public concerns relate to: the intrinsic character of animals (and the need to respect their dignity); effects on animal welfare; and the effectiveness of organizations responsible for regulation and institutional oversight.
- Misgivings about GMA reflect broader mistrust about the institutions responsible (frequently illustrated by reference to the crises over BSE and GM foods).

(From AEBC, 2002)

It is clear that while most people do not reject animal biotechnologies out of hand, there is much concern that they might result in unnatural and risky outcomes, which there is little confidence that governments will be able to either control or prevent.

9.7 Responses to arguments presented for animal biotechnologies

A standard defence of GM technologies is that they introduce no new ethical issues. But this claim has been questioned on several grounds. A distinctive feature of GM is that it can rapidly produce qualitatively major genetic changes, sometimes with highly adverse effects. The diseased state of the Beltsville pigs (9.5.2) revealed an unanticipated capacity for harm which was associated with the pursuit of a purely economic objective – cheaper meat. So, given risks of adverse welfare consequences, it might be argued that it would only be permissible to produce GM animals in cases of genuine 'need'. But how do we define 'need'?

9.7.1 Needs and wants

We might instinctively draw a clear distinction between *needs* (necessary for survival) and, less worthy, *wants*. But, since the Industrial Revolution which began in nineteenth-century Britain, the institutionalization of market demand for 'goods' has created a new type of 'need' – that deemed necessary to drive economic progress. Persuasive advertising often creates this demand, so fuelling the competition underpinning modern market forces. The result is that the pursuit of wealth becomes an aim in itself, irrespective of whether it serves to satisfy real needs. The ethical acceptability of this concept of economic needs is open to question, especially when the genuine survival needs of many people remain unsatisfied (chapter 4).

9.7.2 Trust and moral concerns

But if we envisage a situation in which the technical limitations of animal transgenesis had been overcome, so that scientists could define with precision the procedures necessary to permit the production of GMA which suffered no reduced welfare, would not this allay public fears?

In reality, there must be substantial doubts that public acceptance would be gained easily. This is for two reasons. First, the safe application of any biotechnology would seem to depend on a sophisticated economic and political infrastructure.[56] But even if there were confidence that this currently existed (which many people might question), there could be no guarantee that it would exist in future, or globally. In that case, societies might be wise to only invest in technologies that were robust in the face of unpredictable future changes; and many people might draw the line at technologies that address genuine needs and entail only minimal welfare and/or environmental risks. Second, there is evidence of substantial public concern about the possible alteration of animal natures in inappropriate ways (Box 9.2), whether by accident or design. Typically, advancing social development leads to what philosopher Peter Singer has called 'the expanding circle' of moral concern, which increasingly embraces not just peoples' local concerns, but those of wider social groups and non-humans as well.[57]

9.7.3 The significance of emotional responses

Another important constraint on public acceptance of animal biotechnology is based in the emotions. Emotional reactions or sentiments are often disparaged by those with a scientific outlook because they appear to be subjective, and not capable of rational explanation. But philosopher David Cooper claims that *'Attempts to immunise "reflective moral judgements" against contamination by sentiment are incoherent.'* For if what is ethically objectionable about Jack's actions are that they show a callous disregard for Jill's life and welfare, Jim's emotional outrage on observing the actions might be considered entirely appropriate, and a cool, unemotional response quite inappropriate.

Analogously, as Cooper puts it: *'Much of the vocabulary in which people express their outrage at current and predicted bioengineering of animals registers the fact that it is the people, the bioengineers who practise it, who are the objects of this outrage.'*[58] (Of course, Cooper was not implying that such outrage can ever constitute ethically justifiable grounds for acts of terrorism against researchers: see 8.9.)

9.7.4 Is genetic modification merely an extension of current practices?

Finally, the 'continuity' argument for GMA has also come in for serious criticism. This suggests that modern biotechnology is simply an extension, albeit now performed more efficiently, of the genetic selection which has been practised for hundreds of years. The suggestion is clearly flawed, because in the transfer of genes between species, transgenesis can result in entirely new forms of animal, which could not be produced by traditional breeding practices.

But two problematical issues are also often raised. First, there is the question of where it will all end. The utilitarian justification that permits 'minimal' erosions of animal welfare to secure marginal economic benefits has no logical end-point, since every additional 'cost' is deemed, in itself, minor. As philosopher Jonathan Glover puts it: *'By easy stages we could move to a world which none of us would choose if we could see it as a whole from the start'*.[59] Second, the endless pursuit of human 'happiness', by exploiting animals, might be seen as *'an immature perception of the ends of life, a perception of life without measure, shape, or significance'*. What animal biotechnology implies is that humans have a right to disregard the subjectivity of animals. But the basis on which that right is claimed is by no means clear.[60]

9.8 Ethical analyses

The above account reveals that animal biotechnologies raise a very wide spectrum of concerns, of which animal welfare and rights represent only two. Depending on the application, there are also often important ethical implications for consumers, patients, citizens in society and the biotic environment, and indeed for future generations.

Although there may be significant utilitarian arguments for developing such technologies, for many people, deontological considerations could prove to be a decisive constraint. So, for innovations having such a pervasive impact, explicit justification based on a detailed ethical analysis would seem essential. And for scientists and technologists involved in such research, awareness of others' views (even if rejected) is undoubtedly important.

A form of ethical matrix (Table 9.2) may thus serve as a useful means of structuring the required ethical analyses. For example, in an ethical analysis of xenotransplantation, prospective benefits to patients need to be weighed against risks identified in Table 9.1 and the deontological concerns summarized in 9.7. Or in the case of GM fish designed to grow more rapidly through supplementing the GH genes, the increased efficiency of production needs to be seen in the context of the welfare and environmental constraints identified in 9.5.2, as well as deontological concerns discussed in 9.7.

Table 9.2 An ethical matrix specifying respect for the three principles (well-being, autonomy, and fairness) in the use of genetic technologies applied to animals (genetic modification or cloning)

Respect for:	Well-being	Autonomy	Fairness
Patient/Consumer	Health; Acceptability	Choice	Rights; Affordability
Treated animals	Welfare	Behavioural freedom	Intrinsic value
Biota	Conservation	Biodiversity	Sustainability
Society	Social harmony	Democratic choice	Fair allocation of resources

Notes: For details on use of the matrix see 3.5–3.7. The matrix can be used to help identify the extent to which each of the specified principles might be respected, and to facilitate deliberation on how positive and negative impacts should be weighed.

THE MAIN POINTS

- Reproductive technologies in farm animals, such as AI and MOET, facilitate rapid increases in productivity; however, the procedures involved may reduce animal welfare and thereafter overtax animals' physiological capabilities.
- GM and cloning have many applications in laboratory, farm and companion animals: these include biomedical research, agricultural products, bioreactors (producing food, medicines, or materials) and xenotransplant organs (for human patients).
- Utilitarian justifications for such biotechnologies in terms of gains in efficiency and improved quality of products and services are challenged by concerns over animal welfare and public health (e.g. for xenotransplantation).
- Deontological concerns focus on a perceived instrumental use of animals, which pays insufficient regard to their intrinsic value, and on reservations over the (future) ability of governments to regulate these technologies wisely.
- Sociological research reveals much ambivalence in public attitudes to animals (e.g. pet-owners who eat meat), but there are widespread reservations about challenging the 'natural order', which appear to constitute a significant constraint on the application of many prospective animal biotechnologies.

EXERCISES

These can form the basis of essays or group discussions:

1. Is there a significant ethical difference between the use of MOET in cattle (9.2.2) and the use of ART in women (5.4)?
2. Discuss the concept of the instrinsic value of animals (7.3) in relation to: (*a*) use of GM rats in medical research on cystic fibrosis, and (*b*) use of GM pigs designed to produce lean pork with greater economic efficiency (9.5).
3. Given the range of alternative strategies identified for avoiding or treating heart disease, discuss the ethical acceptability of the prospective use of xenotransplant hearts from pigs (9.5.4).
4. If it were possible to produce GM animals that were less sentient in a way that reduced their capacity to suffer (9.6.1), would this provide ethical justification for using such animals instead of non-GM animals currently used in some painful tests, such as the Draize test for skin and eye irritancy (8.5.2)?
5. Discuss the arguments for and against the claim that there is a need for GMA in agriculture (9.7.1).

FURTHER READING

- *Animal Biotechnology and Ethics* edited by Alan Holland and Andrew Johnson (1998). London, Chapman and Hall. A valuable collection of essays from a variety of perspectives on animal biotechnology.
- *Livestock, Ethics and Quality of Life* edited by John Hodges and I K Han (2000). Wallingford, CAB International. Presents a global perspective on animal biotechnologies.
- *The Second Creation* by Ian Wilmut, Keith Campbell, and Colin Tudge (2000). Chatham, Headline. A readable account of the project to create Dolly the cloned sheep.

USEFUL WEBSITES

- **http:/ /www.aebc.gov.uk/** *Agriculture and Environment Biotechnology Commission*: The site of the UK Government's advisors on social and ethical aspects of non-human biotechnologies. Covers a wide range of issues, including animal biotechnologies.
- **http://www.genewatch.org./** *Genewatch UK*: A prominent NGO which addresses animal and other biotechnologies.
- **http:/ /www.guardian.co.uk/flash/** *Guardian Unlimited*: An interactive 'Guide to Cloning'.

NOTES

1. Zeuner F E (1963) A History of Domesticated Animals. New York, Harper and Row
2. Hodges J (2000) In: Livestock, Ethics and Quality of life. (Eds) Hodges J and Han I K. Wallingford, CAB International, pp. 253–262
3. Gamborg C and Sandoe P (2003) In: Key Issues in Bioethics. (Eds) Reiss M J and Levinson R. London, Routledge Falmer, pp. 133–142
4. Agriculture and Environment Biotechnology Commission (2002) Animals and Biotechnology. London, DTI, p. 25
5. Thibier M and Wagner H-G (2002) World statistics for artificial insemination in cattle. Livest Prod Sci *74*, 203–212
6. Armstrong D T (1993) Recent advances in superovulation of cattle. Theriogenology *39*, 7–24
7. Sreenan J and Diskin M (1992) Breeding the Dairy Herd. Dublin: Teagasc
8. Murray R D and Ward W R (1993) Welfare implications of modern artificial breeding techniques for cattle and sheep. Vet Rec *133*, 283–285
9. Rauw W M, Kanis E, Noordhuizen-Stassen E N, and Grommers F J (1998) Undesirable side effects of selection for high production efficiency in farm animals: a review. Livest Prod Sci *56*, 15–33
10. MAFF Statistics (2000) http://www.maff.gov.org

11. Mepham T B and Crilly R E (1999) Bioethical issues in the generation and use of transgenic farm animals. ATLA *27*, 847–855

12. Wilmut I, Schneike A E, McWhir J, Kind A J, and Campbell K H S (1997) Viable offspring derived from fetal and adult mammalian cells. Nature *385*, 810–813

13. Wilmut I, Campbell K, and Tudge C (2000) The Second Creation. Chatham, Headline

14. Farm Animal Welfare Council (1998) Report on the implications of cloning for the welfare of farmed livestock. Surbiton, FAWC

15. Ministry of Agriculture, Fisheries and Food (1995) Report of the committee to consider the ethical implications of emerging technologies in the breeding of farm animals ('Banner Report'). London, HMSO

16. Reiss M J and Straughan R (1996) Improving Nature? The science and ethics of genetic engineering. Cambridge, Cambridge University Press, p. 185

17. Home Office Annual Statistics on use of animals in scientific procedures (2002) http://www.homeoffice.gov.uk/docs/animalstats.html

18. Home Office Annual Statistics on use of animals in scientific procedures (2004) http://www.homeoffice.gov.uk/docs/animalstats.html

19. Gordon J W (1997) In: Animal Alternatives, Welfare and Ethics. (Eds) van Zutphen L F M and Balls M. Amsterdam, Elsevier, pp. 95–112

20. Mepham T B *et al.* (1998) The use of transgenic animals in the European Union (ECVAM workshop report 28) reprinted in ATLA *26*, 21–43

21. Ibid.

22. Robinson V *et al.* (2003) Refinement and reduction in production of genetically modified mice. Int J Lab Anim Sci Welfare *37* (suppl. 1), S1:1–S1:51

23. Rollin B (2002) In: A Companion to Genethics. (Eds) Burley J and Harris J. Oxford, Blackwell, pp. 70–81

24. Animal Procedures Committee (2001) Report on Biotechnology (para. 122). London, Home Office

25. LaFollette H and Shanks N (1996) Brute Science: dilemmas of animal experimentation. London, Routledge, p. 188

26. Mench J A (1999) In: Transgenic Animals in Agriculture. (Eds) Murray J D, Anderson G B, Oberbauer A M, and McGlouchlin MM. Wallingford, CAB International, pp. 251–268

27. Royal Society (2001) The Use of Genetically Modified Animals. London, Royal Society, p. 14

28. Ibid.

29. Agriculture and Environment Biotechnology Commission (2002) Animals and Biotechnology. London, Department of Trade and Industry, p. 38

30. Dunham R A and Devlin R H (1999) In: Transgenic Animals in Agriculture. (Eds) Murray J D, Anderson G B, Oberbauer A M, and McGlouchlin M M. Wallingford, CAB International, pp. 209–229

31. Royal Society (2001) The Use of Genetically Modified Animals. London, Royal Society, p. 14

32. Ibid., p. 15

33. Agriculture and Environment Biotechnology Commission (2002) Animals and Biotechnology. London, Department of Trade and Industry, p. 15

34. Rutovitz J and Mayer S (2002) Genetically Modified and Cloned Animals. Buxton, Genewatch UK, p. 59

35. Stewart H (2004) Firm that cloned Dolly the sheep faces bankruptcy. The Guardian, 9 March 2003. http://www.guardian.co.uk/

36. Rutovitz J and Mayer S (2002) Genetically Modified and Cloned Animals. Buxton, Genewatch UK, p. 60
37. Nuffield Council on Bioethics (1996) Animal to Human Transplants: the ethics of xenotransplantation. London, NCB
38. Advisory group on the Ethics of Xenotransplantation (1996) Animal Tissue into Humans. London, The Stationery Office
39. Allen J S (1996) Xenotransplantation at the crossroads: prevention versus progress. Nat Med *2*(1), 18–21
40. Langley G and D'Silva J (1998) Animal Organs into Humans. BUAV and CIWF, pp. 52–54
41. Mepham T B and Crilly R E (1997) Transgenic xenotransplantation; utilitarian and deontological caveats. In: Animal Alternatives, Welfare and Ethics. (Eds) van Zutphen L F M and Balls M. Amsterdam, Elsevier, pp. 355–360
42. Agriculture and Environment Biotechnology Commission (2002) Animals and Biotechnology. London, Department of Trade and Industry
43. Rutovitz J and Mayer S (2002) Genetically Modified and Cloned Animals. Buxton, Genewatch UK, p. 73
44. Tudge C (2000) In Mendel's Footnotes. London, Jonathan Cape, p. 232
45. Royal Society (2001) The Use of Genetically Modified Animals. London, Royal Society, p. 17
46. Rollin B E (1989) The Unheeded Cry. Oxford, Oxford University Press, p. 203
47. Holland A (1995) In: Issues in Agricultural Bioethics. (Eds) Mepham T B, Tucker G A, and Wiseman J. Nottingham, University Press, pp. 293–305
48. Seidel G E Jr (1999) In: Transgenic Animals in Agriculture. (Eds) Murray J D, Anderson G B, Oberbauer A M, and McGlouchlin M M. Wallingford, CAB International, pp. 269–282
49. Farm Animal Welfare Council (1998) Report on the Implications of Cloning for the Welfare of Farmed Livestock. London: MAFF
50. Paver M (1998) Legislation and regulation: a view from the UK. In: Animal Biotechnology and Ethics. (Eds) Holland A and Johnson A. London, Chapman and Hall, pp. 276–288
51. EC Directive (1998) 98/44/EC of the European Parliament and of the Council on the legal protection of biotechnological inventions. Official Journal L213, 30 August 1998, pp. 0013–0021
52. Neeteson A-M *et al.* (1999) In: The future developments in farm animal breeding and reproduction and their ethical, legal and consumer implications. EC-ELSA project, 4th Framework Programme. Utrecht, The Netherlands, pp. 87–96
53. Stevenson P (1998) Animal patenting: European law and the ethical implications. In: Animal Biotechnology and Ethics. (Eds) Holland A and Johnson A. London, Chapman and Hall, pp. 288–302
54. Mepham T B (2000) 'Würde der Kreature' and the Common Morality. J Agr Environ Ethic *13*, 65–78
55. Agriculture and Environment Biotechnology Commission (2002) Animals and Biotechnology. London, DTI
56. Holland A and Johnson A (1998) In: Animal Biotechnology and Ethics. (Eds) Holland A and Johnson A. London, Chapman and Hall, pp. 1–9
57. Singer P (1983) The Expanding Circle: ethics and sociobiology. Oxford, Oxford University Press
58. Cooper D (1998) In: Animal Biotechnology and Ethics. (Eds) Holland A and Johnson A. London, Chapman and Hall, pp. 145–155
59. Glover J (1984) 'What sort of people should there be?' Harmondsworth, Penguin, p. 14
60. Holland A and Johnson A (1998) In: Animal Biotechnology and Ethics. (Eds) Holland A and Johnson A. London, Chapman and Hall, pp. 1–9

PART FOUR
Bioethics, plants and the environment

We cannot command nature except by obeying her

Francis Bacon (1620)
Novum Organum

10

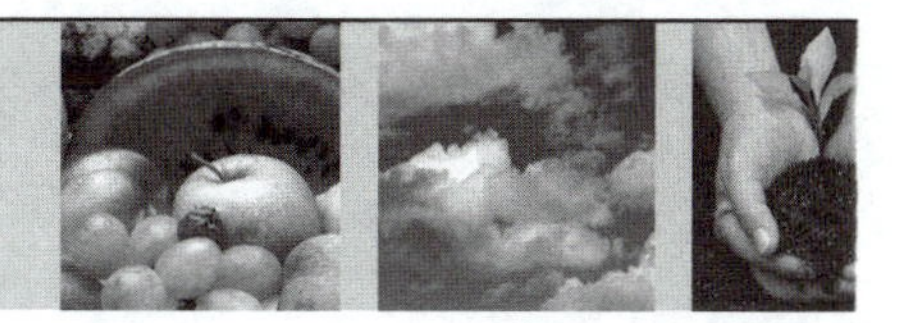

The first generation of genetically modified crops

OBJECTIVES

When you have read and discussed this chapter you should:

- appreciate the benefits which it is claimed the first generation of GM crops will deliver in developed countries
- be aware of the political events associated with the differing public attitudes to GM crops evident in the EU and USA
- appreciate the ethical issues raised by GM crops, as identified by use of the ethical matrix
- understand the rationale for the UK Government's decision on GM herbicide-tolerant crops following the farm-scale evaluations of impacts on biodiversity indices
- appreciate the divergent ethical attitudes to GM crops associated with different worldviews

10.1 Introduction

Biotechnology, in the broadest sense, can be applied to crops in various ways. Traditional plant breeding involves cross-breeding different crop varieties to produce novel offspring, with those possessing desirable characteristics being selected for propagation. **Marker-assisted breeding** involves screening the offspring of crossing experiments to identify desirable genes at an early stage of the breeding programme. In **induced mutation** approaches (involving doses of radiation or treatment with mutagenic chemicals), random changes in DNA which confer novel and desirable characteristics on plants are selected and then bred in traditional programmes. With **tissue culture** selection, a nutrient medium is used to grow millions of individual plants which are exposed to damaging conditions or chemicals (such as herbicides), allowing selection of cells showing high levels of resistance to the hazard applied.

10.1.1 Genetic modification

In **genetic modification** (GM) specific genes are inserted into, or removed from, a plant genome in order to confer on the plant desirable characteristics. Two main features

BOX 10.1 SPECIAL FEATURES OF GENETICALLY MODIFIED CROPS

- the novelty and power of the techniques employed
- the extent of cross-species genetic transfers made possible by GM
- the potential irreversibility of some modifications to the environment and food chain
- the speed with which modifications become possible, and may become established in agricultural practice if economically advantageous
- the complete control of agronomic aspects of plant cultivation made possible
- the extensive scientific and economic control of GM technology which major multinational companies appear to be establishing

distinguish GM technology from conventional plant breeding:

- *precision*, because a few, well-characterized genes are inserted, instead of thousands of unwanted genes with unknown side effects
- the *range* of possible transgenes is much greater, because they may originate from any species (microbial, plant or animal) and consequently might, in principle, have significant effects on human health and/or the environment.

This chapter concentrates on GM approaches, although the boundary between GM and non-GM techniques (such as marker-assisted breeding) is likely to become increasingly blurred.[1] The focus is on so-called 'first generation GM crops', which have principally been designed to modify agronomic practices, e.g. by conferring on crops herbicide tolerance (HT)[a] or insect resistance. The ability of GM to introduce novel changes into crops much more easily and rapidly than conventional techniques is the source of both potential economic benefits and serious concerns, identified by the Nuffield Council on Bioethics as six special features (Box 10.1). The ethical questions raised have received much prominence in recent years, especially in view of the polarized attitudes adopted towards them by a largely sceptical public in Europe, on the one hand, and by a public in North America which appears to have largely accepted them, on the other.

10.1.2 The political and economic background

By 2002, two-thirds of the global area devoted to GM crops was in the USA, which, together with Canada, Argentina, and China, accounts for 99% of the total.[2] The global area planted with GM crops has increased annually by more than 10% since they were introduced in 1996, but from 1998 several EU states (though not the UK) imposed a moratorium, which effectively blocked approvals for applications to cultivate GM crops. In 2003, the US Government launched a legal case with the WTO against the EU, demanding that it abandon its ban on the growing of GM crops (which the USA considers illegal) and requesting payment of £1 billion in compensation for loss of exports over the previous six years. Consequently, it has been suggested that: '*The WTO*

[a] Sometimes, the term 'herbicide resistance' (HR) is used to mean the same thing.

is now facing the biggest case in its history, one that could spark a damaging trade war between the USA and Europe and split the international community.'[3]

In the UK, the Government reached a voluntary agreement with the agricultural biotechnology industry that no commercial growing of GM crops would be undertaken before the results had been assessed of the **farm-scale evaluations** (FSE) designed to measure the impact on biodiversity of the associated agronomic management practices. However, according to the AEBC, in deciding whether or not to commercialize GM crops, the FSE *'cannot be, as widely interpreted, the final piece of the jigsaw'*, since *'additional information and consideration of a wide range of viewpoints must be factors in the eventual decisions'.*[4]

The first GM food to be on sale in the UK was tomato purée, made from tomatoes with higher solid and lower water content than traditional varieties. Priced in supermarkets up to 15% cheaper than non-GM purée, it substantially outsold the latter. The product was labelled as from a GM source, and its marketing attracted little public opposition. However, in 1999 several factors combined to disturb this situation, with modern biotechnology becoming *'the subject of more intense and acrimonious debate in the British media than at any time in its previous 25-year history'.*[5] The trigger event was a television programme in which, on the basis of experiments on rats, a scientist at a Scottish research institute, Árpád Pusztai, expressed concerns about the safety of consuming certain GM potatoes. (Despite his being a leading authority in this field, his views were discredited by the scientific establishment and he was forced to resign from his post.)

But public apprehensions about GM crops had been growing since 1996, when the UK media reported that North American commodity crops (soya and maize) were being imported into the EU, containing unsegregated mixtures of GM and non-GM material. Then, in 1998, a frozen-food company, *Iceland,* decided to exclude all GM ingredients from its own-label products (a practice subsequently followed by all other major supermarkets); the government advisory agency English Nature called for a three-year moratorium on the planting of GM crops in the UK; and after many years of difficult negotiations, the EU introduced regulations on the labelling of GM foods.

The 'media storm' peaked in February 1999, but the issue continued to receive much attention for months afterwards (14.6.2). GM crops were front-page news in most national daily newspapers for weeks, and some launched anti-GM campaigns. Many comments referred to the BSE outbreak as grounds for doubting Government assurances of safety, while Government ministers and the scientific establishment accused the press and broadcasting media of scaremongering. Overall, the general tenor of UK public opinion on GM food shifted from cautious approval in 1997 to widespread scepticism and disapproval in 1999.[6]

Less developed countries In the case of LDC, GM crops are grown chiefly in China, Argentina, and Brazil. In China, in contrast to other countries, GM crop research is funded exclusively by the State. In addition to GM cotton, which accounts for more than 50% of the Chinese cotton harvest, GM tomatoes and paprika are also cultivated, and field trials are at an advanced stage for several other crops.

Elsewhere, LDC governments, e.g. in Africa, have resisted GM crops. For example, Zambia and Zimbabwe refused GM food-aid shipments. It was argued that GM shipments could entail health risks or adverse effects on biodiversity, since imported grain could have been used, as is traditional, as seed. Currently, these countries have no adequate control mechanisms, but independent testing seems essential because of the substantial international differences in both ecology and nutritional status of the population. Refusal to accept GM food-aid shipments might also have been related to concerns about losing their own non-GM status, with adverse effects on exports. Moreover, states providing GM food aid have been accused of fobbing off needy countries with GM products that are difficult to sell on the world market - so trying to force countries in need to rely on GM imports.[7]

Concern has also be expressed over the development of *genetic use restriction technologies* (GURT), sometimes called **terminator technology**, which makes seeds sterile in the second generation. This prevents farmers saving and replanting seed, a practice which is very common in LDC.

10.2 Ethical issues

Clearly, concerns about GM crops cover a wide range of issues. In the following discussion, an ethical matrix (3.5) will be used to frame these issues in a structured manner (Table 10.1). The interest groups identified are: arable farmers, consumers, the biota and the GM crop itself. There may be some surprise at the inclusion of 'the GM crop', implying as it does that plants have 'interests'. However, this view needs to be considered, and perhaps especially by those who would reject it out of hand - even if conscientious deliberation confirms initial reactions. This discussion of ethical issues is confined to use of GM crops in DC. Factors specific to LDC are discussed in 11.7.

With reference to Table 10.1, the specifications given in Table 3.1 (which in some cases were specific to animals) are re-stated in terms applicable to crops, with *welfare* and

Table 10.1 An ethical matrix applied to use of genetically modified crops in agriculture

Respect for:	Well-Being	Autonomy	Fairness
Arable Farmers	Satisfactory income and working conditions	Managerial freedom of action	Fair trade laws and practices
Consumers	Food safety and acceptability. Quality of life	Democratic, informed choice, e.g. of food	Availability of affordable food
The Biota	Conservation	Biodiversity	Sustainability
GM Crop	Flourishment	Adaptability	Intrinsic value

Notes: For arable farmers and consumers, both impacts and duties need to be considered, whereas for the biota and genetically modified (GM) crop, only impacts are involved.

behavioural freedom being replaced by *flourishment* and *adaptability*, respectively. The main emphasis is the HT crops that were the subject of the FSE in the UK in 2001–2003 (10.1.2), but some reference is also made to crops transgenic for internal insecticides, since these have been used extensively in the USA. The aim is to explore the ways in which use of the GM crops has been perceived either to respect or to infringe the principles in the ethical matrix by comparison with non-GM crops. However, since the latter include a wide spectrum of possible management systems, reference will sometimes be made specifically to organic farming (since this has well-defined standards), as well as to the prevalent conventional non-GM systems.

10.3 Arable farmers: well-being, autonomy and fairness

The European Commission has described biotechnology as a whole as *'the next wave of the knowledge-based economy'*, predicting that global markets for sectors where *'biotechnology is likely to constitute a major portion of the new technology applied (excluding agriculture)'* could be worth more than £130 billion. But because of the considerable reservations over GM crops in the EU, this may not be a reliable guide to future economic benefits of agricultural biotechnology.

10.3.1 Farmers' incomes

Although prediction of future farming incomes is complex, economic data from North America might be expected to provide some useful clues. However, the interpretation of these data differs markedly, depending on the sources consulted, so that it is difficult for people who are not expert in agricultural economics to assess the validity of the different arguments. To illustrate the differences, the accounts of two organizations are summarized below.

The UK Soil Association (the main UK organic farming body) has made, among others, the following claims about North American farmers' experiences of GM crops: [8]

- GM seeds typically cost 25–40% more than non-GM seeds, and farmers have to sign an agreement with the biotechnology company prohibiting them from saving seed (or retaining a portion of the harvest and planting next year). As 20–25% of US farmers traditionally save seed, this could be an additional cost.
- Yields of GM crops have often been lower than for non-GM, but because agrochemical prices (e.g. of the herbicide used with HT soya) have recently fallen dramatically, this has offset some of the other increased costs of growing GM HT soya; however, growing *Bacillus thuringiensis (Bt)* maize was less profitable than non-GM crops.
- Overall yield increases had not been realized (except for a small increase for *Bt* maize).
- An eight-year survey by US agricultural economist Charles Benbrook indicated that farmers had not only become more reliant on herbicides, but had suffered severely reduced choice in the way they farmed, so that *'Many farmers have had to spray incrementally more herbicides on GM crops in order to keep up with shifts in weeds towards*

tougher-to-control species, coupled with the emergence of genetic resistance in certain weed populations.' [9]

- For North American exports, GM crops have proved an economic disaster, because exports of US maize and Canadian rape (called canola) to the EU virtually collapsed. It is estimated that GM crops cost the US economy at least US$12 billion net from 1999 to 2001; in fact, farmers who gained in terms of market prices were those who could supply guaranteed 'GM-free' produce.

By contrast, *PG Economics (agricultural economics consultancy)* made, among others, the following comments in response to the UK Government's Environmental Audit Committee report on the FSE GM crops, which included reference to the Soil Association report:[10]

- The Soil Association evidence was biased in that it took evidence from the Canadian Farmers Union, a small body which is unrepresentative of Canadian farmers as a whole.
- The experience of growing GM crops in North America has not been predominantly negative, as evidenced by the facts that in 2003, 81% of the US soyabean crop, 40% of the US maize crop, 84% of the US canola crop, 48% of the Canadian soyabean crop, 58% of the Canadian maize crop, and 68% of the Canadian canola crop planted were GM varieties.
- *'the majority of farmers have positive experiences, e.g. in terms of low tillage cultivation, reduced use of toxic pesticides and more secure yields.'*
- The Canadian Canola Council did not find the issue of HT volunteers to be problematic for farmers, and it can in any case be easily controlled.
- Benbrook makes adjustments and amendments to the (USDA) data on herbicide usage which the United States Department of Agriculture USDA does not support. He also omits to mention the eco-friendly nature of the herbicide glyphosate compared with alternatives used before the introduction of GM crops.
- There is a reasonable body of evidence suggesting that total herbicide use has declined since the introduction of GM crops.

It is apparent not only that the data might be open to different interpretations, but also that outcomes might depend quite critically on economic, political and agronomic variables. In the case of the UK, the difficulty of making economic predictions is emphasized by the fact that even the Government Strategy Unit's report on GM crops did not attempt to attach overall monetary values to the costs and benefits associated with commercialization, since it would have relied too heavily on arbitrary assumptions.[11] In a rare specific example, the report referred to GM HT oilseed rape, which *'might only offer savings for farmers with worse than average weed problems'*. More generally, economic impacts would be influenced by farm type, farm size, the crop concerned, and the geographical region. Perhaps just as significantly, they might be greatly affected by the particular regulatory regime in place. If segregation and labelling were given low priority, farmers might find GM crops conferred economic advantages which might not apply with more rigorous regulations (10.7).

10.3.2 Farmers' managerial freedom

This is also likely to be greatly influenced by the economic and regulatory circumstances. A critical question concerns **coexistence**, i.e. the ability of both GM and non-GM farming systems (and notably those certified as organic systems) to be carried out without detriment to the other. Under EU law, use of gentically modified organisms (GMO) by organic farmers is forbidden, and they are not allowed to sell crops or food as 'organic' if GM material is present. Although no legal threshold has yet been stipulated for the presence of adventitious (accidental) GM material in organic food, the Soil Association is seeking to set this at 0.1% (the alleged practical limit of detection), which is substantially lower than the threshold of 0.9% for which GM labelling is legally required.

For farmers using GM, the North American experience suggests that managerial freedom might be significantly reduced by the terms of the contract signed with the biotechnology company (10.3.3) and by restrictions on seed saving (10.3.1).

10.3.3 Fair trade laws and practices

Respect for this principle might be felt most acutely by farmers who wish to avoid contamination of their crops with GMO, and this is by no means confined to organic farmers. Thus, important decisions need to be made about the legal framework underpinning the coexistence regulations and their implementation, and the appropriate economic compensation due to farmers if their crops became contaminated with GM material through no fault of their own. This raises a number of questions, on which the authors of an AEBC report in 2003 could not agree; for example:

- the legal status of the 0.1% and 0.9% thresholds
- whether either threshold would be achievable in practical terms
- to what degree organic farmers would be required to take measures to reduce cross-contamination
- who would be liable to pay compensation to affected farmers, and how much
- whether insurance policies are the best way of providing financial redress (and who should pay the premiums) *or*
- whether, in the absence of insurance cover, compensation would be most appropriately provided by the biotechnology companies, GM farmers, the Government, or *all* farmers through a small levy on the harvested crops.[12]

The prominent biotechnology company Monsanto has made its position on liability very clear, namely: *'In no event shall Monsanto or any other seller be liable for any incidental, consequential, special or punitive damages.'*[13]

In North America, the legal implications for farmers of patents (9.6.2) on GM crops have had serious consequences for some farmers. A farmer growing a GM crop without paying the technology fee, or saving seed from a GM harvest to sow the following year, can be accused of stealing the biotechnology company's intellectual property. While it is doubtless sometimes intentional, some farmers may *unwittingly* sow GM seed, for example because: (*a*) they purchased it from someone who had breached the contract; (*b*) their fields were contaminated by airborne pollen or seeds; (*c*) it accompanied new livestock, hired machinery, or flood water; or (*d*) it derived from 'volunteers' (unwanted licensed GM plants that, e.g. failed to germinate in the previous season). Some farmers claim they have been wrongfully accused of behaving illegally. A prominent case is that of the Saskatchewan farmer Percy Schmeiser, whose crop of allegedly non-GM rape was found to contain Monsanto's GM HT seeds. Despite denying any guilt, Schmeiser had to pay a legal bill of $600,000, because the judge ruled that the source of the GM material *'is not really significant for the issue of infringement'*.[14]

Another perspective on fair trade is presented by international differences in the regulations governing GM. In the USA more than forty GM crop varieties have been cleared through the federal review process with enhanced agronomic and/or nutritional characteristics or one or more features of pest protection (insect and viruses) and herbicide tolerance. All foods derived from these GM crops are considered to be *substantially equivalent* (10.4.2) to their conventional counterparts, and therefore to require no labelling. It follows that, from the US perspective, their farmers are being financially penalized by unfair trade laws according to WTO regulations (10.1.2).

10.4 Consumers: well-being, autonomy and fairness

The claim that GM crops provide a basis for improving the quality of life is both one of the most contested and the most difficult to define. In this context, 'consumers' are essentially equated with 'citizens', because almost everyone consumes food derived from crops. A simplistic interpretation would identify *well-being* as average personal income, and attempt to derive this from the profits of the nation's biotechnology industry - considered to impact on everyone, whether or not they have any direct contact with the industry.

However, a more perceptive understanding of well-being would include issues such as: job security; job satisfaction; fair treatment; the acceptability of the farming systems adopted; the cultural diversity of village life (which might be strongly influenced by the numbers of economically active farm workers - and perhaps reflected in viable services in the form of shops, schools, and social amenities); and the appearance, and biodiversity, of the countryside. A simplistically monetarist interpretation of well-being fails to do justice to these important criteria. Despite this, politicians and industrialists often assume a direct correlation between 'economic competitiveness' and the nation's 'quality of life' - an assumption embodied in the mission statements of all the UK government's research councils (14.2.2).

10.4.1 Food safety

A more defined interpretation of consumer well-being concerns the safety of GM crops and the food and feed derived from them. Can some GM crops improve our health or could they make us ill? A definitive analysis was provided by the GM Science Review Panel (SRP), part of the UK government's enquiry into GM crops, which considered four issues, viz: (*a*) possible nutritional and toxicological differences in GM food; (*b*) food allergies from GM crops; (*c*) the fate of transgenic DNA; and (*d*) the effect of GM-derived feed in the food chain.[15] In terms of negative effects on health, most potential health consequences could in theory be due to:

- any inherent toxicity in the transgenes and their products
- unintended (mutagenic or pleiotropic - see 6.4.5) effects due to insertion of the new gene into the plant's genome, e.g. over-expression of an endogenous active substance, gene silencing or altered metabolic pathways.

Risks to health of the first type are deemed negligible, because of their very low levels in GM food, rapid hydrolysis, and innocuous nature. Risks of the second type, which in theory might increase the toxicity (including carcinogenicity) or allergenicity of the GM crop or food, raise questions (not unique to GMO) which are not technically simple to address. In conventional toxicology testing, a test substance is fed to animals (8.5.2) at a range of doses, some of which are several orders of magnitude higher than the expected human exposure level, to determine adverse effects (13.2.2). But with GM foods the amount that can be fed is limited by the appetite of the animals and by palatability and nutritional effects.

10.4.2 **Substantial equivalence**

In view of such limitations the concept of **substantial equivalence** has been used to identify differences that might potentially compromise the safety levels established for conventional foods. Introduced in 1993, it is based on the assumption that *'existing organisms used as foods, or as a source of food, can be used as the basis for comparison when assessing the safety for human consumption of a food or food component that has been modified or is new'.* [16] In practice, this means GM foods are examined for certain compositional variables (such as proteins, carbohydrates and vitamins), which, if they are sufficiently similar in the non-GM food, allow the GM food to be designated 'substantially equivalent'.

However, this assumes negligible safety implications of the known genetic and biochemical differences - differences which must exist for the GM material to be patentable. Moreover, the way GM crops are treated in the field may have safety implications. For example, the rationale for using GM HT soya is that the crop can be exposed to high concentrations of a herbicide (such as glyphosate), which kills the weeds but leaves the crop unaffected. Despite the fact that this significantly changes their composition, e.g. with respect to levels of phenolic compounds such as isoflavone, the tests are conducted on beans that have *not* been treated with glyphosate. Some critics have described substantial equivalence as a *'pseudo-scientific concept because it is a commercial and political judgement masquerading as if it were scientific'.*[17]

The Royal Society acknowledges that substantial equivalence may not reveal unexpected effects of genetic modification, such as those leading to the *'presence, perhaps at very low levels, of previously unknown toxins, anti-nutrients, or allergens'.*[18] Consequently, it recommended that, in future, safety assessments make use of new profiling techniques such as micro-array technology for detailed studies of mRNA expression, quantitative two-dimensional gel electrophoresis and mass spectrometry for protein analysis, and metabolic analyses to look at changes in all metabolites and metabolic intermediates. Nevertheless, the report states: *'We believe that at present there is no evidence to suggest that those GM foods that have been approved for use are harmful'* – an opinion some might consider risks confusing 'absence of evidence' with 'evidence of absence' (13.5.1).

10.4.3 Allergenicity

In theory, GM foods have the potential to introduce novel allergens into the diet, with the risk of inducing severe allergic reactions in susceptible consumers. Exposure could have serious consequences for individuals who suffer anaphylactic shock from allergic reactions, which sometimes prove fatal. Such hazards are, of course, not confined to GM food, although it is worth noting that if a troublesome allergen became incorporated into many processed foods, susceptible people might find it very difficult to avoid it. In the UK, the SRP suggested that in view of a relative lack of knowledge about allergenicity, caution should be exercised with *all* new foods.

Two high-profile cases have attracted attention to the hazard. One involved inclusion of a protein from brazil nuts into soyabean to improve nutritional quality. Laboratory tests raised doubts about safety, and the development was stopped before it was cleared for commercialization.[19] In the case of StarLink™ corn, a *Bt* gene was inserted to provide resistance to the corn borer. Although it had been approved only for use in animal feed, the company concerned (now Aventis) submitted a request to the Environmental Protection Agency (EPA) in the USA for its approval in human food. The EPA refused the request on the basis of risks of allergic reactions, but it was subsequently discovered that Kraft taco shells (marketed under the name Taco Bell) tested positive for StarLink. The resulting outcry led to withdrawal of the product, but not before the EPA announced that fourteen people had complained of adverse reactions after eating products containing StarLink.[20]

10.4.4 Antibiotic resistance genes

Introducing a transgene depends on the ability to select the transformed cells that have acquired the GM gene, and antibiotic resistance genes (usually conferring resistance to the antibiotics kanamycin and neomycin) have often been used as markers. Concerns have been raised that these genes might transfer to gut bacteria and thus limit antibiotic use in patients suffering from bacterial infections. Even so, it is recognized that antibiotic resistance is widespread in bacteria, so that the rare gene transfer from a GM food source is unlikely to significantly add to the risks.[21] However, it is generally agreed that caution is called for, and the Royal Society recommends that: *'such genes, if used in future, should be removed at an early stage in development of the GM plant, and where possible, alternative marker systems should be used'.*[22]

10.4.5 Animal feeds

The use of GM animal feeds poses qualitatively different human safety questions. The SRP suggested that because in future an increasing number of fodder crops will be developed to improve nutritional value and digestibility, and reduce pollution, we are likely to see more GM crops exhibiting multiple transgenes: this *'would make economic sense in the production of nutritionally enhanced animal feed, but it raises the question of whether this approach is more risk prone'.*[23] Thus the committee recommended:

- studies on animals under different levels of stress (which might affect gut permeability)
- research on the fate of GM silage
- research on the interactions of multiple transgenes in a single fodder crop.

10.4.6 Democratic informed choice

This operates at two main levels:

- personal decisions over food purchases
- the democratic decision about whether GM technology should be employed, and how it should be regulated (discussed in 14.6).

Food purchase decisions are influenced by several factors, of which cost is only one. It follows that respect for consumer choice depends, at minimum, on reliable labelling indicating how and where the food was produced. The strength of the reaction to GM crops in the EU in the late 1990s appears to have been largely a response to the discovery that GMO were present in food without being labelled.

A lack of official recognition of such concerns was demonstrated by a UK report, published in 1992, on *the ethics of genetic modification and food use*, which discussed the possible use of human genes in food for humans. The authors did not think there was a *'case on ethical grounds for an absolute prohibition of the food use of organisms containing copy genes of human origin'* and merely recommended that official guidelines should *'require notification by those seeking to market a novel food of why a copy gene of human origin had been used rather than an alternative'.*[24] This attitude now appears quite out of line with evidence on public attitudes, which are strongly opposed to GM in general in food, let alone that containing human genes.

Consequently, all GM food now appears to be widely considered as 'ethically sensitive', and EU regulations reflect this by requiring labelling of food and food ingredients containing novel DNA or protein.[25] Originally, this ruling excluded highly processed ingredients and derivatives (e.g. soya oil) which originate from GM crops but could not be scientifically detected. However, public opinion has appeared to demand a much more rigorous definition of GM, one based on the *process* involved rather than the characteristics of the food product, as revealed by surveys indicating that 87% of people want the GM food to be labelled irrespective of whether its products are chemically detectable in the food.[26] Sensitive to these opinions, the EU introduced new Regulations on Traceability and Labelling of GM Food and Feed in 2004. Under these rules:

- the labelling threshold for adventitious or technically unavoidable GM content is reduced to 0.9% for GMO approved in the EU
- a lower threshold of 0.5% applies to GMO assessed as safe, but which are still awaiting final approval by the EU
- labelling is extended to include any food ingredient that is derived from a GM product, even if it does not have any detectable GM content (e.g. highly refined oils)
- labelling is extended to animal feeds.[27]

Even so, not everyone is satisfied. For example, (*a*) small amounts of GM ingredients can still find their way into food without a necessity for labelling, and (*b*) consumers will be unaware of whether meat or milk is from animals fed GM feed.[28]

However, in line with consumer demands, many retailers, manufacturers and caterers (as for example represented in the UK by the Food and Drink Federation and the British Retail Consortium) have introduced *identity preservation systems* to ensure traceability throughout the food chain. Typically, the food chain includes: seed supply, planting, cultivation and harvesting; transportation (e.g. via ports), storage and drying; and processing and ingredient/commodity use. The multiplicity of links in the chain emphasizes the number of critical points at which intentional or unintentional contamination could occur, and underlines the challenge to reliable surveillance and enforcement control. Enforcement is the responsibility of local authority trading standards officers, whose effectiveness is subject to two sorts of constraint: (*a*) the low precision of some of the analytical procedures for detecting GM (especially in processed foods); and (*b*) the limited resources available, many of which might be allocated to work assigned a higher priority.[29]

However, in the UK the Food Standards Agency (FSA) is unsympathetic to the public (and the EU) opinion on GM process labelling, arguing that consumers would have better protection if *GM-free* labelling were introduced alongside the existing regulations. Several criticisms of this proposal have been made, for example:

- the onus would be removed from those using GM to label it
- in view of the permitted levels of GM in unlabelled foods, the situation would be very confusing
- *processes* using GM (which concern most consumers) would not be identified
- the *GM-free* label might be exploited by overpricing or by using the label when no equivalent GM foods exist.[30]

It is clear that respecting consumer choice through labelling is not a straightforward matter. Decisions need to be made about the definition of 'GM' in both qualitative and quantitative terms, the onus for labelling, the appropriate level of regulation, and the penalties for non-compliance. Moreover, choice is not necessarily satisfied by labelling. It is, for example, illusory when:

- alternatives to either GM or GM-free foods are not freely available
- one type (GM or GM-free), though available, is prohibitively expensive for many people

- food purchasers are unable to understand the labelling, either through language difficulties or because of a lack of awareness of the issue.

10.4.7 Affordability

Impacts of GM on food affordability are difficult to assess. Justifications for technology are typically made in economic terms, and it seems clear that these are often valid: e.g. the cost-of-living adjusted prices of cars, electrical foods and audiovisual equipment have fallen greatly over the years. Similarly, food prices in DC now account for a much smaller proportion of disposable income than formerly (although at least part of this apparent gain might be offset by increased costs, e.g. to the environment; see 12.5.3).

However, to date, there do not appear to be clear economic benefits to consumers from use of GM crop technologies. In part, this might be due to the weak link between production costs and shop prices. For example, when GM tomato purée was on sale in UK supermarkets its mark-up price was lower than the non-GM variety, but whether this reflected production costs is uncertain, since it might have been marketed as a 'loss leader' to encourage acceptance of GM food in general.

10.5 The biota: conservation, biodiversity and sustainability

Because all three principles are closely interrelated, they are conveniently considered together. Here we consider some of the direct and indirect effects of GM crops on the biota (3.4).

10.5.1 Invasiveness or persistence

The possibility of an 'alien' species becoming invasive is revealed by non-native plants that have been brought to the UK, such as *Rhododendron pontica* and *Buddleia davidii*. But such exotic plants are not necessarily good models for GM crops. Future GM plants may not be comparable with non-GM crops, because transgenic technology may so fundamentally change their physical and biochemical characteristics as to convert them effectively into new species. There is clearly need for research to understand the traits which, when subjected to GM, are likely to affect the plant's performance in the natural habitat, and when exposed to competition and predation.

10.5.2 Toxicity to wildlife

It is generally agreed that, because insect toxins are rarely species-specific, GM plants designed to produce a toxin can also be toxic to non-target wildlife. The only types of GM insect resistance currently used widely are the *Bt* toxins, which are derived from the bacterium *Bacillus thuringiensis*. More than 100 types of such toxins have been discovered, each specific to certain species of Lepidoptera or Coleoptera. *Bt* sprays have been used for many years by organic farmers (whose regulations do not allow them to use synthetic pesticides), but GM allows a targeted application, from within the plant. A potentially important difference is the fact that the whole bacterium produces a prototoxin, which is only converted to the active form after ingestion by the insect, whereas the GM crops directly express the active toxin over a longer period. This exposes pests to higher selection pressure and could accelerate development of resistance.[31]

Potential impacts on non-target species are illustrated by results of studies examining the effect of Monarch butterfly larvae consuming *Bt* maize pollen. Some laboratory studies suggested a 44% reduction in survival of larvae fed GM pollen by comparison with controls.[32] However, when repeated on a field-scale the adverse effect was not observed – and the issue is often cited by GM proponents as an example the 'scare-mongering' tactics employed by critics. Even so, identifying whether GM crops might have ecologically significant impacts is not straightforward. For example, toxicity may exert *tritrophic* effects, i.e. affect the plant at the first level, the herbivore consuming it at the second level, and the herbivore's predator at the third level. Moreover, a wide range of taxa might come into contact with GM plant-produced toxins in the soil, such as bacteria, fungi, protozoa, nematodes, mites, millipedes, centipedes, woodlice, molluscs, earthworms and a range of soil-dwelling insects.

The GM SRP deduced that: *'Since our understanding of the impacts of GM crops on non-target species will never be complete, in cases where the environmental risks are assessed to be acceptably low, regulators are likely to grant commercial consent with the option of withdrawing consent if monitoring programmes identify significantly harmful impacts.'*[33] As on other issues, such decisions (e.g. on what is 'acceptably low') require ethical judgements, since science alone cannot provide a definitive answer.

10.5.3 The development of resistance

Two important aims in plant breeding are to develop varieties resistant to pests and diseases, and crops tolerant of herbicides. But these strategies have a limited lifespan, because such resistant varieties provide a strong selection target for new pests and pathogens to attack, while new weeds often develop tolerance to the herbicides.[34] By far the commonest forms of GM crops are those designed to resist spraying with proprietary, broad-spectrum herbicides (notably, glyphosate and glufosinate ammonium). The stated objectives are to reduce herbicide applications, replace environmentally damaging chemicals with more benign versions, and reduce the need for mechanical tillage. Whether these objectives have been fulfilled is open to some debate (10.3.1), but here we consider the resistance problems that might occur.

There is now much evidence for emergence of herbicide tolerance in a range of weeds. In Canada, in 1998, volunteer (10.3.3) oilseed rape was reported to show multiple resistance to glyphosate, glufosinate and imidazoles, just three years after the GM HT rape was first sown. It is presumed that this **transgene stacking** is due to sequential crossing of several HT varieties. Some argue that the multiplicity of herbicides available means that gene stacking is not a serious problem for farmers. But the UK Government agency

English Nature, in discussing the Canadian results, stated that: *'the use of transgenic techniques... may pose additional risks to our natural heritage due to potential impacts on ecological food webs (and) the potential for GMO to enable changes in agricultural, forestry and, fisheries management... could be detrimental to wildlife.'*[35]

The strategy adopted in the USA to reduce development in pests of resistance to the insecticidal *Bt* toxin is to plant refuges of toxin-free crops. However, a report in 2004 indicated that pollen-mediated gene flow up to 31 m from *Bt* maize caused low to moderate *Bt* toxin levels in kernels of non-*Bt* maize plants. Most studies on gene flow from GM crops have emphasized potential effects on wild relatives of the crops, organic plantings etc. - with implications for pest resistance being largely ignored. But the authors of the report argued that *Bt* production in seeds of refuge plants could accelerate pest resistance to *Bt* crops - suggesting a need to revise the separation guidelines to reduce gene flow.[36]

10.5.4 The farm-scale evaluations

Following the expression of concerns by several conservation bodies that GM HT crops might exacerbate the already serious decline in farmland wildlife, the UK Government established the FSE (sometimes called *the trials*) to investigate the effects on biodiversity of four GM HT crops, namely, beet (sugar and fodder), maize and oilseed rape (spring and winter sown). The trials were designed to discover the effects on several biodiversity indices of the management of the GM crops by comparison with those observed with non-GM varieties, using a split-plot design. Fieldwork was undertaken between 2000 and 2003, all the results being available by 2004. The trials involved 60 to 75 fields of each crop on farms of varying type throughout Britain. They became the target of protest groups (such as Greenpeace), who on several occasions destroyed the crop; but enough remained unaffected to allow statistically valid data to be obtained.

The trials showed that growing non-GM beet and spring rape was better for many groups of wildlife than the GM varieties (Figure 10.1). Insects such as bees (for beet crops) and butterflies (for beet and spring rape) were recorded more frequently in and around the non-GM crops because the greater prevalence of weeds provided them with food and cover. Similarly, more weed seeds (which are important components of the diet of birds and other animals) were present in the non-GM crops, although some groups of soil insects occurred more frequently in GM beet and spring rape crops. In contrast, GM HT maize was better for many wildlife groups than non-GM maize, with more weeds in and around the crops, and more butterflies and bees at certain times of the year.[37] However, the maize results were complicated by the fact that the herbicide atrazine, used widely on the non-GM crops, was due to be banned for use in the EU, so that only more benign herbicides would be permitted in future.

10.5.5 The UK Government policy on genetically modified crops

On the basis of the FSE results, and following the advice of the UK Advisory Committee on Novel Foods and Processes (ACRE),[38] in 2004 Margaret Beckett, the Minister at the

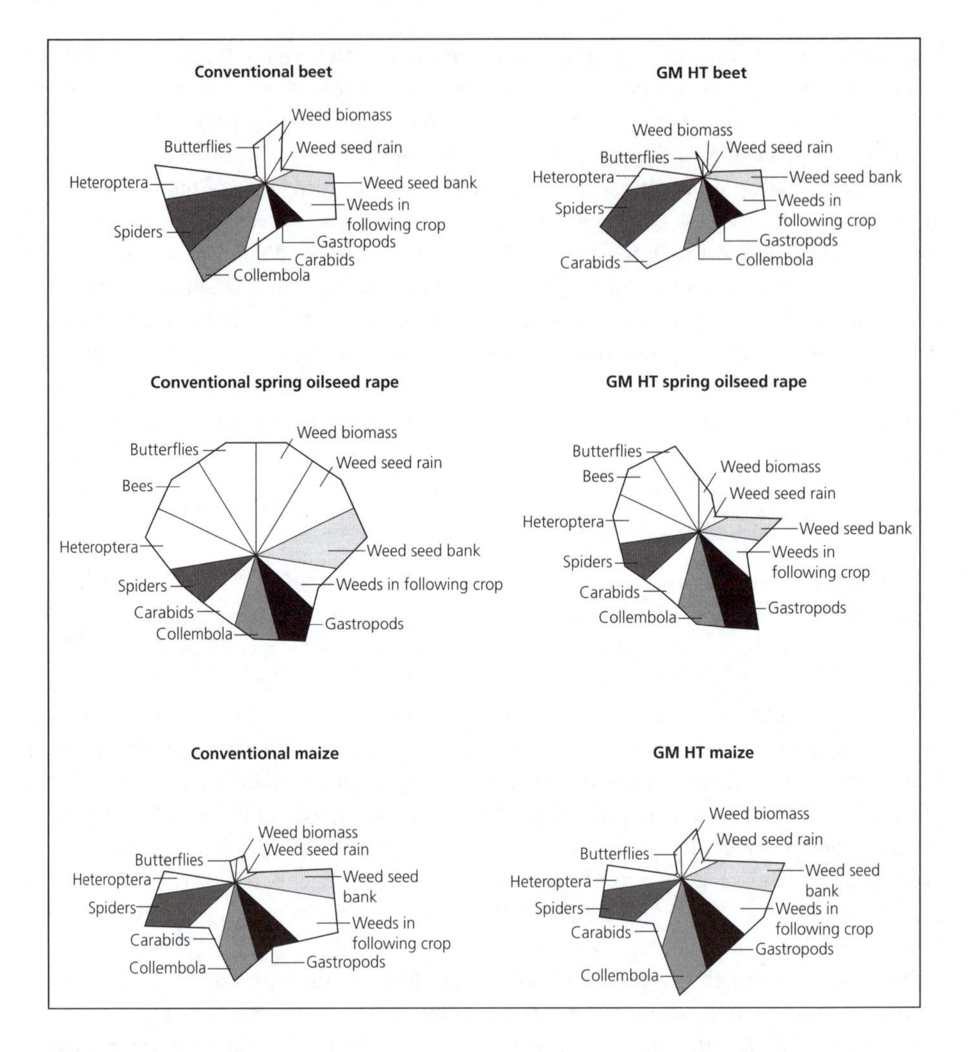

Figure 10.1 Star plots comparing the mean values of major biodiversity indicators across conventional and genetically modified (GM) herbicide tolerant (HT) treatments of beet, maize, and oilseed rape crops in the UK farm-scale evaluations. For each indicator, the length of the star corresponds to the value relative to the maximum value found in any of the six combinations of crop and treatment: for example, the most gastropods were found in GM HT oilseed rape. [From: Department of Trade and Industry (2004) GM Science Review: second report. London, DT1]

Department for the Environment, Food and Rural Affairs (DEFRA), announced:

- that the UK should oppose commercialization of GM beet and oilseed rape anywhere in the EU, when using the management regime tested in the FSE
- agreement should be given in principle to commercial cultivation of GM HT maize, subject to two conditions:
  - maize can only be grown and managed as in the trials

  - further tests must be conducted to monitor changes in herbicide use after the phasing-out of atrazine use.

However, the Minister also stressed the importance attached to ensuring that conditions were established for the co-existence of GM and non-GM (especially, organic) crops. To this end, she stated: *'I will also consult stakeholders on options for providing compensation to non-GM farmers who suffer financial loss though no fault of their own. But I must make it clear that any compensation scheme would need to be funded by the GM sector itself, rather than by Government or producers of non-GM crops.'*[39]

A more extensive Government response to the various lines of evidence made the following points. The Government will:

- protect human health and the environment through robust regulation of GM crops on a case-by-case basis, consistent with the Precautionary Principle
- ensure consumers can choose between GM and non-GM products
- safeguard farmers' interests by putting in place effective and proportionate measures to facilitate the coexistence of GM and non-GM crops
- provide guidance to farmers wishing to establish voluntary GM-free zones
- consider the best ways of providing the information which the public wants and in an open and transparent way.[40]

Collectively, these constitute some very stringent conditions. In fact, the biotech company Bayer decided to drop the intention to market its GM *Chardon LL* maize only a few days after the Government gave approval for planting, claiming that the conditions imposed were too restrictive.

10.6 The genetically modified crops: flourishment, adaptability and intrinsic value

How this 'interest group' is perceived depends very largely on the extent to which intrinsic value (3.6.3) is assigned to the crop; this, in turn, reflects the fundamental attitudes by which individuals position themselves intellectually and emotionally in the world. These attitudes have been described as the basic value orientations - or **worldviews** - which guide our interpretation of the natural world (hereafter described as 'Nature'). The Dutch philosopher Petran Kockelkoren has identified four fundamental attitudes positioned at points along a continuous scale; and it may be instructive for the reader to review each of them to discover which corresponds most accurately to her or his position.[41]

At one end of the scale is the **dominator**. The attitude of the dominator is that Nature, in itself, only has instrumental value (3.6.3) in respect of human culture; i.e. it is mankind that has identified and fashioned whatever is now considered to have value. Dominators believe that Nature is a resource which we are free to exploit fully for human benefit. But Nature often needs to be 'tamed' or 'conquered', as illustrated by the occurrence of natural disasters (such as earthquakes and floods) and disease outbreaks (such as AIDS and severe acute respiratory syndrome (SARS)). This attitude is **anthropocentric**.

At the opposite end of the scale, the **participant** locates intrinsic value in Nature itself, with which human culture is obliged to comply. In conformity with natural law (2.3), we should work with Nature rather than try to dominate it. Life demands respect in and of itself, irrespective of its usefulness. This attitude is **ecocentric**: we oppose Nature at our peril, or at least seriously risk undermining our real well-being. Despite the vaguely mystical associations, it is notable that people whose lives are otherwise based in human cultural activities (such as wealthy pop stars, sporting personalities or business tycoons) often choose to live close to Nature, e.g. in country mansions situated in unspoilt countryside.

Between these extremes are two intermediate positions. Next to the dominator is the **steward**, who accepts a duty of care for Nature, because bound by a sense of obligation either to a Creator and/or to future generations of humanity. While principally anthropocentric, this attitude is tempered by deontological constraints (2.6) which curb the extent to which Nature can be exploited. Between the steward and the participant is the **partner**, who regards non-human life forms as potential allies. Exploitation of Nature is permitted, but only by according it due respect and recognizing the needs of ecological integrity.

To illustrate the impact of such worldviews it is useful to consider the scale employed by the Dutch Society of Biotechnological Engineers, which defined four decision-options pertaining to GM. Applications to humans (with only extreme exceptions) were assumed to be impermissible (i.e. **no**). Applications to animals and plants were, respectively, **no, unless**, and **yes, provided that**: i.e. the burden of proof lay with those proposing the biotechnology in the case of animals, and with those opposing the technology in the case of plants. The response **yes** (subject to standard safety precautions) applied to GM micro-organisms. These decisions have been characterized as largely anthropocentric, but the scheme can be used more generally to give substance to the four worldviews discussed above.

Table 10.2 indicates the decisions on different forms of GM crop, which seem likely to be adopted by the four worldviews.[42] The range of applications has been chosen to illustrate different degrees of modification. Some might be perceived as important and others as trivial, some might pose greater environmental risks than others, and some might primarily benefit consumers, whereas others might be most beneficial for farmers.

Table 10.2 Differentiated ethical evaluations of four genetically modified crop technologies considered likely to be reached by dominators, stewards, partners and participants (for details see 10.6.)

Crop Technology	Dominator	Steward	Partner	Participant
Herbicide tolerance	Yes	Yes, provided	No, unless	No
Biotic stress (disease resistance)	Yes	Yes	Yes, provided	No, unless
Abiotic stress (cold, drought etc.)	Yes	Yes	Yes, provided	No, unless
Aesthetic changes (colour, taste etc.)	Yes	No, unless	No	No

The perception of crops' intrinsic value will necessarily have a bearing on the ethical impacts of each GM technology on the other principles (Table 10.1), but the effects might be interpreted either negatively or positively. For example, GM might be considered to undermine adaptability by the imposition of a new *telos* (9.6.1), whereas the GM-enhanced ability to survive abiotic stresses could be viewed as facilitating the crop's ability to flourish. But such impacts are only likely to be considered ethically relevant by those adopting the participant or partner attitudes.

10.7 The prospects for genetically modified crops in the UK

Of all the prospective GM technologies, those involving the growing of GM crops seem the most likely to find early commercial application in the UK (and EU more generally), as they have in the USA and elsewhere. At the time of writing, the one GM crop approved for growing in the UK has been withdrawn (10.5.5), but the issue seems certain to return to centre stage.

The intense public interest in the possibility of GM crops being grown in the UK led to the Government-sponsored *GM Nation?* debate in 2003 (14.7.1), and to the setting up of the SRP, the reports of which have been cited in this chapter. The Government's Strategy Unit also performed a valuable 'cost–benefit' analysis of the issues, in which it outlined five possible scenarios; and the aim of this section to summarize the report's main points.[43] Figure 10.2 shows how the scenarios are located on a grid encompassing GM regulations on one axis and public acceptability on the other.

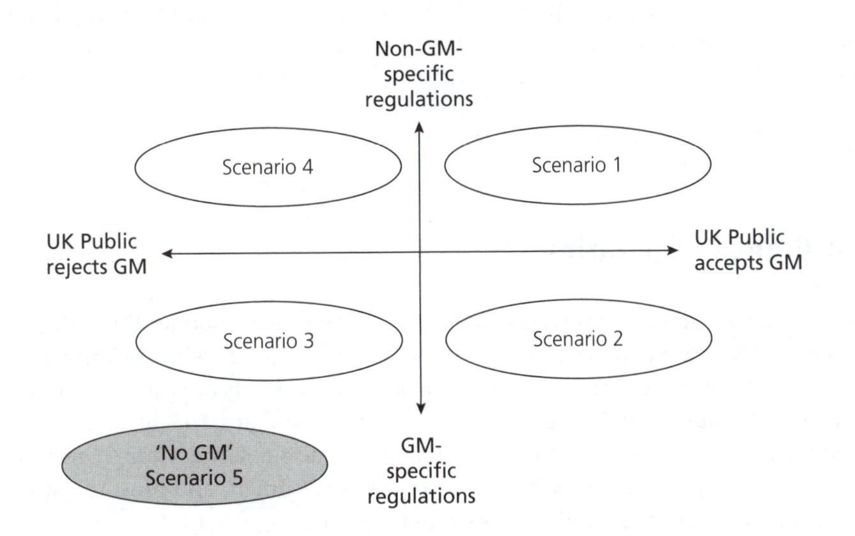

Figure 10.2 Genetically modified (GM) crop scenarios used in the UK. For details, see 10.7. [From: UK Government's Strategy Unit (2003) Field Work: weighing up the costs and benefits of GM crops. London, Cabinet Office, p. 25]

In *Scenario 1* (**part of the fabric**) GM crop technology would increase the range of products available, producing positive economic impacts for farmers, processors and consumers. The combination of consumer acceptance and a cheaper, less stringent regulatory regime, could benefit UK biotechnology - and send a very pro-GM message to the rest of the world. Disadvantages would be that the absence of GM-specific regulations (as in the USA) might carry increased risks of adverse (e.g. environmental) effects, while consumer and producer choices to avoid GM would be severely restricted.

In *Scenario 2* (**separate weave**) consumer and producer choices would be supported by tighter regulations, minimizing adverse effects. However, the more stringent regulatory regime would impose some additional costs, and slow down technological developments. The signal from the UK would be positive and supportive of strong regulations.

In *Scenario 3* (**bare minimum**) considerable regulatory costs would be incurred, so that many GM crops would not reach the market. But consumer choice would be maintained, sales of non-GM crops (such as organic) might be boosted, and the likelihood of negative effects would be reduced. Internationally, mixed messages would be sent.

In *Scenario 4* (**tangled threads**) uncertainty over the regulations and acceptance of GM would confuse industry, retailers, and producers alike, leading to consumer dissatisfaction and a weak ability to cope with adverse impacts. Overall, the outcome would be muddled, with benefits for some but problems for others, and alarm in overseas markets.

In *Scenario 5* (**no GM**), an explicit non-GM policy would have international implications for trade and reduce the types of crop available to consumers and producers, while limiting options in the biotech industry. Some, but not all, regulatory costs would be avoided; consumers and producers wishing to avoid GM would be satisfied; risks of adverse effects would be reduced. The UK might see benefits from having a GM-free environment, e.g. by developing a niche market for national and international consumers.

10.8 Ethical evaluation

Which of the above scenarios seems likely to prevail? And as importantly, which ought to prevail? The use of the ethical matrix (Table 10.1) might help to clarify and, for individuals, quantify the various ethical impacts. But in terms of policy decisions over whether GM crops should be grown, and how they should be regulated, much will depend on the weight given to public opinion, because it seems clear that there is a substantial discrepancy between the views of the Government and most scientists, on the one hand, and of the public, on the other (14.7.1).

A view often advanced is that because of the complexity of the issues the only appropriate approach is one that relies on a 'case-by-case' analysis, which is scientifically evaluated.[44] But it is worth asking whether this does not implicitly adopt a utilitarian stance, because it implies that one cannot make a decision in advance of knowing the outcome in terms of costs and benefits. On the contrary, it might be claimed that one *can* make such a decision in terms of the *objectives*. Particular worldviews (10.6) might constitute intrinsic barriers to using GM to achieve a certain category of aims; and in that sense resort to 'case-by-case' analysis would be inappropriate. For example, if there were significant public concern about power over the global food supply being concentrated into the hands of a very few multinational companies, this might provide grounds for some people to reject GM crops as a whole, despite any economic benefits which might be predicted. Although this could have the most direct adverse effects on consumer autonomy (see Table 10.1), in reality indirect effects might well impact on several other cells of the matrix.

Moreover, the moratorium on growing all GM crops in the EU, which was instituted in 1999, had its justification in an appeal to the Precautionary Principle (PP). This interpretation of the PP (13.4) placed the burden of proof for the absence of harmful effects on the proponents of GM, and included the threat to public acceptability as a risk to be avoided. For if the public acceptability of a risk common to different GM crops was low, and the risk assessment itself was subject to significant uncertainty, a moratorium might be the appropriate regulatory outcome.[45]

On the other hand, proponents of GM crops often dismiss these issues of doubt as minor problems in the face of the great benefits which they could deliver. According to plant scientists Steve Hughes and John Bryant:

> *Undoubtedly there will be mistakes made in the choice of goals for the technology just as there have been for all new technologies, but these should be seen in the light of the decision-making process and its deficiencies, not as a characteristic of the technology itself. The ethical challenge is to find a solution that takes account of the legitimate interests of all parties but does not assign a veto to any one of them.*[46]

THE MAIN POINTS

- GM crops, grown extensively in the USA but not in the EU, permit more precise modification of crops thus enabling both agronomic and nutritional benefits; however, the power and speed of implementing GM raise ethical concerns.
- Ethical issues include: profitability, managerial freedom, and fair trade laws for farmers; consumer safety and choice; and impacts on environmental biodiversity and sustainability.
- Many people's views on appropriate human relationships with the natural world (i.e. whether the latter is appreciated only instrumentally or also for its intrinsic value) influence attitudes to GM.
- In the UK, the *farm-scale evaluations* of the management of GM herbicide tolerant crops provided extensive data on impacts on biodiversity indices, which suggested that growing GM beet and oilseed rape was environmentally detrimental, but that GM maize was beneficial.
- The profitability of GM crops seems to be highly dependent on public attitudes to GM food, longer-term environmental impacts, and the extent to which it will be possible for GM and non-GM (especially organic) forms of agriculture to co-exist.

EXERCISES

These can form the basis of essays or group discussions:

1. Is the co-existence of GM and non-GM (and especially organic) forms of crop production feasible in the UK? Explain the reasons for your opinion (10.3.2).
2. How do you respond to the suggestion that *substantial equivalence* is a 'pseudo-scientific concept' (10.4.1)?
3. According to the AEBC, the results of the *farm-scale evaluations* should not be seen as the 'last piece of the jigsaw'. Which are the other important pieces? (10.5.4)
4. Where do you position yourself on Kockelkoren's scale of fundamental attitudes to Nature? How do you justify this position? (10.6)
5. The UK Government Strategy Unit identified five possible GM crop scenarios. Explain which you consider would be the best (10.7).

FURTHER READING

- *The Doubly Green Revolution: food for all in the twenty-first century* by Gordon Conway (1997). London, Penguin. A reasoned defence of GM technology by the President of the Rockefeller Foundation.
- *Against the Grain: the genetic transformation of global agriculture* by Marc Lappé and Brit Bailey (1999). London, Earthscan. A critique of crop biotechnologies.
- *The Threatened Gene: food, politics, and the loss of genetic diversity* by Cary Fowler and Pat Mooney (1990). Cambridge, Lutterworth. Examines the social effects of genetic erosion.

USEFUL WEBSITES

- **http://www.defra.gov.uk/corporate/ministers/statements** *Defra* (2004) 'UK Government Statement on GM crops policy': the Government's considered judgement on whether the planting of GM HT crops should be allowed.
- **http://www.strategy.gov.uk/output/page3673.asp** *UK Government Strategy Unit* (2003) 'Weighing up the costs and benefits of GM crops': an economic assessment of GM crops.
- **http://www.gmsciencedebate.org.uk/report** *The GM Science Review Panel* (2004) Second report London, Department of Trade and Industry: the sequel to an earlier report, which provides a scientific assessment of GM crops.

NOTES

1. Cabinet Office Strategy Unit (2003) Field Work: weighing up the cost and benefits of GM crops. Cabinet Office, London, p. 34
2. Ibid., p. 22
3. Brown P (2004) US seeks £1bn from Europe over GM ban. London, Guardian. http://www.guardian.co.uk (accessed 24 April 2004)
4. Agricultural and Environment Biotechnology (2001) Crops on Trial. London, Department of Trade and Industry, p. 13
5. Parliamentary Office of Science and Technology (2000) The 'great GM food debate'. London, POST
6. Ibid.
7. Swiss Ethics Committee on Non-Human Gene Technology (2004) Berne, ECNH
8. Soil Association (2002) Seeds of Doubt. Bristol, Soil Association, p. 4
9. Benbrook C (2003) Impacts of genetically engineered crops on pesticide use in the United States: the first eight years. http://www.biotech-info.net/technicalpapers6.html
10. PG Economics (2004) Comments on GM foods: evaluating the farm scale trials. http://www.pgeconomics.co.uk/environment_select_committee_report.htm
11. Cabinet Office Strategy Unit (2003) Field Work: weighing up the cost and benefits of GM crops. Cabinet Office, London, p. 55
12. Agricultural and Environment Biotechnology (2003) GM Crops? Coexistence and Liability. London, Department of Trade and Industry
13. Biotech-Info Network (2002) Pollen movement. http://www.biotech-info.net/pollen_movemet.html
14. Soil Association (2002) Seeds of Doubt. Bristol, Soil Association, p. 48
15. GM Science Review Panel (2003) First Report. London, Department of Trade and Industry, p. 59
16. European Commission (1997) Official Journal. L253. 16.9.97 http://europa.eu.int/
17. Millstone E, Brunner E, and Mayer S (1999) Beyond substantial equivalence. Nature *401*, 525–526
18. Royal Society (2002) Genetically Modified Plants for Food Use and Human Health - an update. London, Royal Society
19. Ferrara J and Dorsey M K (2001) Genetically modified foods: a minefield of safety hazards. In: Redesigning Life? (Ed.) Tokar B. London, Zed, p. 57
20. Greenpeace (2001) Starlink corn contamination chronology. http://www.greenpeaceusa.org/ge/starlink
21. GM Science Review Panel (2003) First Report. London, Department of Trade and Industry, p. 96
22. Royal Society (1998) Genetically Modified Plants for Food Use: Statement. London, Royal Society, p. 13
23. GM Science Review Panel (2003) First Report. London, Department of Trade and Industry, p. 107
24. Ministry of Agriculture, Fisheries and Food (1992) Report of the Committee on the Ethics of Genetic Modification and Food Use (Polkinghorne Report). London, Stationery Office
25. Parliamentary Office of Science and Technology (2002) Labelling of GM foods. Postnote 172. London, POST
26. Consumer Association (2002) Policy Report: GM dilemmas. London, CA
27. European Commission (2003) Regulation 1830/2003. http://europa.eu.int/
28. Friends of the Earth (2004) New GM labelling laws in force. http://www.foe.co.uk/

29. Consumer Association (2002) Policy Report: GM dilemmas. London, CA
30. Ibid.
31. Swiss Ethics Committee on Non-Human Gene Technology (2004) Gene Technology and Developing Countries. Berne, ECNH, p. 25
32. Steinbrecher R A (2001) Ecological consequences of genetic engineering. In: Redesigning Life? (Ed.) Tokar B. London, Zed, p. 85
33. GM Science Review Panel (2003) First Report. London, Department of Trade and Industry, p. 135
34. Steinbrecher R A (2001) Ecological consequences of genetic engineering. In: Redesigning Life? (Ed.) Tokar B. London, Zed, pp. 75–102
35. English Nature (2002) Gene Stacking in Herbicide Tolerant Oilseed Rape: lessons from the North American experience. Peterborough, EN
36. Chilcutt C F and Tabashnik B E (2004) Contamination of refuges by *Bacillus thuringiensis* genes from transgenic maize. P Natl Acad Sci USA, 10 May 2004. http://www.pnas.org/cgi/
37. GM Science Review Panel (2004) Second Report. London, Department of Trade and Industry
38. Advisory Committee on Releases to the Environment (2004) Advice on the implications of the farm-scale evaluations of genetically-modified herbicide-tolerant crops. http://acre.gov.uk/
39. Department of Food, Environment and Rural Affairs (2004) Secretary of State Margaret Beckett's statement on GM policy. http://www.defra.gov.uk/corporate/ministers/statements/mb040309.html
40. Department of Food, Environment and Rural Affairs (2004) The GM Dialogue: Government response. http://www.defra.gov.uk
41. Kockelkoren P J H (1993) Ethical Aspects of Plant Biotechnology. Netherlands Ministry of Agriculture, Nature Management and Fisheries
42. Kockelkoren P and Linskens M (1997) In: The Future of DNA. (Ed.) Wirz J and Lammerts van Bueren E. Dordrecht, Kluwer, pp. 205–217
43. Cabinet Office Strategy Unit (2003) Field Work: weighing up the cost and benefits of GM crops. London, Cabinet Office, pp. 82–103
44. Nuffield Council on Bioethics (1999) Genetically Modified Crops: The Ethical and Social Issues. London, NBC
45. Food Ethics Council (2003) Engineering Nutrition. Brighton, FEC, p. 4
46. Hughes S and Bryant J (2002) GM crops and food: a scientific perspective. In: Bioethics for Scientists. (Eds) Bryant J, Baggott la Velle L, and Searle J. Chichester, Wiley, p. 138

11

Dietary futures

OBJECTIVES

When you have read and discussed this chapter you should:

- appreciate the nature and scale of diet-related disorders in the UK and other developed countries
- understand the aims of functional, and other novel, foods and the legislation which controls their marketing
- be aware of the ethical issues raised by functional foods that relate to consumer interests
- appreciate the prospective roles of GM functional foods
- be able to perform an ethical analysis of Golden Rice (based on an ethical matrix) as a technology aimed at alleviating malnutrition in LDC

11.1 Introduction

Within the bioscience community it is often assumed that the low public acceptance of the first generation of GM foods (10.1.2) in the EU is due to a lack of clear benefits for consumers, either in greatly reduced prices or in markedly increased nutritional quality. According to Robert May, President of the Royal Society, and formerly the Government's Chief Scientist, *'What is needed is a glamorous GM product that would change the image of biotechnology. We need a GM apple that will make you thin and wicked. I suspect that would do the trick.'*[1] The suggestion is that the widespread public resistance to GM is ill-founded and will dissolve in the face of a demonstration that GM foods can confer real benefits on consumers. This 'second generation' of GM foods, many of which are designed to enhance nutritional quality, is one focus of this chapter. The aim will be to consider what problems these GM foods are designed to address, their efficacy in doing so, and ethical issues raised by their production, marketing, and consumption.

But in addition the chapter considers other, non-GM, dietary innovations that also fall into the category designated **novel foods**[2] (Box 11.1), and which are subject to EU legislation. In particular, it considers a category of novel foods called *functional foods*.

Although hunger is not a problem for the vast majority in DC, it is becoming clear that the diet of many people is far from ideal; and in theory novel foods might provide

BOX 11.1 NOVEL FOODS

A food is considered *novel* if it has not previously been used to a significant degree in the EU, and falls into one of the following six categories:

- contains or consists of GMO
- is produced from, but does not contain GMO
- has a new or intentionally modified molecular structure
- consists of or is derived from micro-organisms, fungi, or algae
- consists of or is derived from plants or animals (but excluding those which are obtained by traditional practices and have a history of safe food use)
- has come from a novel production process which causes changes affecting the nutritional value, metabolism, or presence of undesirable substances

a means of improvement. But because deficiency diseases are common in LDC, it is also often argued that certain novel GM foods could significantly contribute to efforts to counter malnutrition in those countries. The second generation of GM foods could thus have some highly beneficial impacts on a global scale. Before assessing these claims, it will first be useful to survey the current state of diet and health in the UK, viewed as a representative DC. An equivalent survey for LDC is provided in 4.3.

11.2 Nutritional standards in the UK

Since, in the time-honoured words of Brillat-Savarin, *'We are what we eat'*,[3] the increased stature and lifespan of people in the UK over the last 100 years suggest that many of the dietary inadequacies of earlier years (often due to ignorance about the importance of macronutrients and micronutrients)[4] have now been largely corrected. Despite this, there is much concern that the nation's diet is far from satisfactory. In the words of a Minister for Public Health: *'The consumption of salt, fat and sugar – especially through processed foods – is still too high; consumption of fruit and vegetables – three portions a day – are well below the recommended level of five portions a day, and there are still areas of deprivation where low income consumers lack opportunities to eat well.'*

The result is that diet-related diseases affect many people who, if they were to eat more wisely, would live much longer and healthier lives. For example, CHD, the country's 'biggest killer', is a preventable condition that causes 110,000 deaths p.a., with 275,000 having heart attacks and 1.4 million suffering from angina. More than million people are estimated to suffer from diabetes, which probably accounts for 6% of all deaths. Medical opinion is now firmly of the view that such conditions, as well as obesity, various cancers, digestive disorders, and dental decay, would be significantly reduced by improved diets (coupled with more exercise and less smoking and alcohol).[5] In Germany, it is estimated that one-third of all costs incurred by the health system are caused by diet-related illness, while in Switzerland 60% of deaths are caused by food-related diseases.[6]

Many of these health problems originate in childhood, and are already established by the time of adolescence. According to a BMA report, *'Obesity has come to be considered a global epidemic and excess body weight is now the most common childhood disorder in Europe.'* Thus, the prevalence of 'overweight' (defined as a body mass index[a] of 25–30) in 7–11-year-olds rose by 60% between 1994 and 1998; and in 1997, in the 16–24 age group, 23% of men and 19% of women were overweight, while a further 6% of men and 8% of women were obese (defined as a BMI of more than 30). But obesity and overweight are not just cosmetic problems. They can cause serious medical conditions, such as CHD, type 2 diabetes, cancer, gallbladder disease, asthma, and impaired fertility, as well as psychological and social problems, such as low self-esteem, loneliness, eating disorders and educational disadvantage.[7]

However, the problem is not only a dietary one. The term **obesogenic** describes the modern environment that encourages both high food-energy intake and inactivity. Features of this environment are excessive sedentary activities, such as computer use and television viewing; lack of exercise because of frequent vehicle use and reduced involvement in sporting activities; and increased consumption of fast foods and carbonated drinks, coupled with 'snacking' and 'binge-eating' patterns.

This obesogenic environment is sharply distinguishable from genetic causes of obesity, and as food marketing is largely directed at adolescents, it is especially difficult for them to make healthy food choices.[8] Many of the trends evident in the UK follow those originating in the USA: they suggest that commerciogenic malnutrition (4.3.1) has become a serious public health problem. For example, in 1997 American children obtained 50% of their dietary energy from added fat and sugar, while only 1% regularly ate diets conforming to official dietary guidelines. This was doubtless greatly influenced by the US$12.7 billion spent in direct advertising of food to children that year.[9] In the UK, up to 99% of advertisements during children's commercial television are for products high in fat and/or sugar and/or salt, and fatty and sugary foods are advertised in proportions up to eleven times higher than those recommended in dietary guidelines.[10]

11.2.1 The modern food system

Food consumption may seem to be an essentially private activity, especially for food eaten at home. Individually selected foods, picked from the vast range available at the supermarket, confirm a strong sense of individual choice, in keeping with our personal likes and dislikes. But in reality, behind the scenes, most food originates in a massive, highly integrated agro-food industry; and the rigours of competitive market economics have resulted in a very few, very powerful companies controlling most of the food chain from 'field to fork'.

In the UK, supermarkets play a major role, with 75% of food sales (nearly £80 billion p.a.) being purchased from the four major retailers. It is widely recognized that supermarkets, through their enormous buying power, increasingly dictate the way agriculture

[a] Body mass index (BMI) = body weight (kg) divided by the square of the height (m)

is practised, both in the UK and abroad. Food and drink are also the basis of a major manufacturing industry. In fact, it is the UK's largest manufacturing sector–employing almost half a million people, buying two-thirds of UK's agricultural produce, and having an annual turnover of £66 billion.[11]

Increasingly, people are eating out, and much food is purchased through fast-food outlets. For example, the US company McDonald's, which now operates in many countries of the world, have based their businesses on *'the quantifiable principles of speed, volume, and low price*... (with) *customers offered a highly circumscribed menu using assembly line procedures for cooking and serving food'* (a phenomenon termed 'McDonaldization').[12] Almost inevitably, fast food has serious consequences for public health.

11.2.2 Government dietary advice

The importance of diet in promoting health is hardly a new discovery, being established since at least the early 1980s, albeit resisted by sections of the agricultural and food industry who feared its adverse impacts on food purchases.[13] The Government's **healthy eating guidelines** have not changed much since then, and were summarized by the Health Education Authority (1997) in eight points, as:

- enjoy your food
- eat a variety of different foods
- eat the right amount to be a healthy weight
- eat plenty of foods rich in starch and fibre
- eat plenty of fruit and vegetables
- don't eat too many foods that contain a lot of fat
- don't have sugary foods and drinks too often
- if you drink alcohol, drink sensibly.

11.3 Functional foods

The deteriorating state of diet-related public health (11.2) is clearly a social problem, which, it might be argued, is ripe for a technological solution. Thus, it is claimed that the knowledge of human nutritional requirements, which was systematically acquired during the last century through a combination of epidemiological studies, animal experimentation, and human dietary intervention, now needs updating to meet the different demands of our modern lifestyles. Moreover, the realization that our individual genetic predispositions are likely to have important implications for optimal diets (explored by the new science of **nutrigenomics**) suggests that food technology might have a critical role in maintaining our future health and well-being. The stage seems set for the introduction of novel foods designed to counter the adverse effects of an outdated dietary regime.

But it would be naïve to ignore the commercial motivations for developing novel foods. Food, viewed simply as nutrients, is a commodity with limited commercial potential. So, as population levels are now fairly stable in most DC, food might not seem to satisfy a central requirement of capitalist economics - the capacity for growth.[14] Whereas the richer people get, the more consumables they are able to buy (whether they be houses, cars, or holidays), the finite capacity of the stomach places a limit on the amount that even the greediest can eat. This has resulted in the necessary 'growth' (from a commercial perspective) being achieved by increasingly *adding value* to food - that is to say, modifying the food so that consumers are prepared to pay more for it. Consequently, food manufacturers are constantly seeking to market new products with higher added value; and there is keen competition between different companies to secure as big a slice of the market as possible. *'The reality of the modern food industry is an unending struggle for profit margins in often unresponsive markets.'*[15]

The search for innovative products has coincided with the growing awareness that many disease conditions are diet-related. This has resulted in the introduction of a new class of foodstuffs called **functional foods** (FF), which have been defined as *'any modified foods or food ingredients that may provide health benefits beyond the nutrients they contain'.* For most FF there is only one active ingredient, such as a bacterial culture or specific fatty acid. Others are fortified with antioxidants (which reduce cancer risks) or additional nutrients designed to improve specific types of performance, which may be physical or mental. Types of FF are illustrated in Box 11.2.

In 1999, the market for FF in the USA accounted for 60% of global sales, while in the EU and Japan the figures were about 17% and 16% respectively. But sales are expected to increase substantially, with the US sales of about US$15 billion in 2000 rising to almost US$50 billion by 2010.[16]

11.3.1 Health claims for functional foods

It is expensive to develop FF: one estimate puts the average development costs at US$10 million per product, with a *market entry time* of two years. This means that only large food companies are able to afford to market FF, and they need to be able to make

BOX 11.2 TYPES OF FUNCTIONAL FOOD

- Disease-risk-reducing foods (nutraceuticals; e.g. containing folic acid to prevent neurological disorders in infants; phytosterols to reduce blood levels of cholesterol; and probiotics to promote a healthy gut microflora)
- Risk-group-specific foods (e.g. gluten-free for coeliacs, and nut-free for allergy sufferers)
- Age-retarding foods (e.g. containing docosahexaenoic acid, claimed to retard senility)
- Physical-performance-enhancing foods (e.g. sports diets designed to provide instant energy and promote electrolyte balance)
- Mental-performance-enhancing foods (e.g. containing choline, claimed to improve short-term memory)
- Mood foods (e.g. claimed to relieve stress through increased choline content)
- Enhanced variants of traditional foods (e.g. with processing to improve bioavailability of nutrients)

claims for substantial consumer benefits if the product is to be a commercial success. Claims are a contentious issue, for several reasons:

- **medical claims** are prohibited in the EU, under Article 2.1 of Directive 2000/13/EC, which states that *'labelling and methods used must not . . . attribute to any foodstuff the property of preventing, treating, or curing human disease, or refer to such properties'*
- **nutrition claims** (under the EU Directive on Nutritional Labelling 89/398/EEC) generally limit claims to those listing the energy and nutrient contents
- **functional claims** cover reference to the physiological role of a nutrient on growth, development and normal body function; but any statement that the nutrient would provide a cure or treatment for, or protection from, a disease is not allowed.

Consequently, there is much discussion in the EU about the desirability of an agreed approach to **health claims**, so that manufacturers can give their products some identity and exclusivity in the highly competitive market. The problem is that FF lie at the interface between foods and drugs, and this may cause some confusion, especially with members of the public who think of these as totally distinct categories. Currently, it is proposed to describe a health claim as: *'any claim that states, suggests or implies that a relationship exists between a food category, a food, or one of its constituents and health'.*[17] Health claims are essentially of two types:

- **enhanced function claims** imply that the food or its constituent/s has a specific beneficial effect beyond that normally obtained from the diet, e.g. 'Calcium may help improve bone density. Food X is rich in calcium.'
- **reduction of disease risk claims** imply the food or its constituent/s reduce a major risk factor in disease, e.g. 'High blood cholesterol is a risk factor in heart disease. Food Y has been shown to reduce blood concentrations of cholesterol.'

However, the difficulty of reaching agreement on the wording of health claims, on food labels and in advertisements, is only one aspect of the problem. Criteria also need to be established to verify the *accuracy* of claims so that consumers are not misled. But this is problematical. As most of us are genetically, environmentally and behaviourally distinctive, we all respond somewhat differently to the food we eat, so that it would be extremely difficult to predict precisely how a single food ingredient is likely to affect us individually. One way around this is to assess the efficacy of FF by reference to **biomarkers** – e.g. the level of blood cholesterol. *Hypercholesterolaemia* is not itself a disease state, and some people with high blood levels may live long, healthy lives; however, it can be taken as a statistically valid marker of increased *risk* of CHD. Consumer acceptance is thus dependent on a high level of trust, since there is usually no way of readily assessing the efficacy of FF in delivering health benefits. Box 11.3 shows two examples of current FF.

Because the EU has so far failed to devise a ruling on the issue, individual Member States have their own codes of practice. In the UK, the *Joint Health Claims Initiative* (JHCI) is administered by a group of representatives drawn from enforcement authorities, industry trade associations and consumer groups.[18] A priority is to ensure that claims made are based on sound evidence and do not exaggerate the health benefits or mislead the consumer. By 2002, the JHCI Code had been endorsed by the FSA,

BOX 11.3 PROMINENT CURRENT FUNCTIONAL FOODS

- **Probiotics**: cultures of live micro-organisms that affect the digestive system beneficially by improving the properties of indigenous microflora. Claimed benefits include: alleviating symptoms of lactose intolerance, playing a role in immunomodulation, and treatment and prevention of viral diarrhoea. Most probiotics for human consumption consist of lactic acid bacteria (e.g. *Yakult*) or bifidobacteria (e.g. *BioPot*).
- **Phytosterol-enriched foods**: foods that lower blood levels of cholesterol, especially low-density-lipoprotein cholesterol (LDLc). By competing with the cholesterol space in mixed micelles (the form of lipid delivery for absorption into mucosal cells), they reduce intestinal absorption of cholesterol by about 50%.

  These plant sterols (e.g. *Flora/Becel ProActive*™) or stanols (e.g. *Benecol*) are often included in spreads. At doses of 1–3 g/day, LDLc levels are lowered by 5–15%.

several food companies, trade associations, local trading standards officers, advertising clearance and regulatory bodies, and consumer interest groups.[19]

Assessment of the safety and efficacy of novel foods is the responsibility in the UK of the Advisory Committee on Novel Foods and Processes (ACNFP), which reports to the FSA (14.3.3). The majority of ACNFP members are scientists with expertise in food safety issues, but members also include two consumer representatives and an ethicist. If necessary, additional scientific expertise is sought from other specialist advisory committees (namely, those on toxicity, mutagenicity, carcinogenicity, environmental releases, microbiology, animal feed, and nutrition: Table 14.2).[20]

11.3.2 The social context of functional food use

Because of the centrality of food in all our lives, its provision is greatly affected by changing lifestyles; and this means that there is likely to be demand for an increasing diversity of foods (Table 11.1).[21] But, in designing novel foods, food manufacturers need to take account of three important features of many foods:

- their authentic taste, smell and appearance
- certain non-nutrient components, such as dietary fibre, which may have beneficial properties for consumer health
- their cultural significance - often associated with factors such as the origin, modes of production and preparation, and rituals of presentation of the food.

In fact, the latter consideration is likely to play a critical role which is often poorly appreciated by food scientists, whose emphasis on *functionalism* is primarily concerned with food's physical qualities. In contrast, *structuralist* attitudes seek deeper and broader causes and meanings of food habits - which have to do with factors such as position in society, class and power relations.[22] In such terms, food choices may be an expression of a political outlook or feminist ideology.

Despite enthusiasm by the food industry, to date, FF have not found wide public support in the UK. According to the Leatherhead Food Research Institute, this is due to

Table 11.1 Impact of socio-demographic trends on the food market (From: National Council for Agricultural Research (Netherlands))

Socio-Demographic Trends	Impacts on Food Market
Stable population size	Stable market size; increased competition; development of niche markets
Ageing population	Demand for: increased nutritional value; smaller portions; traditional taste
Greater ethnic mix	Demand for non-traditional foods
Smaller households	Demand for individual and smaller portions
Awareness of diet–health links	Demand for increased nutritional value; functional foods; high quality; food safety
Both adults in full-time employment	Demand for convenience foods; *one-stop shopping*
Increased leisure time	Demand for *fun foods*
Increased international travel	Demand for travel and exotic foods
Children's increased independence and purchasing power	Demand for snacks; *fun foods*; small portions

(From National Council for Agricultural Research, 2000)

several factors:

- they are expensive
- most health claims are poorly understood by the public
- there are concerns about overdosing
- some contain unfamiliar ingredients.

Furthermore, the blurring of the distinction between food and medicine, and questions about the role of FF in national nutritional policy, have tended to undermine their acceptance by both consumers and health professionals.[23]

11.4 International differences

The terminology used for FF has not been standardized, so readers should be aware that different publications may use terms differently. For example, FF are sometimes equated with nutraceuticals (Box 11.2), while in some North American literature the term *techno-foods* is used to include all modified foods, e.g. those fortified with added vitamins. In the USA, GM foods are often referred to as *bioengineered foods*.

11.4.1 The situation in the USA

The growth of the FF market in the USA, which has set the trend for many global developments, followed the decision of the FDA to authorize certain health claims for techno-foods. When the Nutrition Labeling and Education Act (1990) was introduced, manufacturers began to add to foods various phytochemicals (such as lycopenes, indoles, flavonoids and sterols), which were known to reduce the risks of contracting cancer. As these are naturally occurring substances, the FDA regarded them as GRAS (*generally regarded as safe for human consumption*), and consequently required no specific pre-marketing approval.

Subsequently, the 1994 Dietary Supplement Health and Education Act allowed manufacturers to market FF with **structure/function claims**, indicating potential health benefits. Nutritionist Marion Nestle (who has no connection with the food company of that name) has described how this liberal labelling regime has permitted manufacturers to advertise the speculative health benefits of some components of the food, without reference to the total dietary context. A prominent example was the advertisement for a tomato ketchup which claimed that *'lycopene may help reduce the risk of prostate and cervical cancer'.*[24] (Lycopene is a plant pigment which is naturally present in tomatoes and other fruits and vegetables.) However, it has been claimed that *'82% of the* [sixty] *contributory risk factors for prostatic cancer are unknown'* – and of the known factors, *'nutrition contributes a mere 5%'.*[25]

The US market for FF is booming, fuelled by the almost complete focus on disease and its prevention, coupled with the fact that sections of the US population have never been more health conscious.[26]

11.4.2 The situation in Japan

The concept of FF originated in Japan in 1988, with the launch of a dietary fibre soft drink, which was marketed for 'gut regulation'. Manufacturers may use either of two marketing routes. **Foods for specified health uses** (FOSHU) are defined as those for which 'a functional ingredient has been added for a specific healthful effect and are designed to maintain and promote good health'. Most of the nearly 200 approved FOSHU foods are fermented-milk drinks. The non-FOSHU route simply entails the manufacturer launching the product without making any specific health claim. For example, *The Calcium* is a calcium-fortified sandwich biscuit, which the knowledgeable consumer may judge a valuable product, based on the stated information that one biscuit contains 600 mg calcium. More than 2000 non-FOSHU products have reached the market.[27]

The buoyant growth of the Japanese FF market seems to be due to strong government support in creating a favourable climate for research and a liberal regulatory regime, coupled with a high degree of consumer awareness and acceptance.[28]

11.5 Genetically modified functional foods

Very few GM FF have to date been licensed for commercial use, but some projects are at advanced stages of development. For example, several projects concern the genetic manipulation of oil composition (Table 11.2), often with the aim of facilitating food processing.[29] More consumer-oriented GM FF projects aim to modify the levels of certain

Table 11.2 Approaches to the genetic manipulation of oil composition (From: Dibb and Mayer, 2000)

Crop	Trait	Genes	Potential uses
Oilseed rape	High lauric acid	12 : 0 acyl carrier protein thioesterase gene	Food industry; confectionery
Soybean	High oleic acid	Sense suppression of GmFad2-1 gene which encodes a Δ^{12}-desaturase enzyme	More stable oil for frying, baking, and food manufacturing. Animal feed applications
Oilseed rape	High stearic acid	Garm FatA1, an acyl-acyl carrier protein thioesterase	Shortening, margarine, confectionery and bakery products
Oilseed rape	High gamma-linolenic acid	Gene from borage	Therapeutic/substitute for evening primrose oil

proteins, fatty acids, and carbohydrates in food, or those of micronutrients such as vitamins, minerals, and phytochemicals that are believed to act as anticancer agents. Examples are:

- low-linoleic-acid soyabean oil, which has claimed health benefits through reduction of *trans* fatty acids
- protein-enhanced sweet potatoes, aimed at countering malnutrition
- modified starch content potatoes, designed to reduce fat absorption during frying
- increased flavonol tomatoes, thought to protect against CHD.

As more is understood about the biochemical pathways leading to the synthesis of plant micronutrients, attention is increasingly being turned to producing GM plants with raised levels of beneficial phytochemicals, and/or with reduced levels of anti-nutrients.

11.6 Functional foods and consumer interests

Functional foods might be thought the 'best thing since sliced bread' (which certainly was a 'functional food' of another sort!). But technological enthusiasm needs to be tempered by taking due account of reasonable scepticism; in this respect, consumers' interests are a critical factor. This section concentrates on issues raised by the ethical matrix (Table 10.1) that concern consumer well-being, autonomy and fairness. It is confined to consideration of the situation in DC.

11.6.1 Consumer well-being

The rationale for most FF is that they confer health benefits or benefits that respect well-being by improving physical or mental performance. Two dimensions of this question are:

- the efficacy of the FF in meeting health (or other) claims
- safety in its use.

In both cases it appears impossible to generalize, and bodies like the ACNFP are charged with examining each product on a case-by-case basis. Where long-term use has permitted collection of extensive data, much statistical evidence for efficacy and safety has been obtained. For example, evidence based on biomarkers for both probiotics and phytosterol-enriched spreads (Box 11.3) appears to indicate that they are safe and efficacious. However, for the vast majority of individual users, who will not (wish to) be involved in **post-marketing surveillance studies**, there can be no guarantee of significant benefit.

In some cases, the possibility exists of adverse side effects - a risk which, of course, attaches to the consumption of almost *any* food. For example, concerns have been expressed that consumption of phytosterol-enriched spreads may reduce the absorption of fat-soluble vitamins, amounting to 25% lower blood carotene levels in some studies.[30] This may not cause a problem for most people in DC, but it represents a theoretical constraint on use of the FF. Beyond this specific case there are numerous unknowns which could reduce efficacy or increase risks. How does the bioactivity of the FF change with storage? Is there a possibility of over-dosing if several FF, each with the same physiological effect, are consumed? Do specific states (such as pregnancy, old age, or illness) increase the risks of adverse effects? It is inevitable that for any new FF on the market relatively few data will be available on which to base answers to such questions.

Well-being might also be affected by FF in ways less to do with nutrition, and more with the conviviality and social bonding that is associated with meals. The shared experience of meals with friends or family has much more significance than simply feeding people efficiently.[31] But this could well be undermined if replaced by individually prescribed diets based on FF.

11.6.2 Consumer autonomy

In the context of food purchases this principle is often expressed as **consumer sovereignty**, which has been claimed to comprise: **capability** (to understand the claims made for the product and any risks); **choice** of goods (provided by market competition); and **information** (sufficient to judge how expectations will be satisfied).[32] Thus, consumers should be able to decide, on the evidence presented, whether to purchase and/or consume any particular FF. Ensuring that this is so entails the establishment of several procedures. For example:

- reliable pre-market modelling studies are required, to eliminate any product which fails to meet efficacy or safety criteria
- health (or other) claims need to be approved by the appropriate authorities (e.g. in the UK, according to the JHCI code)
- labelling and advertising of the FF need to conform to advertising standards (e.g. as enforced by the Advertising Standards Authority).

It might be argued that such provisions are excessive - especially if the purchaser/consumer is regarded as a 'free agent', with the right to make personal choices irrespective of other people's opinion. The nineteenth-century philosopher J S Mill (Box 2.1) argued that: *'The only purpose for which power can rightfully be exercised over any member of a civilised community, against his will, is to prevent harm to others. His own good, either physical or moral, is not a sufficient warrant.'*[33] But is Mill's principle now applicable? Certainly, modern governments often overrule it, when, for example, education is made compulsory for all school-age children, or we are compelled to wear seat belts, or drink fluoridated water. On similar grounds, it might well be argued that it is right to prevent people from choosing to consume food about which there are serious doubts on grounds of safety and efficacy, and to encourage people to consume food that confers health benefits.

But even if such information is made available - and fully understood, which could be more problematical - consumer decisions will only be sound if there is a much broader understanding of the relative virtues of the FF and *alternative* ways of achieving the same benefit. Conceivably, government advice to *'consume five portions of fruit per day'* is just as effective in conferring health benefits (because of the content of antioxidants and fibre) as consuming a much more expensive FF. But 'information' is of limited use without 'capability'. Many consumers are unaware of FF, and thus excluded from enjoying their potential benefits. And some, though aware, equate them with *junk food*. It appears that the term 'functional food' is unpopular with both the public and scientists alike.[34]

11.6.3 Consumer fairness

Some FF have a proven record of conferring health benefits, and since much illness is diet-related, a case could be made for those people most at risk, or most in need, having greatest access to them. However, a Swiss study shows that while children and the poor could derive greatest benefit from FF, most sales are to well-educated women aged between 30 and 50 years old - a market targeted by the manufacturers.[35] The explanation is clear: some phytosterol-enriched spreads cost up to five times that of the conventional product, and are thus prohibitively expensive for those on lower incomes.

Does this imply that, in the interests of promoting public health, governments ought to make FF available to needy sections of society at the tax-payers' expense? The proposal might make good sense as a general principle, but justifying it in specific cases seems likely to meet several objections. For example, because FF are usually only one way of addressing health risks, more cost-effective action might be achieved through measures such as health education programmes and providing better sporting facilities.

11.6.4 Functional foods in context

The above analysis has focused on consumer concerns, but a full ethical analysis would need to consider the ethical principles as they affect farmers, the biota and (for some people) the interests of the treated organism. While many people would not consider the intrinsic value of plants (10.6) amounted to a strong argument against using GM crops as FF, the use of GM animals for this purpose (e.g. as bioreactors, see 9.5.3) might well be considered differently.

It is also important in performing ethical evaluations to consider *alternative* courses of action, and not to assume that comparison of the proposed novel approach with the *status quo* exhausts the possibilities. As noted (11.6.3), the prospective benefits of promoting the healthy eating guidelines (11.2) need to be assessed alongside (or possibly as an adjunct) to new technological approaches to dietary improvements in health.

11.6.5 Nutrigenomics

The latest development in nutritional science is nutrigenomics – the study of the relationship between what people eat and how their genes function. Nutrigenomics combines elements of genomics (how genes function) and proteomics (how proteins function) to develop detailed predictions of an individual's susceptibility to develop disease conditions, and the role of diet in suppressing or exacerbating these conditions. It is closely related to the science of **pharmacogenetics**, which may be defined as 'the study of genetic differences between individuals in their response to medicines'.[36]

For example, it is believed that the approximately 500 genes that contribute to the propensity for CHD interact and respond to environmental stimuli as groups and subgroups. These genes may be switched on or off by certain phytochemicals in the diet, such as lycopenes and carotenoids, presenting the possibility of using dietary intervention to prevent the onset of disease or ameliorate its effects.

What are the likely future impacts of nutrigenomics? It is envisaged that by 2007, use of *gene-mapping technology* will enable prediction of an individual's responses to foods (specifically micronutrients such as phytochemicals) based on their genetic profile. Such predictions are then likely to allow individualized therapies and preventive strategies, based on natural foods, supplements and/or FF.

So, by 2010, the information-empowered consumer might be saying: *'I should eat this food with [vitamin] B_{12} because, based on my genetic profile, my body can use B_{12} to turn off the expression of my GSA7 gene, which causes hypercholesterolaemia [excess cholesterol in the blood].'*[37] For centuries, folk wisdom has identified herbs that could act as poisons, potions and panaceas;[38] the science of nutrigenomics now promises to tailor such treatments to individual needs.

11.7 Genetically modified foods from crops in developing countries

Although several second-generation GM crops are designed to add value to foods consumed in DC (11.5), debate over the ethics of GM foods has also achieved prominence in relation to their prospective roles in addressing the problems of global hunger and malnutrition. In 1999, the Nuffield Council on Bioethics stated that *'The moral imperative for making GM crops readily and economically available to developing countries who want them is compelling'*, and recommended a major increase in financial support for appropriate GM crop research in LDC. Moreover, Nuffield suggested that this objective was capable of securing the 'high moral ground' for GM crops, for '*"More food for the*

BOX 11.4 THE GREEN REVOLUTION

- The introduction of conventionally bred, high-yielding varieties of rice and wheat led to increased food production (estimated at 15% between 1980 and 2000).
- Maximization of yield depended on large-scale, monoculture-based, highly mechanised, intensive systems, which required extensive inputs of agrochemicals.
- In areas benefiting from the Green Revolution, food production increased faster than falling prices, so that small surplus farmers were better off and were able to increase employment.
- However, despite increased yields e.g. in Latin America and South Asia, numbers of hungry people often increased.
- With the emphasis on wheat and rice, poor farmers who grew sorghum, millet, and beans also missed out.
- Moreover, hybrid seed varieties made farmers dependent on seed companies, which raised farm costs and benefited bigger farmers and agrochemical companies.
- Ownership of land became progressively concentrated in the hands of fewer rich farmers, increasing the gap between the rich and poor.
- There have also been adverse environmental impacts, e.g. as biodiversity has been reduced with adoption of modern seed varieties and with high use of agrochemicals.

hungry", unlike "tomatoes with longer shelf-life", is a strong ethical counterweight to set against the concerns of the opponents of GM crops.'[39]

Drawing analogies with the *Green Revolution* of the 1960s (Box 11.4), Nuffield argued that the yield increases which GM crops could deliver were necessary in the face of the global slowdowns in both the growth of yields and the growth of areas devoted to arable production, together with the unlikelihood of any change in the priorities of the major biotechnology companies (which are focused on DC markets). Hence, Nuffield argued that a major increase in support for GM crop research was needed, involving public sector scientists in LDC and international agricultural research institutes. Gordon Conway, President of the Rockefeller Foundation, conscious of the deficiencies of the original Green Revolution, has suggested the need for a new **Doubly Green Revolution**, which stresses conservation as well as productivity, with GM crops making a significant contribution.[40]

Among the strongest critics of the Nuffield report was the NGO Christian Aid, who argued that:

- GM crops were irrelevant to ending hunger
- the new technology put too much power into too few hands
- too little was done to help small farmers grow food in sustainable and organic ways.

Rather than help solve the problem of global hunger, Christian Aid argued that *'GM crops are taking us down a dangerous farm-track creating classic preconditions for hunger and famine.'*[41]

In view of the highly polarized debate and the introduction of several technical developments, Nuffield again reported on the potential role of GM crops in LDC in

2003, stating as their *'main conclusion... that possible costs, benefits, and risks associated with particular GM crops can only be assessed on a case by case basis'.* [42] In addition to the nature of the transgene and the target crop, important factors to be taken into account include:

- the prevalence of specific climatic conditions
- the presence of wild relatives of the crop
- the availability of water for irrigation
- the level of infrastructure in place
- the extent to which commercial fertilizers or pesticides are used
- the proportion of the farm produce which is sold
- the relative proportion of crop production directed to domestic and/or export markets
- access to export markets
- the effects of competition from subsidized agricultural products from DC
- the nature of the regulation of biotechnology.

Taken together, these factors might be thought likely to limit the appropriate use of GM crops in LDC quite substantially.

Nuffield's analysis was informed by several case studies, examples of which are summarized briefly in Box 11.5. Most of the examples cited are at an early developmental stage, so that benefits are often speculative. Nuffield identifies several risks of introducing GM crops, such as possible adverse effects on the labour market, environmental impacts, and the particular hazards associated with GM biopharmaceutical crops which may contaminate other crops or be accidentally eaten as food crops by humans or farm animals. However, the report re-affirms the conclusion of the 1999 report that *'there is an ethical obligation to explore the potential benefits [of GM crops] responsibly, in order to contribute to the reduction of poverty, and to improve food security and profitable agriculture in developing countries'.* [43] This conclusion, applied particularly to Golden Rice, is discussed below (11.8).

11.7.1 The question of justice

In a world of such vast disparities between the standards of living of the haves and the have-nots (chapter 4), it is crucial to assess the ethics of new technologies in terms of their impacts on global justice. A lucid analysis of the situation has been presented by the Swiss Ethics Committee on Non-Human Gene Technology (ECNH), and the following account summarizes their report.[44] GM technology can be said to impact on four fundamental deontological principles (which are implicit in the ethical matrix):

- the right to food, often called **food security**
- **food sovereignty**, referring to the freedom to make individual choices about what is eaten or grown

BOX 11.5 SOME CASE STUDIES OF GENETICALLY MODIFIED CROPS FOR LESS DEVELOPED COUNTRIES

- *Abiotic stress-resistant rice*. Transfer of a set of genes controlling expression of the sugar trehalose into a variety of Indian rice, which accounts for 80% of rice grown worldwide, allows survival of the plant under prolonged drought conditions. It is estimated that the GM variety has the potential to increase yields under poor conditions by up to 20%.
- *Increased yields of rice by dwarfing*. As demonstrated by the Green Revolution, shorter plants make more nutrients available for grain production. The same strategy has been adopted in rice by introduction of a gene from the common weed *Arabidopsis thaliana*.
- *Improved resistance to viruses in sweet potato*. Yields of this important African subsistence crop are often greatly reduced by viruses and weevils. GM varieties resistant to the feathery mottle virus are currently undergoing field trials, and are anticipated to show 18–25% increases in yield. However, the crop is attacked by three main viruses so that protection may be limited.
- *Improved resistance to diseases in bananas*. Bananas, which are important elements of food security in many LDC, are prey to numerous nematodes, viruses, and fungal diseases: one fungal disease alone (*black sigatoka*) sometimes reduces yields by up to 70%. A current research goal is to sequence the genome of inedible wild bananas which are resistant to black sigtoka, in order to identify the genes conferring resistance and introduce them into the edible bananas.
- *Biopharmaceuticals*. GM crops can be used in the manner of bioreactors (9.5.3) to produce substances such as vaccines. Examples are GM tomatoes modified to produce a vaccine against hepatitis, and GM potatoes expressing vaccines against rotavirus and against the *Escherichia coli* bacterium, which causes diarrhoea.

(From Nuffield, 2003)

- the obligation to provide a sustainable lifestyle (of which protection of **biodiversity** is an integral element)
- the right to **social peace** (entailing fair rules to resolve conflicts and ensure co-existence of different viewpoints).

Each of these is discussed briefly, in turn, below.

Food security It seems clear that GM technology has the potential to improve the yields and quality of conventional crops, and could thus play a key role in promoting food security in LDC. However, the fact that GM products emerging from the market are largely the result of privately funded research conducted by a small number of DC multinational seed companies means that research priorities have not, to date, been directed to LDC needs.

The belief that the problems of malnutrition can be solved by a *technological fix* would seem to be misguided, and sometimes the potential role of GM has been exaggerated. This is because food security depends on the interaction of a number of factors, such as political circumstances, the social and commercial infrastructure, the mechanisms by which communities function, and global economic developments. Moreover, non-GM approaches can often yield greater benefits. For example, studies in Africa, Asia, and

Latin America showed that substantial increases in yield per hectare (100% for millet, 5–30% for irrigated rice, and 20–200% for maize) were achieved by switching to integrated farming systems. Consequently, the ECNH believes that it is currently impossible to assess the contribution GM crops could make to food security, and that *'it would be wrong to promote gene technology research alone as a means of ensuring sufficient food and a healthy diet'*.

Food sovereignty The principle is relevant at different levels, such as: (*a*) the ability of countries or communities to regulate food supply according to choice; (*b*) the right of an individual to decide what to eat; and (*c*) the farmer's right to make independent decisions on what to grow and market.

However, the patenting of GM seeds by multinational seed companies has serious consequences for LDC farmers. *'If farmers are not fully free to make their own choice of seed, this constitutes a dependency and hence a constraint on food sovereignty. The more patents are granted to quasi-monopolies, the greater the dependency becomes.'* The ECNH suggests that LDC have good reasons to set their own safety standards, because different farming methods, climates, and ecological conditions result in context-specific risks. Moreover, the close relationship which exists between food, nutrition, and personality (i.e. the anthropological significance of food), rather than the special nature of the GM technology, dictates that monopoly situations should be avoided. To respect this principle, farmers' and breeders' privileges should be fully guaranteed to safeguard against unfair monopolies.

Biodiversity A decline in biodiversity in LDC has been a serious concern for many years, and has been exacerbated by industrialization and Green Revolution technology (Box 11.4) independently of any possible impacts of GM crops. GM crops thus need to be assessed, among other things, according to the extent to which they help or hinder sustainable biodiversity.

But recent history has shown that when there has been an expansion of industrial agricultural production in LDC, the associated increase in monocultures is linked to additional genetic erosion. There are growing signs that reduced agro-biodiversity is making arable agriculture vulnerable to both climatic factors and pests, and such concerns are given weight by the fact that only a few commercial crops have so far attracted the interest of GM seed companies. Although it is difficult at present to assess the extent to which GM crops have contributed to reduced biodiversity, the majority of ECNH members consider that the current use of GM crops in LDC is contributing to reduced biodiversity.

Social peace The conflicts evident in DC between supporters and opponents of GM technology are reflected within LDC societies, and between LDC and DC states. Social peace will only exist when all social groups are willing and able to resolve such conflicts peacefully, and this is of course a necessary condition, because conflicts pose serious threats to economic recovery and sustainability.

A prerequisite for social peace through acceptance of new technologies which affect the whole of society so profoundly is the availability of a broad basis of information (of both the pros and cons entailed) and conditions in which governments earn public trust. But such conditions are undermined in cases where access to information is blocked by corrupt journalists and politicians; or, worse, when conflicts are resolved militarily rather than by consent. In many LDC, agriculture is not just about food production, but is enshrined in cultural norms which often have religious or mythical connotations. According to the ECNH, it would seem to follow that social peace will only be achieved over conflicts involving GM crops if provisions are made to ensure: *'adequate freedom of the press/media; collaboration with cultural and religious bodies; and a decline in corrupt permit practices'.*

11.8 Golden Rice

One GM crop which is being developed to serve the needs of LDC is Golden Rice (GR), which will act as a useful case study to conclude this chapter. This crop, designed to counter vitamin A deficiency (VAD) in LDC, is perhaps the most prominent example of a GM FF. But it has also become a *cause célèbre* – for supporters it embodies the true promise of GM; while for opponents it is viewed as a cynical, and flawed, attempt to claim 'the high moral ground'. GR is transgenic for several genes, which are collectively designed to: (*a*) increase production of beta-carotene (the main precursor of vitamin A); and (*b*) enhance its iron content.[45] The product was so named because the polished rice grains are turned yellow by the beta-carotene.

11.8.1 Vitamin A deficiency

WHO estimates that, globally, VAD affects 100–140 million children, with up to 500,000 becoming blind every year, of whom half die within a year of losing their sight. More than 40% of the children affected live in South and South-East Asia.[46] Although in 1990, the World Summit for Children set a target to *'virtually eliminate VAD and its consequences by 2000'*, little progress has been made. In fact, the overall target of reducing deaths from preventable causes was only reached in five of the fifty-five countries with an *under-5-years mortality rate*,[b] of 100 or more in 1990.[47]

There are three main strategies for reducing VAD:

- **supplementation:** this involves giving high-dose vitamin A capsules twice per year. A level of 70% supplementation was achieved in forty-three countries in 1999, compared with only eleven in 1996
- **fortification:** consists of adding vitamin A (often together with iron and iodine) to foods (such as fats, oils, sugar, cereal flours, maize, milk and dairy products) before sale
- **dietary diversification:** because meat, milk and dairy products, and leafy green vegetables are rich sources of vitamin A, encouragement of livestock farming and horticulture is likely to be a sound strategy.[48]

GR is a prospective fourth strategy for tackling VAD, which may be classed as **bio-fortification**. Development of GR, by Ingo Potrykus, Peter Beyer, and colleagues, has been carried out in the public sector at the Swiss Federal Institute of Technology in

[b] The *under-5-years mortality rate* is expressed 'per 1000 live births'.

Zurich. Failing initially to obtain commercial support, the project won financial backing from the Rockefeller Foundation, the EU and the Swiss government. Although the research team had intended the project to serve the *public good*, especially the poor in LDC, it came as a surprise to discover that about seventy patents had been used in the initial development process, which would have incurred the payment of excessive royalties. To overcome this problem the research team transferred exclusive rights for GR to the biotechnology company Syngenta, so that the technology immediately became a *private good*. Other biotechnology companies are also supporting the product, e.g. Monsanto is providing royalty-free licences for all technology that can help further development of GR.

Under the terms of the deal, the GR project is legally divided into two strands. In the *commercial strand*, Syngenta hope to sell GR (principally in DC) as a premium FF, based on the antioxidant properties of beta-carotene - which is believed to have anti-cancer properties. In the other strand, a *humanitarian board* will guide development of GR as a 'pro-poor' technology, being distributed free of charge to LDC farmers earning less than US$10,000 p.a. from the rice. Syngenta claims that *'GR has the potential to provide massive benefit countering vitamin A deficiency-related diseases'.*[49]

11.8.2 An ethical analysis of Golden Rice

An ethical analysis of GR is usefully structured by the ethical matrix (Table 10.1). This section aims to highlight only some important features pertaining to consumers' well-being, autonomy and fairness. Other issues are similar to those identified in 10.3–10.6, which would usefully be amplified by consideration of the ECNH report (11.7.1). But it needs to be appreciated that the usual sharp distinction between producer and consumer, justifiably assumed for DC, is often invalid in LDC, where many small farmers grow food for their own consumption.

Respect for consumers' well-being concerns the benefits and costs of GR, the former principally in terms of health benefits (in relieving VAD or reducing cancer risks) and the latter encompassing any health risks associated with GR use together with financial costs. The health benefits in LDC have been hotly contested, and depend on various contrasting assumptions which are made. Thus, the developers of GR claim that only 100 g GR would supply the full daily requirement of Vitamin A, whereas Greenpeace claims that 3 kg would be required. Nuffield notes that questions about the efficacy and efficiency of GR require further scientific research, and recommends that *'reliable empirical data from nutritional and bioavailability studies be obtained as a priority'.*[50] So large is the discrepancy between the two estimates that it seems impossible to assess the nutritional impact of GR in the absence of more details on its efficacy in preventing VAD, its cost, and the level of consumer acceptability (because the yellow colour could either be an attraction or a deterrent).[c]

In purely utilitarian terms, it would also be important to assess the relative benefits and costs of the three alternative strategies (11.8.1). Such assessments are intrinsically

[c] It was reported in early 2005 that a new strain of GR contained twenty times as much beta-carotene as its predecessor.

complex, and might prove to be highly dependent on speculative economic projections. For such reasons, the UK government was unable to perform any definitive financial cost–benefit analyses of GM crops which might be grown in the UK (10.3.1).

The prospective health benefits of GR when used as an FF in DC are also unpredictable. Essentially, this is because, as in countering VAD, the benefits and efficacy of GR and of alternative dietary strategies are very difficult to quantify.

Respect for consumers' autonomy centres on the informed choices that consumers are able to make concerning GR and alternatives. While accurate and informative labelling might be considered crucial, it may not be sufficient, or always easy to achieve, even with the best intentions. Labelling should be easy to understand by the least informed potential consumer (i.e. the least educated in nutritional and health matters), but this task can often become a compromise between accuracy and brevity, especially when poorly understood quantitative factors are critically significant.

A significant risk is that consumers might be deceived into paying more for a perceived, but unrealistic, health benefit – which might be better, and/or more economically, delivered by an alternative food or habit. The case for GR rests on the claim that it provides a cheap, effective means of countering VAD, which is not likely to be achieved otherwise.

Respect for consumer fairness is usually translated as *affordability*, whereby an injustice is perpetrated if a health-giving food is unavailable to those too poor to buy it. The fact that this is a common situation in LDC in no way mitigates the impacts of infringements of the principle in this case. As noted, where producer and consumer are the same person calculation of costs and benefits becomes more complicated; they are even more complex when attempts are made to factor in research and development costs for GR in comparison with other VAD strategies.

For example, in a comparison of the cost-effectiveness in supplying vitamin A by the three strategies listed in 11.8.1, it was reported that *'where combined with fortification of other foods'* GR would be the cheapest strategy over the period 2007–2116.[51] However, these estimates were based on projections of *future* spending on research, breeding, and advertising, and did not count the irredeemable US$2 million *already* spent in development of GR. If the latter (so-called **sunk costs**) are included in the calculation, GR is no longer a cheaper option than food fortification.[52] So if the calculation had been performed ten years earlier, it would have been more rational to opt for food fortification as the best strategy to address VAD.

11.8.3 Public or private goods?

One of the more problematical aspects of modern biotechnology is that it is largely controlled by a very few, very wealthy companies – and this has ethical implications for both consumers and farmers. Up to twenty-five years ago much research in the bio-sciences was carried out in universities and government research institutes, which were largely funded from the public purse via taxes. But the increasing costs of research due to an exponential increase in scientific activity (14.1.1) meant that governments were forced to limit their research budgets in order to meet other pressing demands, such as health, education, and social services.

From the mid-1970s, private industry has played an increasing role in funding research, so that commercial objectives have become an important feature of research in biotechnology. Because most of the expertise was originally, and still to a significant extant remains, in universities, new relationships were forged between the academic and commercial worlds, giving rise to the present-day *university–industrial complex* based on biotechnology.[53] But the fact that, traditionally, the ethos of academic research has been markedly different from that of industrial research (15.2.1) means that tensions have arisen over the right way to manage this relationship. They are well illustrated by the case of GR.

Proponents of GR, such as Nuffield and the Rockefeller Foundation, argue that public research institutions need to enter partnerships with industry (so called **public–private partnerships**) because the poor will not otherwise receive benefit: public funding would simply not be adequate. But they also express concern that the privatization of research will have a detrimental effect on public research by starving the latter of resources.

Patents (see Box 4.2) are a major impediment to research for the public good, because they amount to a legal tool to make knowledge act like private property by granting the licence holder an exclusive monopoly, on the basis that this will stimulate inventions in the interests of the greater 'public good'. The discovery that development of GR had involved about seventy patents (11.8.1) led Potrykus and Beyer to independently sign a contract with Syngenta in return for exclusive monopoly control in DC, while allowing free access to LDC farmers.

Critics of the GR project argue that even if public–private partnerships do produce some beneficial pro-poor biotechnologies, it is questionable whether they will offset the anti-poor effects of other GM crops that are developed commercially. Critics also suspect that under the *doubly-green theme* (11.7), so-called 'pro-poor' GM crops are simply being used as *loss-leaders* for other products which will sell at a premium - such as the GR to be marketed in DC as an anti-cancer FF.[54]

Government views. However, the doubly-green theme is currently strongly supported by most Western governments. Since the 1970s, global commercial investment in biotechnology research has accounted for 80% of the total spending, and governments have also promoted it as a means of pump-priming commercial activity. Economic realities may thus suggest that pro-poor technology will only happen at all if it hangs onto the coat tails of commercially oriented Western technology. This is a view represented by the argument that the EU is behaving selfishly by resisting GM foods they can afford to do without, so depriving LDC of an essential lifeline.

THE MAIN POINTS

- Poor diets are responsible for much illness and premature death in DC, with obesity being strongly related to heart disease, diabetes, cancer and digestive disorders.
- Functional foods, subject to novel foods regulations, are designed to enhance body or mental function and/or reduce risks of disease.
- In a highly competitive market, food manufacturers consider that functional foods present important opportunities, especially when advances in nutrigenomics allow the formulation of personalized diets.
- Ethical issues for functional foods focus on impacts on consumers in terms of respect for: safe and effective use, responsible marketing (e.g. avoiding deceptive health claims), and affordability, especially for the most needy.
- Golden Rice is a prominent GM functional food, which it is claimed could relieve vitamin A deficiency in LDC and provide protection from cancer in DC. Critics argue that alternative strategies are more effective.

EXERCISES

These can form the basis of essays or group discussions:

1. Assess the claim that the introduction of GM novel foods that confer real benefits on consumers is a sound strategy for promoting the acceptance of modern biotechnology (11.1).
2. Discuss the pros and cons of stricter controls on the advertising of foods alleged to contribute to diet-related health disorders when consumed in excess (11.2).
3. Discuss the arguments advanced for the introduction of functional foods as an effective way to improve public health (11.3).
4. Assess the validity of the arguments advanced by: (*a*) the Nuffield Council and (*b*) the Swiss Ethics Committee on Non-Human Gene Technology concerning the use of GM crops in LDC (11.7).
5. Evaluate the arguments advanced for and against the use of Golden Rice as a strategy for ameliorating vitamin A deficiency in LDC (11.8).

FURTHER READING

- *The Functional Foods Revolution: healthy people, healthy profits?* by Michael Heasman and Julian Mellentin (2001). London, Earthscan. A readable account of recent developments, but without mention of GM foods.
- *The Use of Genetically Modified Crops in Developing Countries*. Nuffield Council on Bioethics (2003). London, NCB. Argues a moral case for GM technology in LDC.

- *Engineering Nutrition: GM crops for global justice?* Food Ethics Council (2003). Brighton, FEC. A critique of the arguments advanced by Nuffield.

USEFUL WEBSITES

- **http://www.foodstandards.gov.uk/healthiereating/** *Food Standards Agency*: the official UK government site on food standards.
- **http://www.jhci.org.uk/about.htm** *Joint Health Claims Initiative* (2004): explains the UK Code of Practice on Health Claims.
- **http://www.onepersongenetics.com/articles.html** *Institute for the Future* (2003): a useful perspective on the future direction of food science and technology.

Notes

1. McKie R (2000) Why the West must swallow GM foods. The Observer, 23 January 2000
2. Mepham B (2001) Novel Foods. In: The Concise Encyclopedia of the Ethics of New Technologies. (Ed.) Chadwick R. San Diego, Academic Press, pp. 299–313
3. Brillat-Savarin J-A (1970) [1825] The Physiology of Taste. London, Penguin
4. Drummond J C and Wilbraham A (1957) The Englishman's Food (revised edition). London, Jonathan Cape
5. Blears H (2003) Speech at Westminster Health Forum, 3 April 2003. http://www.doh.gov.uk
6. Chadwick R, Henson S, Moseley B, Koenen G, Liakopoulos M, Midden C, Palou A, Rechkemmer G, Schröder D, and von Wright A (2003) Functional Foods. Berlin, Springer-Verlag, p. 150
7. British Medical Association (2003) Adolescent Health. London, BMA, pp. 4–6
8. Ibid., pp. 4–6
9. Nestle M (2002) Food Politics. Berkeley, University of California Press
10. Sustain (2003) TV dinners. http://www.sustainweb.org
11. Food Ethics Council (2001) After FMD: aiming for a values-driven agriculture. Southwell, FEC
12. Gofton L (1996) Bread to biotechnology: cultural aspects of food ethics. In: Food Ethics. (Ed.) Mepham B. London, Routledge, p. 130
13. Heasman M and Mellentin J (2001) The Functional Foods Revolution. London, Earthscan, pp. 66–77
14. Hoogendijk W (1991) The Economic Revolution. London, Green Print
15. Heasman M and Mellentin J (2001) The Function Foods Revolution. London, Earthscan, p. 226
16. Chadwick R, Henson S, Moseley B, Koenen G, Liakopoulos M, Midden C, Palou A, Rechkemmer G, Schröder D, and von Wright A (2003) Functional Foods. Berlin, Springer-Verlag, p. 62
17. Ibid., p. 127
18. Joint Health Claims Initiative (2004) http://www.jhci.org.uk/about.htm

19. Chadwick R, Henson S, Moseley B, Koenen G, Liakopoulos M, Midden C, Palou A, Rechkemmer G, Schröder D, and von Wright A (2003) Functional Foods. Berlin, Springer-Verlag, p. 131
20. Advisory Committee on Novel Foods and Processes (2003) Annual Report 2002, London, Food Standards Agency
21. National Council for Agricultural Research (Netherlands) (2000) Functional Foods. The Hague, NRLO, p. 15
22. Atkins P and Bowler I (2001). Food in Society: economy, culture, geography. New York, Arnold, pp. 5–7
23. Dibb S and Mayer S (2000) Biotech – The Next Generation. London, The Food Commission/ Buxton, Genewatch UK, p. 8
24. Nestle M (2002) Food Politics. Berkeley, University of California Press, pp. 334–336
25. Heasman M and Mellentin J (2001) The Functional Foods Revolution. London, Earthscan, p. 92
26. Ibid., p. 150
27. Ibid., pp. 134–137
28. Ibid., p. 147
29. Dibb S and Mayer S (2000) Biotech – The Next Generation. London, The Food Commission/ Buxton, Genewatch UK, p. 25
30. Chadwick R, Henson S, Moseley B, Koenen G, Liakopoulos M, Midden C, Palou A, Rechkemmer G, Schröder D, and von Wright A (2003) Functional Foods. Berlin, Springer-Verlag, p. 196
31. Telfer E (1996) Food for Thought: philosophy and food. London, Routledge
32. Hamilton R (1996) Consumer sovereignty in food marketing. In: Food Ethics. (Ed.) Mepham B. London, Routledge, p. 140
33. Mill J S (1910) Utilitarianism, Liberty and Representative Government. London, J M Dent, p. 73
34. Chadwick R, Henson S, Moseley B, Koenen G, Liakopoulos M, Midden C, Palou A, Rechkemmer G, Schröder D, and von Wright A (2003) Functional Foods. Berlin, Springer-Verlag, p. 96
35. Ibid., p. 154
36. Nuffield Council on Bioethics (2003) Pharmacogenetics: ethical issues. London, NCB
37. Institute for the Future (2003) The nutrigenomics revolution. http://www.onepersongenetics.com/articles.html
38. Stockwell C (1988) Nature's Pharmacy. London, Century
39. Nuffield Council on Bioethics (1999) Genetically Modified Crops: the ethical and social issues. London, NCB, p. xv
40. Conway G (1997) The Doubly Green Revolution. London, Penguin
41. Christian Aid (1999) Selling Suicide. London, Christian Aid
42. Nuffield Council on Bioethics (2003) The Use of Genetically Modified Crops in Developing Countries. London, NCB, p. xiv
43. Ibid.
44. Swiss Ethics Committee on Non-Human Gene Technology (2004) Gene Technology and Developing Countries. Berne, ECNH
45. Dibb S and Mayer S (2000) Biotech – The Next Generation. London, The Food Commission/ Buxton, Genewatch UK, pp. 11–13
46. UN Food and Agriculture Organization (2003) The State of Food Insecurity in the World. Rome, FAO
47. Black R E, Morris S S, and Bryce J (2003) Where and why are ten million children dying every year? The Lancet *361*, 2226–2234
48. Food Ethics Council (2003) Engineering Nutrition: GM crops for global justice? Brighton, FEC, pp. 14–15

49. Syngenta (2000) Golden rice background. www.whybiotech.com.html/con678.html. (accessed 5 January 2004)

50. Nuffield Council of Bioethics (2003) The Use of Genetically Modified Crops in Developing Countries. London, NCB, p. 53

51. Dawe D, Robertson R, and Unnevehr L (2002) Golden rice: what role it could play in alleviation of vitamin A deficiency. Food Policy *27*, 541–560

52. Food Ethics Council (2003) Engineering Nutrition: GM crops for global justice? Brighton, FEC, pp. 17

53. Kenney M (1986) Biotechnology: the university industrial complex. New Haven: Yale University Press

54. Food Ethics Council (2003) Engineering Nutrition: GM crops for global justice? Brighton, FEC, pp. 19–21

12

Environmental sustainability

OBJECTIVES

When you have read and discussed this chapter you should:

- appreciate the distinction between instrumental (anthropocentric) approaches to the environment and biocentric and ecocentric approaches
- understand the nature and scale of anthropogenic effects on the biosphere, e.g. as reduced biodiversity and environmental pollution
- appreciate the concept of sustainability (and its limitations) and the ways sustainability is affected by different forms of agriculture
- be aware of the discussions concerning the *limits to growth* and international debates over issues such as global climate change
- be able to form an ethical position on ways of achieving global environmental sustainability

12.1 Introduction

Most chapters in this book attempt to deal concisely with a wide range of issues, but this might be considered the most ambitious of all in addressing the 'big picture'. This is because the remit, both in space (nothing less than the biosphere as a whole) and time (because sustainability encompasses the whole of the foreseeable future), is so extensive, and because so many relevant facts are uncertain. We are essentially concerned with issues of ecology, which may be defined as *'the totality and pattern of relations between organisms and the environment'.*[1] Our daunting task is to consider how we should act, individually and collectively, in the light of what we know about the likely future of life on Earth. However, in order to examine the issues from different perspectives, some of the discussion will focus on global concerns and some on matters specifically relating to the UK.

12.1.1 Environmental ethics

We might usefully start with some definitions. The two key areas of concern are environmental ethics and sustainability.

- Environmental ethics *'debates how to balance the claims of the present and the future, human and non-human, sentient and non-sentient, individuals and wholes. It investigates the prospects for a sustainable relationship between economic and ecological systems, and pursues the implications of this relationship with respect to social justice and political institutions.'*[2]

- The meaning of sustainability is perhaps too contested to provide a simple definition, but when applied to the human population, it can be said, as a working hypothesis, to *'involve the persistence through time of the diversity of human communities and ethical ideals of human flourishing, the dynamically balanced development of economic enterprise, and the preservation and regeneration of ecological systems and resources that sustain that development'.*[3]

The science of ecology and more general concerns for the environment, both locally and as the biosphere, are relatively recent developments. They have emerged outside mainstream biology because, as holistic approaches, they owe little to the dominant reductionist programmes of most biosciences (typified by experimental physiology; 8.2). The reasons for their late arrival are twofold, which might be described as *attitudinal* and *practical*.

The changed attitudes have been explained as a weakening of the ancient view, evident in the works of Aristotle and the Bible, that God had provided the natural world expressly for human use. According to philosopher John Passmore, *'Christianity encouraged certain special attitudes to nature: that it exists primarily as a resource . . . that man has a right to use it as he will, that it is not sacred, that man's relationships with it are not governed by moral principles.'*[4] But just as the view that animals had been created solely to serve human ends (7.1.1) became increasingly untenable with a growing appreciation of Darwinism, so attitudes to the natural world have become encompassed by an expanding circle of moral awareness, as slaves, women, and animals – all previously excluded from the moral community – have, progressively, been granted ethical standing (3.4). (However, challenging Passmore, an alternative view rejects this analysis and considers that stewardship is a more appropriate interpretation of Christian scripture.)[5]

In parallel with such changes, the practical concerns have become increasingly recognized, as adverse effects of human activity (**anthropogenic effects**) on the wholesomeness, viability and resilience of our environment have become more evident. A major problem with discussing the environment is that its sheer size makes it difficult to appreciate global consequences, when each of us only experiences a very small sector. Younger people are also limited by lack of direct experience of 'what has been lost', so that what their elders often consider a marked deterioration might pass as 'the norm' for the younger generation. Even so, few educated people can be unaware of the serious official concerns over issues like the loss of species diversity, climate change, the effect of chemical pollutants on living systems, and the uneven distribution of the Earth's resources between the world's people. The main aim of this chapter is to examine the basis of these concerns, and proposals for how they might be addressed ethically.

However, we need first to consider some rather more philosophical questions. Little would be gained by rushing into solutions to problems without first seriously examining what the problems are – or, at least, surveying the different ideas held by those who have thought deeply about their nature (Box 12.1).

BOX 12.1 INFLUENTIAL FIGURES IN ENVIRONMENTAL ETHICS

- **Aldo Leopold** (1887–1948): US conservationist and pioneer environmental ethicist. In *Sand County Almanac* he claimed 'a thing is right when it tends to preserve the integrity, stability, and beauty of the biotic community. It is wrong when it tends otherwise.'
- **Rachel Carson** (1907–1964): US marine biologist. In *Silent Spring* she questioned the use of chemical pesticides in agriculture and aroused worldwide concern over their impacts on the environment.
- **Fritz Schumacher** (1911–1977): German-born economist who settled in the UK and originated the concept of Intermediate Technology for LDC. In *Small is Beautiful* he claimed that pursuit of profit and progress had resulted in gross inefficiency, environmental pollution and inhumane working conditions.
- **Arne Naess** (1912–): Norwegian philosopher who founded the *deep ecology movement*, highlighting concerns for the Third World, the more distant human future, and the needs of non-human beings.
- **James Lovelock** (1919–): British scientist and inventor who in 1979 introduced the concept of *Gaia*, which conceives of the Earth as akin to a living organism.

12.2 Environmental values

One important question concerns the nature of the value we attach to the 'natural world'. We have already noted the distinction between instrumental and intrinsic value (3.6.3), especially as it relates to animals (7.2.3). Parallel considerations also apply to the environment, both in terms of its living (biotic) and its abiotic components. For example, in the case of plants, we can assign:

- **instrumental value** to them for their usefulness to us, or
- intrinsic value to them as individual organisms (**biocentrism**), or
- intrinsic value to the species or ecosystem to which they belong, including the abiotic components (**ecocentrism**).

Differences of worldview often have significant implications for the ways it is considered right to treat non-human beings. The most ecocentric attitude in Kockelkoren's scheme (10.6) is named *participant*, indicating the belief that humans and non-humans are both integral to a dynamic, mutually beneficial relationship. The most anthropocentric attitude is that of the *dominator*. There is clearly no objective way of deciding between these options – neither can be said to be 'incorrect' – but the decision each of us makes on the matter is often based on deep inner convictions, perhaps influenced by tacit knowledge (1.4.4).

We can clarify the issue by confining attention to plants, assumed here to be non-sentient beings, while acknowledging that we can never be absolutely certain at what level of development sentience first appears. Everyone would agree that plants have instrumental value in numerous ways. For example, they act as sources of food, materials, and combustible material; they collectively provide habitats for animals (e.g. insects and birds) as well as acting as a carbon sink; and they appeal, individually and collectively, to our aesthetic sensibilities (as flowers, fields and forests). But is this the only type of value plants possess? Certainly, all such values are compatible with an anthropocentric view that locates their value entirely in human benefits. According to this view, if the Earth was depopulated of people, no single plant would have any value.

12.2.1 Biocentrism and ecocentrism

One challenge to this view is biocentrism, which claims that plants have intrinsic value, and because each plant has a 'good of its own' it makes a call on our ethical duties, requiring us to treat it with respect for its well-being, and not just as a 'thing'. Putting it starkly, an instrumentalist might derive ethically acceptable amusement from mirthfully cutting the heads off a bed of roses in full bloom, but a biocentrist would consider the act deeply unethical because it would offend the plants' flourishment and intrinsic value (10.6).

In contrast, ecocentrists locate intrinsic value not in individual plants but in biosystems, e.g. as species or ecosystems. According to philosopher Lawrence Johnson, because species and ecosystems are both characterized by a persistent state of low entropy, self-regulation via homeostatic control mechanisms, organic unity, and self-identity, it follows that they have interests, not just in surviving, but in flourishing.[6] Philosopher Holmes Rolston III similarly ascribes 'value in itself' to ecosystems, regarding species extinction as a form of 'super-killing'.[7] A claimed advantage of ecocentrism over biocentrism is that it does not depend on a sharp distinction between biotic and abiotic elements of ecosystems, which cannot readily be drawn. It might also be argued that ecocentrism is not an alternative, but additional, to biocentrism. Perhaps the richest account of the ecocentric view is the concept of *deep ecology* developed by Arne Naess (Box 12.1).[8]

Both biocentric and ecocentric approaches have important implications for attitudes to wild animals and plants. For some people, nature should be left to itself – thereby drawing a sharp distinction between nature and culture. For others, human intervention is required, e.g. to preserve threatened species or control wild populations, especially where imbalances are considered a result of anthropogenic activity.[9] In any event, both biocentrism and ecocentrism demand a major change in the way humanity customarily perceives itself and its relationship to the non-human world. Such considerations are reflected in the amendment to the Swiss Federal Constitution, which now requires, with respect to proposed genetic modifications, that account be taken of the *'dignity and integrity of living beings'* – a condition that applies to both animals and plants.[10]

As observed by philosopher Kate Rawles, *'Anthropocentric approaches to environmental concerns ultimately leave the non-human world hostage to changing human needs and interests. Were humans to (lose) interest in it as a source of wonder, beauty, information, recreation... anthropocentric environmentalists would have reason to conserve only those parts of the non-human world needed for human survival in a literal sense.'*[11] It is when we consider the anthropogenic effects of industrial activity on the global environment over the last 200 years that the short-sightedness of that approach becomes evident.

12.3 Anthropogenic impacts on the biosphere

Life existed for 3.5 billion years on Earth without humans. The basic **biometabolism** of all life (consisting of inputs of food, water, and oxygen, and the discharge of organic wastes) began to be augmented by what might be called **technometabolism** with the emergence of human culture, approximately 65,000 human generations ago (HGA). But during the vast majority of the latter period humans lived in a state of 'aboriginal sustainability', modifying the environment only with the aid of hand tools and selective use of fire. Agriculture emerged about 400 HGA, but with the appearance of cities about 200 HGA there was a marked increase in natural resource use; while the greatest step towards **transitional unsustainability** occurred only about 8 HGA. The Industrial Revolution, which began in mid-eighteenth-century Britain, introduced technologies powered by hydrocarbon fossil fuels (coal and oil) and increased energy use by a factor of 10,000.[12]

The exponential growth of the human population that was made possible by increased industrial and agricultural output, together with increased lifespans, has not, however, been remotely uniform across the world (chapter 4). While there is a strong ethical case for international action to improve the lives of the *have-nots*, the ambitions of most *haves* to obtain yet more material possessions seems undiminished (4.5.1). The impetus for growth is thus fuelled both by the desperation of the poor and by the acquisitiveness of the rich. Yet, as noted by Hermann Daly and John Cobb: '*Further growth beyond the present scale is overwhelmingly likely to increase costs more rapidly than it increases benefits, thus ushering in a new era of "uneconomic growth" that impoverishes rather than enriches.*'[13] What are the reasons for this pessimism? The following sections identify some of the principal concerns.

12.3.1 The birth of capitalism

Before the Industrial Revolution, energy was largely derived from biotic sources which had transformed solar radiation. The people who had real power were thus those who owned large land areas (trapping solar energy in crops for food and trees for fuel) and/or could command the muscle power of large numbers of people and/or animals (as could monarchs and aristocrats). Changes in the social order were largely achieved through military conflict or conquest, and sanctions formulated by religious leaders often imposed a strict order on societies.

With the exploitation of fossil fuels, power accrued instead to those who had oil wells, coal mines or other valued resources on their land, or could add value to these natural resources through manufacturing. Thus capitalism was born. In the eighteenth century, philosopher Adam Smith laid down the principles of capitalism,[14] and claimed to show how it was possible to pursue private gain in ways that would further the interests, not just of the individual, but also of society. In a now famous phrase, Smith argued that the combination of self-interest, private property and competition among sellers in markets would lead producers *'as by an invisible hand'* to an end that they did not foresee – the well-being of society as a whole.

However, twenty-first-century capitalism operates in a very different social climate. Many commercial companies are now very large international corporations, who trade globally and by vertical integration of their operations exert considerable power over both primary producers and consumers. In a highly competitive market, profitability is often greatly influenced by success in advertising and other forms of product promotion. Automation has displaced many people from their jobs on the factory floor, and information technology has internationalized markets and labour forces.

Because a central tenet of capitalism is growth, there is a built-in drive to secure an increased share of the market by expanding output while cutting unit costs. Coupled with increased travel for business and recreation, this has resulted in massive increases in energy consumption in DC over recent years. The two principal adverse consequences are the inequitable depletion of non-renewable reserves, and the environmental pollution resulting from their combustion: i.e. there are problems with both **sources** and **sinks**. Moreover, it would be naïve to assume that because of recent discoveries of new reserves of fossil fuels we need not be concerned about 'running out of oil'. The discovery, production and processing stages all incur costs which it may prove uneconomic to pursue.

Industrial activity has had several adverse effects on the biosphere, which are considered below under two headings: biodiversity (12.3.2) and pollution (12.3.3). These correspond to the specifications of two principles in the ethical matrix (see 3.5), namely, respect for biotic autonomy (preservation of biodiversity) and biotic well-being (protection from pollution), while respect for biotic fairness is discussed in terms of sustainability (12.4). However, there is clearly much overlap between these principles.

12.3.2 **Biodiversity**

Biodiversity concerns the total variety of life forms on Earth, encompassing diversity at the level of genes, species and ecosystems. The principal focus here is species diversity. About 1.5 million species of plants and animals are known to biologists, which is estimated to be only about one-tenth of the total number existing. Striking facts are that a single oak tree may be home to 400 species of insect, while a single tropical forest tree may house 1000 species.[15]

Over the last 500 million years there have been five major species extinctions, so that probably 99% of all species that have ever existed are now extinct. These losses have largely been made good. Over the last 100 million years, species formation, following the changes in continental land masses, and the emergence of the tropical rain forests have been important factors. But species extinction is currently proceeding rapidly due to anthropogenic activities. Before the appearance of humans, one species became extinct every 1000 years. Now, the *annual rate* of extinction is estimated to be between 1000 and 27,000.[16] A study reported in 2004 predicted that by 2050, more than *one million* species of plants and animals (i.e. 15–37% of species in the sample studied) will be lost,[17] a discovery the research leader, ecologist Chris Thomas, described as *'terrifying'*.[18]

There are five main causes:[19]

- *hunting and harvesting*: e.g. overhunting of animals for their fur, shells, bones, tusks, and skins
- *the pressure of population growth*: illustrated by the decreased time taken to double the global population, from 150 years in the eighteenth century to 35 years now (when more than 6 billion people have food, housing, clothing, medical, and social needs)
- *habitat destruction and fragmentation*: e.g. the conversion of tropical forests into ranches (e.g. to feed the demand for hamburgers), which is especially threatening to the habitats of large land mammals such as puma and jaguar
- *the introduction of exotic species*: which often carry diseases that eliminate indigenous species
- *pollution, and atmospheric and climatic change*: e.g. the adverse consequences of the **greenhouse effect** (12.3.3).

In turn, the greenhouse effect is due to: increased industrial activity; the growth of cities (4.2); and increased transport, by land, air and sea.

Traditionally, economic analysis has failed to take account of the value of biodiversity, and in the drive to increase agricultural productivity much has been lost. For many in LDC the outlook is bleak. For example, a UK Government report notes: *'70% of the world's rural poor rely on livestock, many of which are adapted to local conditions and diseases [but] with a third of breeds threatened by extinction, there is a major risk to the food and financial security of the poorest families.'*[20]

The critical question is whether current practices will precipitate the *sixth* major species extinction in the Earth's history – and hence the end of human life.[21]

12.3.3 Environmental pollution

Industrial activity often produces toxic wastes, which pollute air, land and water, and threaten human and non-human life.

Aerial pollution. Notably, carbon dioxide produced in the combustion of fossil fuels is contributing to the greenhouse effect – which is also caused by gases such as methane and chlorofluorocarbons (CFC). By trapping the Earth's heat, this leads to global warming, which it is anticipated will cause an increase of between 2 and 5°C by the year 2070. This could have serious consequences for public health, agriculture and the area of habitable land, because melting of the polar ice caps is expected to cause widespread flooding elsewhere. At the Milan Global Change Conference in 2003 it was noted that 1998 and 2002 had been the two hottest years on record. Many greenhouse gases reached their highest concentrations ever in 2002; and the ozone hole nearly reached the all-time record size in September 2002.[22] The wider implications are examined by the Intergovernmental Panel on Climate Change (IPCC).[23]

Acid rain (with a pH less than 5.65) is another example of aerial pollution: during one storm in Scotland the rain was pH 2.4 (the acidity of vinegar!). Carbon monoxide depletes scavenger hydroxyl radicals in the atmosphere, which normally react with, and destroy, the greenhouse gas methane.[24]

Chemical pollution. Anthropogenic emissions of lead, cadmium, zinc and copper exceed natural emissions by factors of eighteen, nine, seven and three, respectively.[25] At high levels, pollutants undermine the pollution-absorption mechanisms themselves. For example, run off or dumping of pesticides into water courses can kill the organisms that clean up the organic wastes. Water pollution can be due to sewage, infectious agents, organic and other chemicals and minerals, radioactive substances, and heat. Relatively indestructible substances (such as the insecticide dichlorodiphenyltrichloroethane (DDT)) become magnified in the food pyramid, exposing the ultimate consumers (such as people) to concentrations much greater than those experienced by organisms (like algae) at the base of the pyramid.[26]

Industrial processes have also resulted in the production of a large number of synthetic chemicals. Between 30,000 and 100,000 are on the UK market, with several hundred added annually. But less than 5% are specifically approved for uses such as food additives, pesticides or pharmaceuticals – and the rest can be used freely unless specifically prohibited. In 2003, the Royal Commission on Environmental Pollution was highly critical of current controls, and made fifty-four recommendations proposing new ways of risk assessment and management.[27] Toxic chemical can be classed in categories such as *fast poisons*, *slow poisons* and *carcinogens*,[28] but many assay procedures rely on animal tests, which often have limitations (8.3.2).

Among the most alarming pollutants are those affecting food, both because they are rarely detectable and because we consume them voluntarily. It is hardly surprising that food scares often have dramatic economic effects, even when the biohazard is minor. The most notable recent example is the outbreak of BSE in cattle, principally in the 1980s–90s in the UK, which later proved responsible for infecting humans with new variant vCJD, a fatal condition affecting younger people (Box 13.2).

But perhaps even more insidious are pollutants that have drastic effects on human intelligence, the very characteristic on which our distinctively human nature depends. Yet evidence is accumulating that lead pollution, iron deficiency and hormone-disrupting chemicals such as polychlorinated biphenyls (PCB) act synergistically to depress IQ. In some African cities lead pollution may impair intelligence in 90% of the population, and even in the USA as many as 17% of children may be affected.[29]

Ecological footprints. The vast differences in the standards of living between the *haves* and the *have-nots* are underpinned by equivalent disparities in consumption and pollution rates (4.5), which are graphically illustrated by comparing the areas of land devoted to supporting us – the so-called *ecological footprints*. (You can discover an estimate of your own footprint by completing a web-questionnaire.[30]) From a global perspective, using this yardstick, we seem likely soon to exceed to Earth's total land area. The World Wildlife Fund (WWF) has estimated that to support the global population in 2050, at the predicted lifestyle and consumption patterns, will require *'almost three planets'*.[31] The calculation is controversial, since e.g. the footprint includes the area of forest needed to absorb carbon dioxide emissions, but makes no allowance for a shift to renewable energy sources. But the general point is well made: it is *obvious* that the world cannot continue to support the entire population in the style to which middle-class Europeans and North Americans are accustomed.

It is clear that even this brief survey of anthropogenic impacts on the environment paints a disturbing picture. What hopes are there for a sustainable future?

12.4 Sustainability

A landmark in ideas about sustainability was the publication in 1987 of *Our Common Future*, the report of the World Commission on Environment and Development,[32] often called, after its chairwoman, the *Brundtland Report* (BrR). The BrR linked sustainability with development in the phrase *sustainable development*, which it defined as *'meeting the needs of the present without compromising the ability of future generations to meet their own needs'*. Five areas for sustainable development were identified:

- economic growth
- equity (a fair allocation of resources to sustain growth)
- more democratic systems
- adoption of lifestyles within the planet's ecological means (aimed at DC)
- population levels within the productive potential of ecosystems (aimed at LDC).

Despite its commendable objectives, the BrR has been subjected to much criticism from those who claim that development was too often equated with *growth*, which is certain to ultimately prove unsustainable. It is now generally recognized, at least by most governments, that economic development cannot be allowed to go on degrading the environment – which must be protected in ways that preserve the essential ecosystems and provide for the well-being of future generations. A newer, widely used, definition of sustainable development is: *'Improving the quality of life while living within the carrying capacity of supporting ecosystems.'*[33]

12.4.1 Environmental sustainability

It remains the case that sustainability is a buzz word that is often used cosmetically. For example, a US Act describes agricultural sustainability as *'making the most efficient use of non-renewable resources'*.[34] In contrast, Herman Daly, economist at the World Bank, argues that a programme of strong sustainability is required. He claims that sustainable rates of use for different types of resource should satisfy the following criteria:

- *Renewable resource* (e.g. soil, water, forests, fish): no greater than the rate of regeneration (e.g. restoration of fish stocks)
- *Non-renewable resource* (e.g. fossil fuels, mineral ores): no greater than the rate at which a renewable resource can be sustainably substituted for it (e.g. profits of the oil industry invested in solar panels or tree-planting)

- *Pollutant*: no greater than the rate at which it can be recycled, absorbed or made harmless (e.g. sewage effluent discharged only at rates at which the natural ecosystem can absorb the nutrients).

Because the knowledge base on which to calculate the relative rates is often unreliable, there is frequently a need to apply the Precautionary Principle (13.4).

The relationship between environmental impacts and human activity can be expressed concisely by the equation:

$$I = P \times A \times T$$

where I is environmental impact, P is population, A is affluence, and T is technology. That is to say, a nation's total environmental impact is the product of its population, the average level of affluence (expressed as levels of consumption and pollution) and its technology (i.e. the efficiency with which the economy uses natural resources and deals with wastes).[35] It is clear that different countries have different capacities to reduce the impact, namely: LDC by curbing population growth; DC by reducing affluence; and emerging economies, such as China, by use of less polluting technologies.

12.4.2 International action on sustainability

The aim of the UN Environment Program is to *'provide leadership and encourage partnership in caring for the environment by inspiring, informing, and enabling nations and peoples to improve their quality of life without compromising that of future generations'*. An important milestone was the Rio Declaration of 1992, which agreed a global action plan for sustainable development, and required countries to draw up national plans. However, the limited success of this approach is illustrated by the debates over climate change.

At the Kyoto summit in 1997, 100 countries (mainly DC) agreed to reduce their greenhouse gas emissions to 5.2% below the 1990 levels within fifteen years. Each country was assigned a quota based on its previous emission levels, so that the more damage the country had caused in the past, the more it was allowed to do in future. This, then, was a very modest (and arguably unfair) target, because in 1990 the IPCC suggested that a 60–80% cut will be necessary to achieve sustainability. Even so, the USA decided not to ratify the treaty, considering it too stringent. The average US citizen emits 19.5 tonnes of carbon dioxide p.a. (which compares with 0.8 tonnes which the average Indian emits)[36] - although it is generally recognized that before greater efficiency in energy use can be achieved, development in LDC will entail some increase in carbon emissions. With the failure of the US Government to ratify the Protocol, it was not until February 2005, with the Russian Government's agreement to sign up, that the Treaty could finally come into force. On the same day, the DEFRA Minister, Margaret Beckett announced that the EU and its member states had agreed to a burden-sharing agreement of an 8% reduction in greenhouse gases by 2012, and that *'as part of this agreement the UK has taken on a reduction of 12.5%'*.[37]

In a major speech in 2004, Prime Minister Blair stressed the critical importance of cutting greenhouse gas emissions, calling this *'the world's greatest environmental challenge'*. The UK Government's aim is to exceed the Kyoto guidelines by reducing carbon dioxide emissions to 20% below the 1990 level by 2010, and, even more ambitiously, to reduce them by 60% by 2050. This is to be achieved by improving energy efficiency and promoting use of low-carbon technologies, such as fuel cells and carbon sequestration. Mr Blair resolved to use the UK's forthcoming presidency of the EU as an opportunity to influence those nations failing to make progress on the Kyoto targets.[38] Even so, earlier in 2004, the Government's own advisors, the Sustainable Development Commission, had expressed disappointment at the Government's progress on greenhouse gas emissions.[39]

One response to individuals, companies and nations who fail to abide by commonly accepted ethical principles affecting the environment is to implement the **polluter pays principle**, whereby the costs are imposed on the polluter of cleaning up the environment to restore its original state. In some cases it is possible, though rarely straightforward, to assign a financial cost to the damage, but for an issue like global warming this is virtually impossible. However, a somewhat cynical alternative is the trading of **emission rights**, whereby countries with higher emission levels (i.e. DC) can purchase a quota from those with lower rates (i.e. LDC), allowing them to continue to pollute while bearing a financial penalty benefiting LDC. If this allowed DC to gradually reduce emissions while allowing LDC to increase theirs modestly, a process entailing contraction and convergence might achieve an equitable outcome, possibly by mid-century.[40]

12.5 Agricultural sustainability

Agriculture is a prime example of a complex of technologies for which sustainability is of critical importance. It also has significant effects on the environment, illustrated by the fact that in the UK agriculture accounts for about 70% of total land use. So in this section, in order to explore these issues in more detail, we will 'turn up the magnification' to identify some relevant features of UK agriculture, which has experienced several recent crises of confidence.

12.5.1 The Common Agricultural Policy

First, a little historical perspective is in order. To aid reconstruction following the traumas of the 1939–45 World War, countries in Europe sought to cooperate on both political and trade issues. The Treaty of Rome (1957) established the European Economic Community, which made provision for a Common Agricultural Policy (CAP), the main objectives of which were to:

- increase agricultural productivity
- ensure fair living standards for agricultural workers
- stabilize markets
- guarantee a secure supply of food
- make products available to consumers at reasonable prices.

This involved several mechanisms, such as paying subsidies to farmers, protecting markets from foreign imports by imposing levies and quotas, and maintaining prices by taking produce out of circulation if surplus had been produced (resulting, for example, in the notorious *butter mountains* and *milk lakes*).

The general effect of the CAP (which costs the average UK family of four people £16 per week[41]) has been to increase output, thus ensuring food security. However, because farmers have been encouraged to maximize output, there have been several adverse environmental and public health consequences. For example:

- use of agrochemicals (e.g. fertilizers, herbicides, and pesticides) has increased
- hedges have been removed to aid efficient mechanical ploughing and cropping
- fossil fuel energy use has been increased as inputs and products are transported large distances
- animal welfare has been reduced by keeping many of them (especially chickens and pigs) at high density in strictly controlled environments and often feeding them antibiotics to maximize yields and growth.

The effects on biodiversity have been serious. For example, in the UK more than 100 species were lost in the last century,[42] and there has been a 40% decline in the populations of twenty farmland bird species since 1980.[43] According to the Countryside Agency, *'There have been major losses of downland, heathland, flower rich meadows, hedgerows and hedgerow trees, [and] ponds.'*[44]

Following establishment of the EU in 1993 several reforms have been made to the CAP (e.g. the imposition of quotas, and introduction of *set-aside*, preventing use of land for major crops), but many farmers are now experiencing serious financial hardship which is still encouraging short-term cost-cutting practices. A significant factor is the great power exerted by the four major supermarkets, which together have 75% of the total UK market. The keen competition between supermarkets (11.2.1) means that suppliers (farmers) are often financially squeezed.

12.5.2 Modern threats to agricultural sustainability

Cost-cutting exercises, such as feeding animal remains to cattle, have had disastrous public health impacts (as with BSE) and economic consequences (as with BSE and foot and mouth disease (FMD)). The official enquiry into the BSE outbreak[45] concluded that the government department responsible for both agriculture and food safety, the Ministry of Agriculture, Fisheries and Food (MAFF) had a 'conflict of interests'. In seeking to protect farmers' jobs, it ignored the serious potential public health risks of eating BSE-infected meat - an act of negligence which ultimately proved fatal when the link between BSE and vCJD became established (Box 13.2).

To add insult to injury, the UK experienced the world's worst outbreak of FMD in 2002. Although not a contagious disease for humans, it had massive economic effects. In the wake of FMD, MAFF was abolished and replaced by a new government department, the Department for the Environment, Food and Rural Affairs (Defra), incorporating staff from MAFF and the former environment department. It seemed clear that the DEFRA Minister, Margaret Beckett, wished to promote 'greener' attitudes,[46] and to this end the Government established a Policy Commission on the Future of Food and

BOX 12.2 THE CURRY REPORT: MAIN RECOMMENDATIONS

- Early radical reform of the CAP, which is currently dividing farmers from their markets, suppressing innovation, and destroying economic and environmental value.
- Re-targeting public funds towards environmental and rural development goals instead of subsidizing production. Rewards should be provided for farmers who deliver an attractive, healthy countryside, and make the environment a strong selling point for the industry.
- Measures to strengthen the food supply chain and promote collaboration between farmers – so reducing inefficiencies, improving competitiveness, and securing a fairer return for the primary producer.
- A new drive on research and technology transfer – including a network of demonstration farms.
- Honest food labelling to empower consumers, and help them make their consciences felt at the supermarket checkout.
- A comprehensive nutrition strategy to encourage a healthier diet for all.
- Promotion of local food – reducing wasteful 'food-miles'.
- Simpler, easy-to-use, free advice services for farmers, to help them assess future options.

Farming, under the chairmanship of Don Curry (Box 12.2). The central theme of its report was the need for 'reconnection': of farmers with their markets and the rest of the food chain; of the food chain with a healthy and attractive countryside; and of consumers with what they eat and where it has come from.[47]

12.5.3 **External costs of UK agriculture**

But the challenge is formidable, as suggested by the following assessment. Most environmental damage incurs costs which are not paid by those causing it. These **external costs** are paid by society at large, e.g. through money spent in purifying water and in medical costs for treating people made ill by diseases contracted from eating contaminated food. An estimate of such external costs of UK agriculture for the period 1990–96 amounted to £1.15 billion–£3.91 billion p.a. This included costs of damage to:

- drinking water (e.g. due to pesticides, fertilizers, pollution and eutrophication)
- air (due to carbon dioxide, ammonia and nitrous oxide)
- soil (due to erosion and loss of organic matter)
- biodiversity and wildlife
- hedgerows and bee colonies
- human health (due to pesticides, nitrates, micro-organisms, antibiotics and BSE/vCJD).[48]

The wide range shows the difficulty of achieving precision in such estimates, but the mean figure of £2.34 billion p.a. amounted to 89% of average net farm income, or £208 per hectare of arable land and permanent pasture p.a. – suggesting the costs were enormous even if the lowest estimate was the most accurate.

12.5.4 Organic farming

The above observations suggest the need for farmers to adopt more sustainable systems of farming. The most prominent system of sustainable farming is **organic farming**. This holistic system largely avoids use of synthetic fertilizers, pesticides, growth regulators and livestock feed additives. Instead, it relies on crop rotations, crop residues, animal manures, legumes, green manures, organic wastes, forms of biological pest control to maintain soil productivity, and tillage, to supply plant nutrients and to control insects, weeds and other pests. All use of GMO is prohibited. General principles are laid down by the International Federation of Organic Agricultural Movements[49] and administered in the UK by the Soil Association.[50] To use the 'organic' title, farmers must be certified by an official government body and their farms are inspected regularly to prove conformity with regulations.

Environmental benefits of organic farming are evident in biodiversity data, as shown by a comprehensive analysis of seventy-six studies conducted in Europe, Canada, New Zealand and the USA. Biodiversity comparisons of organic and conventional systems were made for groups of organisms ranging from bacteria and plants to earthworms, beetles, mammals and birds. Of ninety-nine separate comparisons of groups of organisms, sixty-six showed that organic farming benefited wildlife, twenty-five produced mixed results (or revealed no difference between the farming methods), and only eight suggested that organic farming was detrimental by comparison with the conventional methods used. Almost all the negative impacts of organic farming related to earthworms and beetles, but even in these cases a greater number of studies showed positive impacts.[51]

It was deduced that the biodiversity benefits were due to use of fewer pesticides and inorganic fertilizers, and to adoption of wildlife-friendly management of habitats where there were no crops. Such strategies included mixing livestock and arable farming, and not weeding close to hedges.

There is disagreement about whether food quality and human health due to consumption of organic food are better than from conventionally produced foods. A literature survey conducted by the Soil Association suggested that organic food had greater contents of vitamin C, minerals, and certain phytonutrients, and (because they are banned) lower levels of pesticides, antibiotics, nitrates and food additives.[52] However, the FSA denies there are any significant nutritional or food safety benefits.[53] Even so, the UK government supports organic farming, largely because of its environmental benefits. The Organic Action Plan aims to increase UK producers' share of the organic market from 44% in 2004 (i.e. 56% was imported) to 70%, in line with the proportion for conventionally produced food. In 2004, the UK was one of three countries in the world (with the USA and Germany) where sales of organic food exceed £1 billion p.a.[54]

Natural and social capital. But organic farming also delivers benefits in terms of increased employment and enhanced **natural capital**. The latter refers to the stocks of plants and animals, along with the atmosphere, water, and minerals, that together comprise ecosystems[55] – which is clearly distinguishable from financial capital. Natural capital is a largely undervalued resource, the value of which is rapidly depreciating through anthropogenic activities (12.1.1). According to ecologist Robert Goodland, *'We must now recognise that natural capital has become scarce for the first time'*, estimating that, if one were to try to put a figure on it, natural capital could be *'worth more than twice as much as global GNP (US$33 trillion compared with US$15 trillion)'.*[56]

Organic farming also develops, and depends on, **social capital**, a term that refers to *'the structure of relations between actors and among actors that encourages productive activities'.*[57] When we think of the networks of social relationships which underpin all successful collaborative activities, built on trust, friendship and fairness, we appreciate just how important social capital is. In his book *Agri-Culture*, Jules Pretty has amply demonstrated the importance of reconnecting people, land and nature in the quest to re-establish sustainable agricultural systems worldwide.[58]

A question of values Of course, these benefits come at a cost, both because yields from organic systems are usually lower than in conventional, intensive, systems, and because the administration of the organic system entails financial costs which are passed on to the consumer as higher food prices. So some people dismiss organic farming as 'a con', arguing that the food doesn't taste any better, and is no healthier than cheaper, conventionally produced food. If those are the only things that matter, there might be something in this argument. But for others, it could be said to miss the point. A crucial question for each of us is how much we, personally, are prepared to pay for the longer-term environmental, animal welfare and social benefits of organic farming. If such benefits are valued sufficiently, they constitute a justifiable motive for buying organic food, thereby, through *ethical consumerism* (14.8), helping to bring about a desired food system.

Problems with organic farming Even so, some deficiencies of the organic system have been identified. For example, the ambiguity of the term 'organic' means that organic farmers are allowed to use some non-organic pesticides, such as Bordeaux mixture, based on copper – which is toxic even at quite low concentrations, as also is the permitted nicotine. Science writer Colin Tudge points out (in his highly readable book *So Shall We Reap*), that *'some of the rules have a strangely arbitrary, and indeed a muddled feel to them'.*[59]

While strongly advocating the principles of organic farming, especially for animals, Tudge also argues that, because of the higher yields, in an 'enlightened agriculture', *'there is a case for very intensive farms dedicated solely to crops'.*[60] Nevertheless, the general philosophy of organic farming, which to date has had only minimal research devoted to its practical application, appears to be attracting more scientific interest. In 2004, an article in the leading science journal *Nature* highlighted the issue by noting that *'elements of the organic philosophy are starting to be deployed in mainstream agriculture'.*[61] As for so many ethical questions, it may be that ready-made solutions to problems are not currently available, and there is a need to choose the appropriate *trajectories of change.*

12.5.5 **Integrated farm management**

Other approaches to integrated farm management (IFM) are less rigorous than organic farming, but also place emphasis on providing environmental benefits. For example, integrated pest management (IPM) looks at each crop and pest situation, as a whole, and attempts to devise a programme that integrates the various control factors optimally. In its recent forms it combines modern technologies, the application of synthetic, yet selective, pesticides and breeding for pest resistance (including GM), with natural methods of control, including use of natural predators and parasites. The outcome can be efficient pest control that is cheaper than conventional pesticide use.[62]

In the UK, a prominent organization is LEAF (Linking Environment and Farming), which emphasizes integrated farming and sustainable approaches to producing healthy crops and livestock in an environmentally sustainable way.[63] This approach to farming involves:

- using crop rotations to sustain soil fertility
- managing hedgerows to provide a variety of habitats and food sources for wildlife
- using natural plant nutrients, supplemented with fertilizers where appropriate
- re-cycling on-farm wastes and conserving energy
- observing high animal welfare standards.

LEAF promotes its approach to both farmers and the public through approximately sixty demonstration farms and innovation centres, at various locations throughout the UK. Farmers growing to appropriate (inspected) standards are able to market a wide range of fruit, vegetables, and meat under the LEAF Marque logo, but no premium is charged for such produce.

12.5.6 **Biotechnology and agricultural sustainability**

In theory, agricultural biotechnology might contribute to sustainability in several ways. For example, by increasing yields, the total area of land necessary for production could be reduced, leaving larger areas free to develop biodiversity. Again, in theory, the ability to create new species by GM could directly enhance diversity in environmentally beneficial ways. Thus, it has been claimed that: *'GM technology offers opportunities for increased resource capture efficiency by crops, that are not available through conventional genetic improvement.'*[64] However, others express doubts that such possibilities will be realized if commercial forces remain the main driver for their application. Often, the claim is made that current GM technologies might threaten sustainability by encouraging greater application of herbicides to HT crops and the development of resistance to pesticides expressed by GM crops (10.5.3). We have seen that proponents and critics of GM crops have interpreted the North American data in different ways (10.3.1), and it may be some years before a reliable cost–benefit analysis is available.

By contrast, the use of biotechnology as a tool in some forms of bioremediation might prove more widely acceptable. For example, GM micro-organisms might be used in treatment of contaminated land, as sensors to detect pollution, or directly in soil to digest pollutants. GM plants can also be used in the bioremediation of pollutants (**phytoremediation**) by absorbing the pollutants before destruction. There are a number of potential benefits of this approach, particularly for soils contaminated with metals – which applies to half the contaminated soils in the UK. But certain risks need to be addressed, e.g. the possibility of gene flow from GM to native populations, and the difficulties of disposing of harvested GM crops in which pollutants have become concentrated.[65]

12.5.7 Sustainability and fairness

Of all the principles identified in the ethical matrix (10.2), respect for consumer fairness might be considered the most critical, not only in terms of rectifying current inequalities (chapter 4), but also with reference to future generations (**intergenerational justice**). There might be general agreement, from the perspectives of both instrumental and intrinsic value (12.2) that the environment should be protected and its diversity preserved. But what of the trade-offs? Adoption of organic systems is likely to reduce yields, thereby removing the cushion which provides food security, while also increasing food prices. In DC, food prices are a small proportion of the *average* disposable income, but for poorer people this proportion is much higher. In the UK, the poorest 10% of the population spend 26% of income on food and drink, compared with the 6% spent by the richest 10% of the population (who also eat a much healthier diet).

However, setting the external costs (12.5.3) against the supermarket checkout price tells a different story; and if the adverse environmental impacts of the CAP (12.5.1) are also factored in, the cheap food argument might appear weak. In whatever ways the problem is interpreted, it seems certain that political decisions will have a crucial influence, albeit subject to the influence of ethical consumerism (14.8).

But the global perspective adds a further dimension to consideration of fairness. The reality is that the environmental sustainability of the Earth is a **public good**, which DC and LDC alike have a common interest in ensuring. But the poor contribute least to the environmental problems while being most vulnerable to their ill effects. They have to live in the most degraded and ecologically fragile areas, while being the least able to cope with the harm inflicted on their health and livelihoods. For it is the consumption patterns of people in DC which contribute most to environmental degradation (4.5), while bearing few of its costs.[66] So it is appropriate at this point to return to global concerns.

12.6 The limits to growth

It is clear that current rates of resource use and waste production are unsustainable. Sooner or later, the sources of the raw materials for production of goods (fossil fuels, minerals, etc.) and the sinks to absorb the wastes (greenhouse gases, toxic chemicals, etc.) will be exhausted. So an important question is: 'How long will it be, globally, before we reach the limits to growth?' The consequences of continuing along the current path of resource use were graphically illustrated by the computer modelling studies of Dennis Meadows and colleagues in the 1990s. Making various assumptions about future population trends, resource availability, industrial output, food supplies and pollution levels, they produced several predictions for the state of the world up to 2100. According to the *standard run*, with society proceeding along its historical path as long as possible without major policy changes, food supplies and industrial output would begin to level off and then decline after about 2030, with population and pollution peaking a little later, before beginning to decline. Thus, this scenario foretells the Malthusian solution (Figure 4.3).

In contrast, according to the most optimistic scenario ('scenario 10'), food supplies, population (at about 7.7 billion), and industrial output would level off about 2040, with pollution levels peaking at about the same time and declining thereafter (Figure 12.1). However, this seemingly optimistic outcome would entail some radical changes. It is based on the assumptions that population growth and industrial output will be drastically reduced from current levels, that additional resources become available, and that new technologies will be developed to conserve resources, protect agricultural land, increase land yields, and abate pollution.[67] But according to Meadows' informal surveys of audiences around the world, most people doubt that scenario 10 will be achieved directly. Instead, they consider global production will temporarily overshoot the maximum carrying capacity that is sustainable over the long term. Meadows claims that *'challenging the existing exponential growth paradigm requires thinking outside conventional norms . . . which will require a change in our ethics, culture, and values'.*

Is this altogether too pessimistic? In favour of continued human survival, is the fact that we now have powerful technologies and detailed scientific awareness to counter adverse effects. But loaded against us are the facts that humans are 1000 times more populous than any other land mammal of similar size in the whole history of the world, that collectively we consume 20–40% of the solar energy captured by plants, and that the tropical rain forests, which house a rich source of biodiversity, are being relentlessly destroyed.

For those with strong stomachs, philosopher John Leslie has provided a detailed analysis of the ways it all might end – ranging from nuclear war, overheating due to the greenhouse effect, poisoning by pollution or pandemic disease, volcanic eruptions, collisions with asteroids or planets, disasters due to genetic engineering, nanotechnologies or computer malfunction – through to an unwillingness to rear children and invasion by extraterrestrials![68] We might reasonably conclude there is little we can do about some of these. But for others, prudence – if not intergenerational justice – might suggest the need to adopt a precautionary approach (13.4).

Technological risks. Threats to human survival are, of course, all the more dangerous if they are unrecognized, or realized too late to do anything about. Bill Joy (co-founder of Sun Microsystems) suggests that use of genetics, nanotechnology and robotics (GNR) technologies may simply develop *uncontrollably* if driven by commercial competition, because this could result in an 'arms race' with catastrophic consequences. In his view, *'If we could agree, as a species, what we wanted, where we were headed, and why, then we might understand what we can and should relinquish'*, a view shared by cosmologist (and

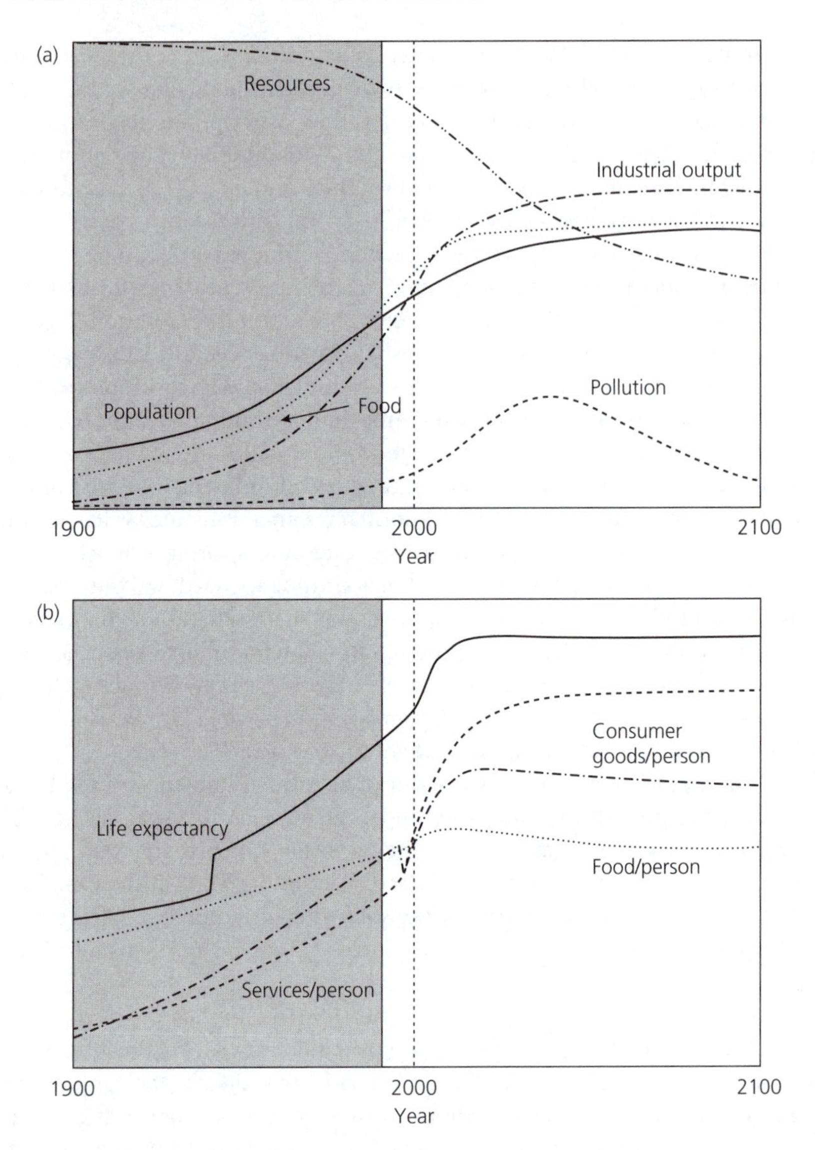

Figure 12.1 Scenario 10: stabilized population and industry with technologies to reduce emissions, erosion, and resource use adopted in 1995 (compare with Figure 4.3). (a) State of the world; (b) material standard of living. [From: Meadows D H, Meadows D L, and Randers J (1992) Beyond the Limits. London, Earthscan, p. 138]

Astronomer Royal) Martin Rees.[69] These are powerful endorsements of the need for global ethics.

But getting an international agreement on a moratorium on dangerous technologies is unlikely to be easy. Even if agreed, complete enforcement would be difficult to achieve. As Rees put it: *'In view of the failure to control drug smuggling or homicides, it is unrealistic to expect that when the genie is out of the bottle, we can ever be fully secure against bioerror and bioterror.'* There would thus seem to be a strong case for saying that we need to adopt an entirely different worldview if we are to secure a sustainable future for humanity.

12.7 Gaia: a different mind set

Indeed, environmentalist James Lovelock has proposed that we regard the biosphere in a quite novel way to encourage the adoption of a more prudent mind set. Most people have been brought up to believe that the Earth is a mass of rock with an almost limitless supply of the resources needed to support the global population, and an almost infinite capacity to absorb the wastes we produce. Lovelock's inspiration was to appreciate that this is a flawed attitude. Just as progress in biology has been associated with re-examination of the ways key ideas have been socially constructed (Box 1.1), so the 'lifeless' model of Earth now, on closer examination, proves seriously limited.

Lovelock's **Gaia** hypothesis (named after an ancient Greek Earth goddess) proposes that the Earth's crust, together with the oceans, atmosphere and biosphere, constitute a single system, which is self-sustaining and self-regulating.[70] Those features essential for the maintenance of life, such as atmospheric pressure and oxygen content, mean that the surface temperature and the salinity of the oceans are closely regulated. For Lovelock, the planet is a super-organism, the activity of which is governed by its **geophysiology**, which encompasses its ecology, geochemistry and geophysics.

The concept of Gaia has undergone evolution since its first appearance, and in the 1988 version the imaginary planet **Daisyworld** was introduced, to explain his thesis.[71] Daisyworld is populated by two kinds of plant only - white and black daisies, which in accordance with Darwinian theory serve to maintain the planet's temperature at an essentially constant level despite a major increase over time in the heat radiated by the planet's 'sun' (Figure 12.2). Initially, when the planet's surface temperature was low, only black daisies could grow, but by absorbing heat they warmed the planet sufficiently to support continued life. With the passage of time, as solar radiation increased, black daisies were displaced by white daisies which, by reflecting heat, cooled the planet's surface, preventing it from overheating. Thus, the biological response to changing solar radiation achieved homeostasis. The example is simplistic, but it conveys sufficiently the idea of the way in which species diversity can provide the necessary 'geophysiological' response to support life.

The name Gaia has not been altogether satisfactory, because although intended as a scientific theory, naming it after a Greek goddess raised the hackles of the scientific establishment, who were disdainful of the allusion to a pagan myth. To avoid that association (although wishing to retain its holistic meaning), Lovelock launched the *medical* model of Gaia - encouraging the idea of the damaged Earth as a patient for whom humanity is the only available doctor. For philosopher Mary Midgley, Gaia performs a vital role by providing *'a more realistic picture of how the earth works, a picture that will correct the delusive idea that we are either engineers who can redesign our planet or*

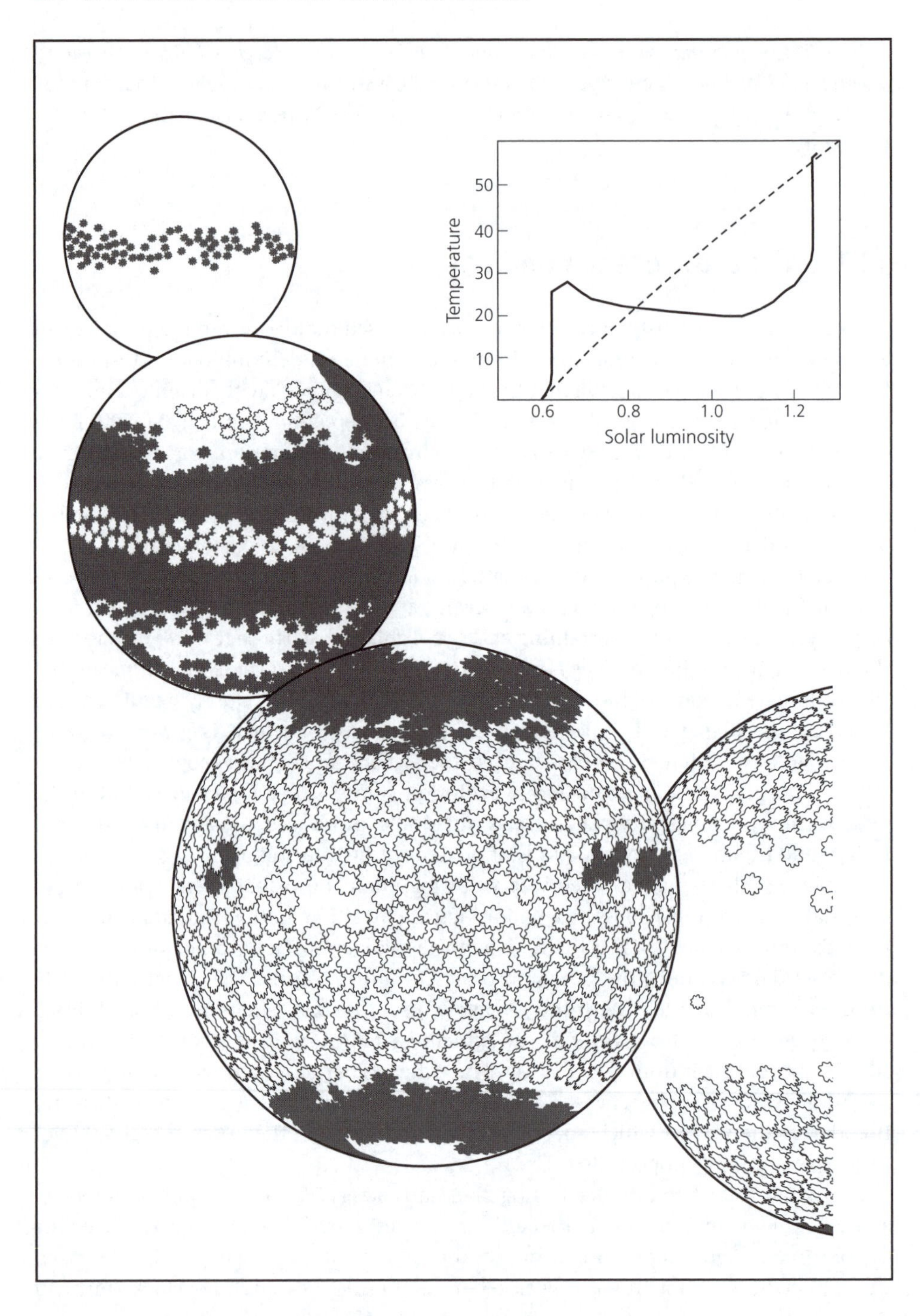

Figure 12.2 Gaia: the evolution of the climate in Daisyworld. [From Lovelock J (1991) Gaia: the practical science of planetary medicine. London, Gaia Books Ltd, pp. 66–67]

chance passengers who can detach ourselves from it when they please'.[72] Gaia would also seem to fit readily into the biocentric and ecocentric worldviews described (12.2.1).

12.8 Sustainability and ethics

In considering how we should act in respect of the future, two facts seem inescapable:

- the current vast disparity between the living conditions of the *haves* and the *have-nots* is ethically indefensible
- globally, current practices are unsustainable, because: (*a*) non-renewable resources are being used too rapidly; (*b*) environmental pollution and degradation are having irreversible adverse effects; while (*c*) population growth is exacerbating both these problems.

Taking ethical responsibilities seriously thus presents people in DC with two sorts of problem: first, the current disparities between rich and poor; and second, the challenge of leaving a world fit for our descendants. Chapter 4 considered the first of these responsibilities, although it is clearly closely related to the second, if only because the slow pace of progress means that the plight of the *have-nots* is likely to demand our concern for years to come.

What, then, are our ethical duties to future generations? To address this question, it might be thought that we need to have some sort of time horizon in mind. According to philosopher John Passmore, we can only reasonably look forward to the immediately following generations, because we cannot know what circumstances will be like in the distant future.[73] But, even from the perspective of limited forms of altruism (1.9.1), it might be imagined that people would have an interest in sustaining a healthy environment at least for their grandchildren's sake – and why stop there? We may not be able to *plan* the future for our descendants, but sustainability presumably operates on a broader canvas.

What is apparent, though, is the extreme difficulty of relying on ethical reasoning based on utilitarian principles. The problems with utilitarianism even in a limited timeframe were discussed in 2.5.3, and they would seem to become insuperable when looking into a distant and uncertain future. Yet this uncertainty can hardly absolve us from the duty of paying due regard to our descendants' well-being. So the relevant theories are almost inevitably of a deontological nature (2.6). In addressing these issues, Edith Brown Weiss has described what she calls the principles of *common patrimony*, as follows:

- **conservation of options**: each generation should leave to its successors a 'robust planet' that will support a variety of life choices, including those not foreseeable at the time
- **conservation of quality**: each generation should leave the planet in as good a shape as it found it

- **conservation of access**: which aims to balance justice requirements within and between generations.[74]

This framework appears to provide a strong basis for an ethic of sustainability for two reasons: (*a*) it rests on a common denominator which is shared by religious and philosophical ideas belonging to several cultures; (*b*) it stresses concern for the protection and preservation the essential features of human habitats.[75]

12.8.1 Idealism and realism

It is not unnatural to think that the scale of the problems humanity faces (only some of which are highlighted in this chapter) means that no individual can hope to influence the future for the better in any significant way. But such a counsel of despair in the past would have prevented the genuine moral progress which society has achieved. After all, slavery *was* abolished, women *did* get the vote, apartheid *was* ended in South Africa, and the welfare of many animals on farms *has* improved. Despite the faltering steps, the efforts of those who sought moral advance have brought real improvements for many.

Alongside the two pressing philosophical questions 'What can I know?' and 'What should I do?', the illustrious eighteenth-century philosopher Immanuel Kant (Box 2.2) posed another: 'What can I hope for?'. And the significance of the virtue of 'hope' should not be overlooked. But for hope to be a justifiable virtue it needs to be underpinned by efforts to realize the ends hoped for. As philosopher Peter Singer puts it: *'We have to take the first step. We must reinstate the idea of living an ethical life as a realistic and viable alternative to the present dominance of materialist self-interest'*. And if a *'critical mass'* of people can share in such objectives, this will offer us a *'real chance of improving the world before it is too late'*.[76]

If we take the considerations discussed in this chapter together, they seem to have two undeniable consequences. Weiss's common patrimony requires **resource redistribution** from the haves to the have-nots; and people in DC need to move towards more **frugal patterns of consumption**. Of course, such conclusions fly in the face of the 'growth ethic' which underpins capitalist societies, so that the whole tenor of Western society militates against this approach. However, in practical terms those who are motivated to promote a sustainable environment have several courses of action at their disposal. For example:

- *at the most personal level*, people can opt for ethical consumerism (14.8), i.e. consume food produced in sustainable systems, use recyclable materials, travel short distances by foot or bicycle, and use public transport for longer distances
- *in choosing a career*, people can opt for jobs which support sustainable lifestyles (e.g. in manufacturing or service industries or through research activities) or promote sustainable thinking (e.g. through education)
- *in exercising democratic rights*, people can vote and campaign for sustainable political programmes, or personally engage in political activity (e.g. for NGO) with such aims.

Concern for future generations whom we shall never meet is perhaps the purist form of altruism imaginable. But if Aristotle was right, it is in the pursuit of virtues such as this that people discover a truly happy life.

THE MAIN POINTS

- Environmental ethics came to prominence with the realizations that the Earth's resources and ability to absorb wastes are limited and that non-human life has intrinsic value.
- Current rates of use of non-renewable resources and impacts on the biosphere of pollutants (e.g. greenhouse gases) pose serious threats to sustainability, which need to be addressed urgently.
- DC could make their greatest contribution by curbing affluence, LDC by reducing population growth rates, and newly industrializing countries by improving the efficiency of their industrial technologies.
- Agricultural production systems need critical attention – by use of holistic, sustainable practices such as organic farming and/or by appropriate use of modern biotechnologies.
- Failure to secure international agreement on limiting greenhouse gas emissions suggests that widespread adoption of a new mind set (e.g. along the lines of Gaia) may be necessary to implement political changes ensuring global sustainability.

EXERCISES

These can form the basis of essays or group discussions:

1. How do you assess the value of biocentric and ecocentric approaches to the biosphere in relation to the anthropocentric approach? (12.2)
2. Can the concept of sustainability ever be practically meaningful? (12.4)
3. Is the USA Government's decision not to sign the Kyoto Protocol ethically defensible and how, if at all, should other developed countries respond? (12.4.2)
4. Given the alleged lower yield potential of organic farming systems, and the difficulties of the co-existence of GM and non-GM forms of arable agriculture, how ethically defensible is it to promote organic farming when hundreds of millions of people are malnourished? (12.5)
5. It is claimed that achieving sustainability will entail *resource redistribution* and *more frugal patterns of consumption*. Discuss the possibility of such objectives being achieved. Would coercive measures (such as the Chinese population policy – see 4.10.3) ever be ethically defensible? (12.8)

FURTHER READING

The following are good introductions to environmental ethics, agricultural sustainability and the political dimensions of sustainability, respectively.

- *Environmental Ethics: an introduction with readings* by John Benson (2000). London, Routledge. A text designed for undergraduate students.
- *So Shall We Reap* by Colin Tudge (2003). London, Penguin. A highly readable account of the history, and future prospects, of agriculture.
- *The Politics of the Real World: meeting the new century* edited by Michael Jacobs (1996). London, Earthscan. The combined message of more than thirty of the UK's leading voluntary and campaigning organizations.

USEFUL WEBSITES

Among useful sites are:

- **http://www.unep.org/** *United Nations Environment Program*: publishes numerous reports on environmental issues.
- **http://www.biodiv.org/** *Convention on Biological Diversity*: the site of the major international body promoting biodiversity.
- **http://www/defra.gov.uk/farm/sustain/** *Department for the Environment, Food and Rural Affairs*: the UK Government's site on sustainable farming and food.

NOTES

1. Odum E P (1975) Ecology (2nd edition). New York, Holt, Rinehart and Winston, pp. 4–5
2. Holland A (1999) In: Cambridge Dictionary of Philosophy (2nd edition). (Ed.) Audi R. Cambridge, Cambridge University Press, p. 269
3. Carpenter S R (1998) Sustainability. In: Encyclopedia of Applied Ethics. (Ed.) Chadwick R. San Diego, Academic Press. Vol. 4, pp. 275–293
4. Passmore J (1974) Man's Responsibility for Nature. London, Duckworth, p. 20
5. Bruce D and Bruce A (Eds) (1998) Engineering Genesis. London, Earthscan, pp. 86–87
6. Johnson L E (1991) A Morally Deep World. Cambridge, Cambridge University Press, pp. 216–217
7. Rawles K (1998) Biocentrism. In: Encyclopedia of Applied Ethics. (Ed.) Chadwick R. San Diego, Academic Press. Vol. 1, pp. 275–283
8. Naess A (1989) Economy, Community and Lifestyle. Cambridge, Cambridge University Press
9. Attfield R (1995) Ethics and the environment: the global perspective. In: Introducing Applied Ethics. (Ed.) Almond B. Oxford, Blackwell, pp. 331–342
10. Swiss Ethics Committee on Non-Human Gene Technology (2001) La dignité de l'Animal. Berne, ECNH
11. Rawles K (1998) Biocentrism. In: Encyclopedia of Applied Ethics. (Ed.) Chadwick R. San Diego, Academic Press. Vol. 1, pp. 275–283
12. Ruckelshaus W (1989) Towards a sustainable world. Sci Am, September, pp. 166–176
13. Daly H E and Cobb J B (1990) For the Common Good. London, Green Print, p. 2
14. Smith A (ca. 1910) [1776] An Enquiry into the Nature and Causes of the Wealth of Nations. London, George Routledge

15. Stork N N E (1995) Biodiversity and bioethics. In: Issues in Agricultural Bioethics. (Eds) Mepham T B, Tucker G A, and Wiseman J. Nottingham, University Press, pp. 205–244
16. Carpenter S R (1998) Sustainability. In: Encyclopedia of Applied Ethics. (Ed.) Chadwick R. San Diego, Academic Press. Vol. 4, pp. 275–293
17. Thomas C D *et al.* (2004) Extinction risk from climate change. Nature *427*, 145–148
18. Brown P (2004) An unnatural disaster. The Guardian, 8 January 2004. http://www.guardian.co.uk/
19. Lee K (1998) Biodiversity. In: Encyclopedia of Applied Ethics. (Ed.) Chadwick R. San Diego, Academic Press. Vol. 1, pp. 285–304
20. Department for International Development (2001) Biodiversity: a critical issue for the world's poorest. London, DfID
21. Leakey R and Lewin R (1998) The Sixth Extinction: biodiversity and its survival. London, Phoenix
22. Department of the Environment, Food and Rural Affairs (2003) Climate change. http://www.defra.gov.uk (accessed 23 January 2004)
23. Intergovernmental Panel on Climate Change (2004) http://www.ipcc.ch/
24. Meadows D H, Meadows D L, and Randers J (1992) Beyond the Limits. London, Earthscan
25. Goudie A (1990) The Human Impact on the Natural Environment (3rd edition). Oxford, Blackwell, p. 293
26. Ibid., p. 189
27. Royal Commission on Environmental Pollution (2003) Chemicals in Products. London, RCEP
28. Rodericks J V (1992) Calculated Risks. Cambridge, Cambridge University Press
29. Williams C (2003) The environmental threat to human intelligence. ESRC Global environmental change programme. www.susx.ac.uk/units/pubs/briefing
30. Ecological footprint quiz (2004) http://www.earthday.net/footprint/info.asp
31. Rees M (2003) Our Final Century. London, Heinemann, p. 102
32. World Commission on Environment and Development (1987) Our Common Future. Oxford and New York, Oxford University Press
33. Jacobs M (1996) The Politics of the Real World. London, Earthscan, p. 26
34. Cherfas J (1996) Sustainable foods systems. In: Food Ethics. (Ed.) Mepham B. London, Routledge, p. 35
35. Meadows D H, Meadows D L, and Randers J (1992) Beyond the Limits. London, Earthscan, p. 100
36. Bruges J (2000) The Little Earth Book. Bristol, Alastair Sawday, p. 43
37. DEFRA (2005) Climate change: Kyoto Protocol in force (16.2.05) http://www.defra.gov.news/
38. Blair T (2004) The climate change speech. http://www.number-10.gov.uk/output/p6333.asp
39. Sustainable Development Commission (2004) Shows promise, but must try harder. http://www.sd-commission.gov.uk/events
40. Bruges J (2000) The Little Earth Book. Bristol, Alastair Sawday, pp. 50–51
41. Consumer Association (2001) http://www.which.net/
42. Department of the Environment, Food and Rural Affairs (2001) UK Biodiversity Action Plan. www.defra.gov.uk
43. Department of the Environment, Trade and the Regions (2000) A better quality of life: a strategy for sustainable development for the UK. www.detr.gov.uk
44. Countryside Agency (2003) http://www.countryside.gov.uk/
45. BSE Inquiry (2002) http://www.bseinquiry.gov.uk/

46. Department of the Environment, Food and Rural Affairs (2002) The Strategy for Sustainable Farming and Food: Facing the Future. London, Defra
47. Policy Commission on the Future of Farming and Food (Curry Report) (2002) London, Cabinet Office
48. Pretty J N, Brett C, Gee D, Hine R E, Mason C F, Morison J I L, Raven H, Rayment M, and van der Bijl G (2000) An assessment of the total external costs of UK agriculture. Agr Syst *65*, 113–136
49. Woodward L (1997) Health, Sustainability and the Global Economy: the organic dilemma. Newbury, Elm Farm Research Centre, p. iii
50. Soil Association (2004) www.soilassociation.org
51. Hole D G, Perkins A J, Wilson J P, Alexander I H, Grice P V, and Evans A D (2004) Does organic farming benefit biodiversity? Biol Conserv *122*, 113–130
52. Soil Association (2002) Organic Farming, Food Quality and Human Health. Bristol, Soil Association
53. Krebs J (2003) Is organic food better for you? www.foodstandards.gov.uk
54. Department of the Environment, Food and Rural Affairs (2004) Action plan to develop organic food and farming in England. London, Defra
55. Daly H E and Cobb J B (1990) For the Common Good. London, Green Print, p. 72
56. Goodland R (2000) The urgency of environmental sustainability. In: Where Next? reflections on the human future. (Ed.) Poore D. Paris, UNESCO, p. 83–94
57. Pretty J (1998) The Living Land. London, Earthscan, p. 8
58. Pretty J (2002) Agri-Culture: Reconnecting people, land and nature. London, Earthscan
59. Tudge C (2003) So Shall We Reap. London, Penguin, p. 337
60. Ibid., p. 360
61. Macilwain J (2004) Organic: is it the future of farming? Nature *428*, 792–793
62. Conway G (1997) The Doubly Green Revolution. London, Penguin, p. 215
63. LEAF: Linking Environment and Farming (2004) www.leafuk.org
64. Agriculture and Environment Biotechnology Commission (2002): evidence from Institute of Arable Crops Research (2002). London, DTI
65. Agriculture and Environment Biotechnology Commission (2003) Looking ahead: an AEBC horizon scan. London, DTI
66. Department for International Development (2000) Eliminating world poverty. London, DfID, pp. 78–79
67. Meadows D H, Meadows D L, and Randers J (1992) Beyond the Limits. London, Earthscan, p. 199
68. Leslie J (1996) The End of the World. London, Routledge
69. Rees M (2003) Our Final Century. London, Heinemann, p. 87
70. Lovelock J E (1979) Gaia: a new look at life on Earth. Oxford, Oxford University Press
71. Lovelock J E (1988) The Ages of Gaia. Oxford, Oxford University Press
72. Midgley M (2001) Science and Poetry. London, Routledge, p. 172
73. Passmore J (1974) Man's Responsibility for Nature. London, Duckworth
74. Carpenter S R (1998) Sustainability. In: Encyclopedia of Applied Ethics. (Ed.) Chadwick R. San Diego, Academic Press. Vol. 4, p. 290
75. Ibid.
76. Singer P (1994) How Are We to Live? London, Mandarin, p. 235

PART FIVE
Bioethics in practice

It is not enough that you should understand about applied science in order that your work may increase man's blessings. Concern for man himself and his fate must always form the chief interest of all technical endeavors, concern for the great unsolved problems of the organisation of labor and the distribution of goods – in order that the creations of our mind shall be a blessing and not a curse to mankind. Never forget this in the midst of your diagrams and equations.

Albert Einstein (1938)
Addressing students at the California Institute of Technology

13

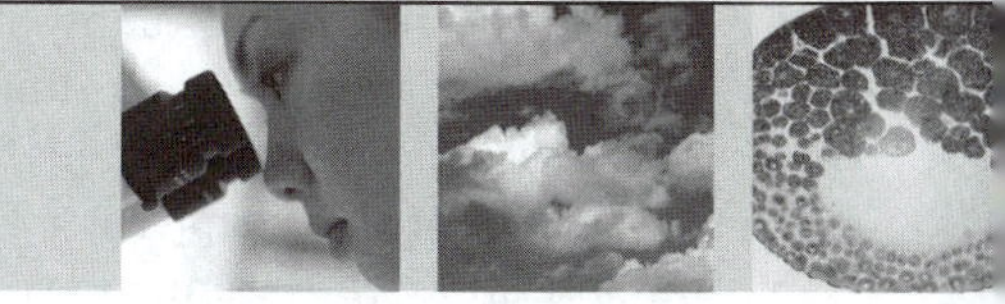

Risk, precaution and trust

OBJECTIVES

When you have read and discussed this chapter you should:

- appreciate the nature of probabilistic concepts of biorisks
- be aware of the character and limitations of risk assessment and risk management
- appreciate the significance of broader dimensions of risk, such as those relating to voluntariness, the distribution of risks and benefits, unfamiliarity and the nature of the *risk society*
- understand the importance of the Precautionary Principle and the different ways in which it is interpreted
- appreciate the importance of trust, the extent to which scientists and politicians are vulnerable to mistrust, and recent measures taken to re-build the trust of a sceptical public

13.1 Introduction

We have seen that utilitarian reasoning (2.5) entails weighing the expected consequences of our actions in a cost–benefit analysis. Bentham's hedonic calculus has been refined in various ways, e.g. to refer to preferences rather than pleasure, and to incorporate rules rather than basing decisions on individual acts; but the aim remains to maximize good by choosing to perform actions with consequences that are, on balance, the most desirable.

However, a limitation of this approach lies in the uncertainty that always accompanies the anticipation of future events. We can never be absolutely sure that things will turn out as expected. And if we are driven by a desire for success or are convinced, without adequate justification, that our optimistic predictions are correct, things may well come unstuck. Uncertainty becomes a problem if it results in loss, injury or damage – which for simplicity can be lumped together as 'harms'. The possibility of consequential harm is defined as *risk*, and it is this that is of most concern in the current context. So, realistically, most utilitarian ethical reasoning is not based on cost–benefit analysis, but on *risk*–benefit analysis.

Consequently, in the ethical matrix (3.5), the risks associated with respect for well-being add complications to assessment of the effects on the interest groups identified. Risk also affects the other principles of the matrix. For example, respect for autonomy, expressed as *consumer choice*, will be compromised if there is uncertainty about whether food labelled *GM-free* actually satisfies the legal definition; or if, when expressed as *biodiversity*, there is uncertainty about whether particular pesticide management regimes significantly affect wildlife. Similar uncertainties arise in respect of fairness. It follows that risk and uncertainty almost inevitably play a significant role in bioethical analysis.

The aim of this chapter is to analyse the concept of risk and how it is dealt with. Almost always, we want to avoid incurring risks and diminish their impacts if they result in harm. (Exceptionally, people who voluntarily take significant risks, e.g. in rock climbing or motor racing, do so for the anticipated sense of achievement or excitement.) The proactive attitude to risk avoidance is often termed *precaution*, a word which has found widespread use in the term **Precautionary Principle** (PP).

13.2 Risk

In the normal course of life we are constantly exposed to one or other **hazard**, which is defined as 'a theoretical possibility of harm'. An electric light socket is a hazard, because by unscrewing the bulb and poking your fingers in you could get a nasty shock; but given reasonable, intelligent behaviour few of us lose much sleep through fear of electrocution. Similarly, substances or systems can be hazardous by virtue of properties such as toxicity, flammability, explosivity or radioactivity. By contrast, **risk** implies a significant, statistically calculable, measure of the likelihood of harm, whether or not the actual calculation has been performed.

Defining risk as the *probability of harm* sounds straightforward enough, and an initial reaction might be that the risks associated with any scientific or technological practice should be quite capable of being calculated using standard statistical and probability theory. This is because *'objective assessment of risk is based on known rates or a logically deduced probability'*.[1] Much useful information has indeed been derived using such an approach. For example, during 1988 in the UK the risk of death due to smoking forty cigarettes per day was known to be 1 : 200; due to contracting influenza 1 : 5000; due to a road accident 1 : 8000, and due to being struck by lightening 1 : 10,000,000.[2]

But the other main feature of risk is the *consequences* of any harm occurring, so that it becomes reasonable to define risk, in its simplest terms, as:

$$\text{risk} = \text{probability} \times \text{consequences}$$

Low probabilities do not necessarily suggest low risks, because if the consequences are dire (such as deaths occurring on a large scale) the risk could still be perceived as great.

Despite the apparent simplicity of the above definition, it is often not possible to calculate risks accurately. This is because, as risk analyst Jerry Ravetz has pointed out, risk assessments based on statistical and historical data are only accurate and adequate when four conditions apply:

- the risk situation is simple, with a few separable causes
- there is a strong historical database for the various empirical probabilities
- the sorts of harm are limited in variety and can be measured on a common standard

- calculated measures of intensity can be used for comparison with related similar risks to facilitate design or management.[3]

However, science in general, and the biosciences and biotechnology in particular, frequently encounter situations where one or more of these four conditions do *not* apply; and this is hardly surprising because, when working at the frontiers of knowledge, there are bound to be areas of uncertainty. For example, there was uncertainty about the effects of CFC on the ozone layer when these chemicals were first introduced.[4] In such cases, there is a need for effective strategies for managing risks.

13.2.1 Characteristics of risk

There are clearly many features of risks affecting the seriousness with which they are treated. Thus, the significance of the consequences of a risk being realized is reflected in its:

- **severity**: this can range from slight (e.g. a mild headache), through serious (e.g. food poisoning), to a distressing and debilitating disease that ends in death (e.g. vCJD)
- **duration**: this can range from brief (e.g. minutes, as in some cases when a laboratory animal suffers reduced welfare during an experiment), through prolonged (e.g. where farm animals are housed in confined conditions throughout their lives), to permanent (e.g. when an animal species becomes extinct)
- **reversibility**: some conditions are temporary and will revert to normal after a period; others (e.g. probably, the effects of global climate change) are irreversible
- **compensatability**: some harms can be offset by other practices, or at least mitigated; others cannot be.

It is clear that coping with risk is rarely a straightforward matter. We can identify four components of the whole process:

- hazard identification
- risk assessment
- risk management
- risk communication.

Addressing the first of these is inherently difficult because hazards can be difficult to identify before they have caused problems. But guidelines such as the Hazard Analysis and Critical Control Point System (HACCP) (13.3) and Control of Substances Hazardous to Health regulations (COSHH) seek to encourage a proactive attitude to hazard identification. The other three steps are informed by some general theories, which are now discussed.

13.2.2 Risk assessment

The precise problems encountered depend on the nature of the risk. For illustrative purposes, the risks associated with a hypothetical toxic substance (*Tx*) in food will be discussed. The actual risks will clearly depend on the level of toxicity, the accuracy of the estimates made to calculate the risk to which consumers are exposed, and the evaluation of the numbers and types of people exposed to the substance. These different steps of the assessment process may be summarized as:

- hazard evaluation
- dose–response evaluation
- human exposure evaluation.

Hazard evaluation of *Tx* depends on collection of epidemiological and experimental toxicity data and the answers to questions such as 'What forms of toxicity does *Tx* cause?' and 'How certain are we that humans will be vulnerable to them?'. For example, *Tx* might have neurotoxic, mutagenic or carcinogenic effects.

The dose–response evaluation usually entails use of animals, such as laboratory rodents; and if toxicity data are available for several species, those showing the highest risk estimates are generally used to estimate risks to humans. However, there are often difficulties in interpreting the data. For example, it may not be clear whether one should extrapolate a dose–response curve linearly on the assumption that even very low doses may be toxic, or assume that at low doses there is a threshold level below which no significant toxicity occurs. In the latter case a *no-effect level* (NOEL) can be established.

But because people differ substantially from other animals (such animals are not good CAM: see 8.3.2), an arbitrary safety factor usually needs to be built into the evaluation to make it applicable to human consumption of *Tx*. In the 1950s scientists at the FDA proposed the use of a safety factor of 100 in employing animal data for calculating acceptable levels of a substance in the human diet. This was based on the product of two safety margins of a factor of 10, the first to allow for the differences in response between humans and test animals, and the second to allow for the fact that different people might vary in their sensitivity to the substance. For example, if a chronic toxicity study on *Tx*, using rats, showed a NOEL of 100 mg/kg body weight per day, the threshold level for safe consumption of *Tx* by humans would be assumed to be 1 mg/kg body weight per day – a value which has traditionally been designated the **acceptable daily intake** (ADI). It is important to note that, while in theory this *might* be a generous margin of error, it is largely an arbitrary 'rule of thumb', which has proved totally inadequate in certain cases, such as the drug thalidomide (8.3.2).[5]

The human exposure evaluation aims to assess the level of health-risk people are exposed to by consuming *Tx*. This is inherently difficult, because if it were possible to consume *Tx* in a number of different foods, or if someone had a particular appetite for a food containing *Tx*, individuals with unusual dietary habits could exceed the ADI based on average consumption patterns. So, depending on the nature of *Tx* and its occurrence in food, some groups of individuals might be especially vulnerable.

13.2.3 Risk management

This term refers to practical measures made to minimize risks, consistent with their nature and severity, and the associated policy decisions. It is clear that achieving zero risk in any activity is, in both theoretical and practical terms, virtually impossible. And even if a dangerous substance is banned, it may be used illegally. The discussion about interpretation of dose–response curves (13.2.2) also emphasizes the impossibility of defining the precise nature of risks. For example, even if human death is taken as a defined measure of how risky a process is, it will hardly ever be possible to ascribe death rates with 100% certainty to a supposed risk factor, because people die from many causes, not all of which can be ascertained accurately.

In the case of our example of food constituent *Tx*, in the UK responsibility for the choice of management options lies with the FSA, which receives advice from its several advisory committees, such as the committees on toxicity, mutagenicity and carcinogenicity (Table 14.2). Risks may be broadly distinguished as of two types. For those that are *voluntary*, with well-understood and accepted risks (such as eating certain types of shellfish), the government's role is confined to providing sufficient information to enable people to make informed choices for themselves and for those in their care, such as children. For *involuntary* risks, where people have no meaningful choice and there may be wider and more serious public health or societal concerns, the Government has an obligation to intervene to protect the public.

In some cases a ban is imposed, as occurred in the case of addition of the amino acid tryptophan to foods and food supplements. (Tryptophan supplementation has been used as a treatment for depression and insomnia but, in 1989 in the USA, 1500 people taking supplements became ill and nearly forty died from a rare blood and neurological syndrome.)[6] Inadequate regulation can lead to a loss of consumer confidence in a product, whereas if too heavy-handed an approach is adopted excessive burdens might be imposed across society, which could also erode public confidence. Consequently, the FSA's stated aim is to achieve risk policies based on *'consistency, transparency, targeting, proportionality, and accountability'*.[7]

Risk management entails making decisions about whether to ban or regulate a substance or process, and how its possible adverse effects should be advertised, monitored and controlled. For example, suitable procedures need to be established to inspect premises where food is manufactured to ensure regulations are complied with: indeed, a rigorous system of inspection might require monitoring at all stages of the food chain (10.4.6).

However, the question of whether or not to ban *Tx* cannot be based on a scientific assessment *alone*. If, according to the best evidence available (usually from animal experiments, with their recognized limitations), it is estimated that an ADI of 1 mg/kg body weight per day might cause the deaths from cancer of, say, twenty-five people over a period of fifty years, a decision has to be made as to what significance to attach to this estimate. Given that thousands of people die annually in UK road accidents, is not a figure of 0.5 deaths p.a. such a drop in the ocean that it can be ignored, or should the substance be banned as likely to cause *any* unnecessary deaths? Or if 1000 deaths in fifty years were estimated to be caused by average consumption of *Tx*, would this be 'acceptable'?

Difficulties that can be experienced due to imprecision in risk assessments are illustrated by a US National Academy of Sciences expert panel report, in the 1980s, on the hazards of using saccharine as a food sweetener. The estimates suggested that on the basis of a daily ingestion of 120 mg over seventy years by the whole US population, a minimum of 0.22

BOX 13.1 HAZARD ANALYSIS AND CRITICAL CONTROL POINT SYSTEM (HACCP)

This is internationally accepted as the system of choice for food safety management. It is a preventive system based on the following seven principles:

- identify any hazards that must be prevented, eliminated or reduced
- identify critical control points (CCP) at the steps at which control is essential
- establish critical limits at CCP
- establish procedures to monitor the CCP
- establish corrective action to be taken if a CCP is not under control
- establish procedures to verify whether the above procedures are working effectively
- establish documents and records to demonstrate the effective application of the above measures.

The principles aim to focus attention on the identification and control of microbiological, chemical and physical food safety hazards. For example, for meat plants, HACCP focuses on control measures that can reduce contamination of meat from microbiological hazards such as *Salmonella, E. coli* 0157, and *Campylobacter*, which can be carried by healthy animals, and cannot be detected by sight or smell.

deaths [sic] from bladder cancer would be caused, while at the other extreme it was unlikely that more than 1,144,000 extra deaths would be caused – an uncertainty range covering six orders of magnitude![8] However, the difficulties are rarely so extreme, and they do not preclude establishing sensible risk-management procedures such as HACCP[9] (Box 13.1).

13.2.4 Risk communication

For completeness, reference is made here to risk communication, as one of the elements of the whole risk process identified in 13.2.1. But it is too complex an issue to summarize in a few words, and much of the remainder of this chapter (and chapter 14) has a bearing on this issue, not least because risk communication is a two-way process. For the present, it is sufficient to note that governments are increasingly aware of the importance of effective communication of risks to the public, a need emphasized by the outbreak of BSE in the UK (Box 13.2). A key factor in communication is **transparency**, which entails explaining the evidence and reasoning on which risk decisions are made. Because technical issues are often discussed in terms which are incomprehensible to the lay public, there is a need for use of appropriate language, the development of skills in writing and speaking in user-friendly ways, and a willingness to be open to questioning and criticism at all stages (14.6.1).

13.3 Broader conceptions of risk

Much of the above discussion applies particularly to a **probabilistic** conception of risk, e.g. the type of risk which can be assessed from the known properties of biological materials and data on how living processes are affected. It is true to say that for a long time, and for

BOX 13.2 ADVICE OF THE SOUTHWOOD COMMITTEE ON BSE

- A scientific committee chaired by Sir Richard Southwood was appointed to advise the UK Government on the risk to humans who consumed beef from BSE-infected cattle.
- The committee assessed the risk as 'remote', but made two 'precautionary' recommendations: (*a*) that sick animals be removed from the food chain, and (*b*) that bovine offal should not be used in baby food.
- The subsequent BSE Inquiry noted that the use of the word 'remote' implied that further precautionary measures were unnecessary, which led to lax implementation of safety measures.
- But as the recommendation was made to ban bovine material in baby food, it must also have been considered a *possibility* that adults might be adversely affected by infected beef.
- Thus, on any reasonable interpretation of the PP there was a *prima facie* case (see 3.2) for also banning beef for consumption by adults.

some people even now (notably, some scientists and prominent politicians), this is the only sort of risk that is meaningful; people who do not fully accept this 'scientific' basis of risk are often thought irrational, e.g. *'Government officials and politicians often express frustration at the "inaccurate and emotional" reactions of the public in the GM debate.'*[10]

But the latter attitude may be thought to be both patronizing and inaccurate, because people often do not limit their conception of risk to the probabilistic kind. For them, the acceptability of imposed risks is, in addition, subject to a number of factors which have less to do with the actual consequences of harm occurring, and much more to do with the context of the risk. These factors include:

- **voluntariness**: some risks are willingly taken, with a view to potential gains (so we might consider that people who takes such risks 'are asking for it'), whereas others are inflicted on people, animals or the environment without permission or even forewarning
- **distribution of risks and benefits**: although, overall, benefits may be delivered, either they, or accompanying risks, may be unfairly distributed
- **trust** in those who make the critical decisions – which may be low where the decision-makers are perceived to have a vested interest in minimizing concerns about the risk
- **unfamiliarity**: novel circumstances may cause apprehension or fear (*neophobia*)
- **threats to a perceived natural order**: a risk may be perceived to be 'a step too far', or to move on to a 'slippery slope'
- **level of compensation** paid to those who suffer ill effects.

The validity of some of these considerations has been acknowledged by John Krebs, then Chief Executive of the Natural Environment Research Council (NERC), and subsequently of the FSA, as follows:

Disagreeing with a scientific estimate of risk is not necessarily irrational: the evidence on analysis may be incomplete, the scientist may have a vested interest in selecting particular bits of evidence, or there may be more than one particular interpretation of the facts. Interestingly, surveys show that within the EU, people in those countries with the best understanding of how science works also tend to be the most sceptical about the ability of science to resolve everyday problems.[11]

In reality, many supposedly scientific judgements about risk are unavoidably subjective, because they depend on (often unstated) assumptions about the nature, magnitude and significance of matters over which there is uncertainty. This means that what are called the **framing assumptions**, i.e. how the problem is framed before attempting to assess the associated risks, can have a major impact on the risk assessment that is arrived at. For example, in the UK insect-resistant crops are treated as *novel organisms*, whereas in the USA they are treated under pesticide legislation – which clearly makes a difference to how the risk assessment is framed. And this will be true even if there is no disagreement about the 'scientific facts' – which is often not the case. For sociologist of science Alan Irwin: *'Despite the rhetoric to the contrary, therefore, scientific analyses must reflect the ideological and institutional assumptions of the "experts" who conduct them, although these assumptions are not necessarily consciously made, and indeed their existence may be strongly denied by those who hold them.'*[12]

Many of the issues included in these broader conceptions of risk have pronounced psychological links, so that an important dimension of risk is the way it is perceived. **Risk perception** varies with individuals' personalities and psychological predispositions, and strategies have been developed to try to understand and predict how people will respond to different types of hazard. The commonest approach employs the so-called **psychometric paradigm** to produce quantitative representations or **cognitive maps** of risk attitudes and perceptions. For example, the most dreaded risks are those which are perceived as unknown, unobservable, uncontrollable, novel, involuntary and inequitable, and whose impacts are potentially catastrophic and a high risk to future generations.[13] To some degree, risk perception can also be correlated with political outlook,[14] and undoubtedly it can be amplified by the ways risks are treated by the news media (14.6.2).[15]

Although in this book, ethics has been interpreted in a way that includes all issues concerned with 'how we should act', the popular conception of ethics is often limited to just those non-scientific factors listed above. And in that sense, it would be true to say that many people now perceive risk as not just a scientific issue, but one also demanding attention to ethics. Indeed, if the common morality (3.1.1) fairly represents a general consensus on what is of value in society, risk might be appropriately perceived as the probability of harm due to failure to respect the principles identified in the ethical matrix (3.5).

13.3.1 The risk society

In recent years, sociologist Ulrich Beck has shown how developments in modern industrialized society (often called the *postmodern* society) have become dominated by perceptions of risk. According to Beck, risk is now an important principle shaping social and institutional structures, and replacing wealth and privilege as the focus of political concerns. He argues that in this **risk society**, the values people assign to different aspects of the world, the trust they place in those who wield power over their lives, and their concept of personal rights, all come to assume at least as much importance as the technological and economic factors which have traditionally underpinned their perception of risk.[16]

Moreover, many risks that are now thought to threaten life on Earth are beyond our power to avoid or alter. The fact is that, as individuals, we can do little to affect the likely outcomes of, for example, climate change (12.3.3), the presence in the environment of hormone-disrupting chemicals, the proliferation of weapons of mass destruction, or the disruptive effects of computer viruses. Perhaps, unsurprisingly, public anxieties over each of these threats to our security are combining to produce, in many people, a state of mind characterized by fatalism, disillusion and mistrust. And reassurances made by governments or industry are often unconvincing because they are seen as cynical attempts to protect their own interests.[17]

Even so, many current biotechnological trajectories appear to have a momentum which it could be difficult to stem. If large investments of human resources, financial capital and reputation have been made in a project, the decision to see it through to completion becomes almost irreversible; and this is especially so if the only conception of risk permitted consideration is one based on probabilistic risk assessments. There will always be a strong motive to capitalize on *sunk costs* (11.8.2). Basing risk assessments on the broader ethical issues identified may thus be an inevitable feature of many people's worldview in the *risk society*.

13.4 The Precautionary Principle

We have seen that probabilistic risk assessments have a number of limitations. For example, in the case of a toxic food constituent:

- the framing of the assessment is sometimes arbitrary
- the dose–response assessment and other aspects of risk evaluation are often imprecise, e.g. because of the difficulty of extrapolating from animals to humans
- the human exposure evaluation may be uncertain
- the permitted ADI is subjective, because based on an arbitrary decision as to what level of risk is acceptable.

So it would be reasonable to conclude that basing risk management on scientific risk assessment alone is inherently unscientific! Moreover, if the putative risk became apparent only after a *delay*, this combination of limitations could, in theory, produce not only an inaccurate risk assessment, but also serious, irreversible, unforeseen consequences. The possibility of such consequences was first appreciated in respect of risks to the environment, of which emissions of greenhouse gases and their putative impact on global warming are a prime example (12.3.3). The reality is that predictions of the impacts of industrial activity on the global environment are characterized by **uncertainty**. Uncertainty is the product of two sorts of limits:

- the *current limits* of actual scientific knowledge (the **unknown**) – due to the lack of appropriate research (e.g. as a result of inadequate funding and/or suitably qualified scientists)

- the *in principle limits* (the **unknowable**) – due to factors such as:
  - the intrinsic difficulties of performing research in the 'open system' of the biosphere
  - the logistical difficulties of testing for the **cocktail effect** – in which, for example, chemicals thought to be safe when tested individually may interact to produce harmful mixtures.

It should be appreciated that: '*This negative assessment of the ability of environmental science to deliver the safety assurances that society requires is not just made by a few dissenters: it is the current mainstream opinion.*'[18]

It is perhaps not surprising then that growing awareness of these problems has led to international action. At the UN Conference on the Environment and Development held in 1992 in Rio de Janeiro, a set of principles were formulated that form the so-called *Rio Declaration.* Of these, Principle 15 states:

In order to protect the environment, the precautionary approach shall be widely applied by States according to their capabilities. Where there are threats of serious or irreversible damage, lack of full scientific certainty shall not be used as a reason for postponing cost-effective measures to prevent environmental degradation.[19]

Often, this has since been called the *Precautionary Principle* (PP), and it has become enshrined in much environmental and health policy, from strategies for sustainable development, to the EU Directive on GMO releases, to the Cartagena Protocol on Biosafety, to the Convention on Biodiversity.

Despite frequent reference to *the* PP, it has no official definition, and is consequently open to different interpretations. The often-quoted Rio Declaration is, in fact, a fairly weak version, in that it does not commit the parties concerned to regulatory action. By contrast, the London Declaration on Protection of the North Sea (1987) states that substances will be regulated '*when there is reason to assume that certain damage or harmful effects on the living resources of the sea are likely to be caused by such substances, even when there is no scientific evidence to prove a causal link between emissions and effects*'. Perhaps not surprisingly, different interest groups have tended to adopt the interpretation that best suits their ideological or political worldviews.

13.4.1 International differences in interpreting the Precautionary Principle

In many respects the different interpretations of the PP are epitomized by the contrasting attitudes to the licensing of GM crops adopted by the USA and the EU respectively (10.1.2).

In the USA, the Animal and Plant Health Inspection Service (APHIS), which is responsible for regulating the planting, distribution, and harvesting of GM crops, allows companies to field test GM varieties of six common crops (maize, cotton, potatoes, tomatoes, soyabeans and tobacco) without seeking approval from the agency. All that is required is notification thirty days in advance of planting and certification that the tests comply with approved standards. If APHIS considers that a GM plant poses no significant risks to other plants in the environment, and is as safe as traditional varieties, it can be granted 'non-regulated status', allowing its widespread commercialization.

With respect to food safety, the FDA announced in 1999 that *'it has not found it necessary to conduct comprehensive scientific reviews of foods derived from bioengineered* [i.e. GM] *plants based on the attributes of these products but... expects developers to consult the agency on safety and regulatory questions.'* The FDA has established a notification procedure whereby *'any person may notify the* [FDA]... *that a particular substance is GRAS'*, i.e. *'generally recognised as safe by acknowledged experts'.*[20]

In the EU, the regulation of GM food is much more stringent, and this undoubtedly reflects public concerns (10.1.2). The EU's initial stance was to proceed by a step-by-step approach, progressively relaxing regulations as experience was gained. In practice, as GM products neared the commercial stage numerous objections raised by Member States (especially Denmark and Austria) led to a tightening of the precautionary measures, and subsequently objections to GM became widespread. The result was the introduction of a **virtual moratorium** on the approval process for GM crops in the EU from 1998 onwards. (Spain has been growing GM maize since before 1998.) In 2004, the approval process was re-started, but many objections are being lodged, which are delaying approvals.

In response to the *virtual moratorium*, and to pressure from the USA concerning the EU's regulatory delays (10.1.2), the European Commission introduced a **Communication on the Precautionary Principle** in 2000.[21] This sought to anticipate objections from the USA and to lay down guidelines emphasizing the systematic, objective and scientific nature of the EU's approach. The Communication recognizes that decision-makers are constantly faced with the dilemma of balancing two opposing factors:

- the freedom and rights of individuals, industry and organizations
- the need to reduce the risks of adverse effects to the environment, human, animal and plant life, and to do so in ways that are *'proportionate, non-discriminatory, transparent, and coherent'.*

The Communication describes the application of the PP as a three-stage process, consisting of:

(a) a trigger stage

(b) a decision stage, and

(c) an implementation stage.

Priority is given, at all stages, to the importance of scientific evidence – although it is stressed that stage (*b*) entails political judgements.

However, policy analyst Susan Carr argues that **value judgements** (i.e. ethical considerations) are important at all three stages, and that more explicit statement of these values would strengthen the EU's position. For example, the Communication claims that at stage (*a*) scientific risk assessment can decide whether or not use of the PP should be triggered. But in the case of impacts of GM crops on the environment, a value judgement first has to be made as to what counts as harm for the purposes of this assessment. Initially, in the EU's assessment of risks of herbicide-tolerant GM crops, indirect effects (e.g. of management practices on biodiversity) were excluded. However, due to pressure from Denmark, France, and the UK it became EU policy to include these – and they formed the basis of the *farm-scale evaluations* in the UK (10.5.4). Analogously, the appropriate baseline for comparison of environmental impacts is a value judgement. Conventional agriculture is the usual comparator, but Austria uses organic agriculture as the baseline, which could be seen as prudent in achieving more sustainable practices.[22]

Carr suggests that the Communication's emphasis on scientific and objective aspects ignores the fact that the PP needs to be viewed as complementary to science. Rather than uncertainty being always viewed as temporary, so that with more 'evidence' it can be integrated into science-based risk assessment, it might be wiser to recognize that values underlie many scientific controversies. In that case, by permitting the introduction only of socially acceptable technologies, a stronger version of the PP could prove as economically sound as it was politically prudent.

In 2004 the EU introduced new legislation on traceability and labelling of GM food products (10.4.6). Such measures were designed to improve consumer confidence in the EU food regulatory system, which in the wake of crises such as BSE and dioxins in food had been low for many years. However, in the USA, government officials, farm groups, biotechnology companies, and food processors and manufacturers expressed deep concern over the impacts of the new EU labelling and traceability requirements, considering them *'costly, unworkable, unenforceable, and discriminatory against US agricultural products'*.[23]

13.5 Different formulations of the Precautionary Principle

If ethics is to conform to the disinterested, rational principles which it has been assigned (3.2), special pleading can hold no weight. We thus need to examine the ethical bases of the different interpretations advanced for the PP. According to its critics, the PP:

- lacks uniform interpretation: one study found fourteen different interpretations, while the Cartagena Protocol adopts it in an ambiguous manner
- marginalizes the role of scientists, e.g. by requiring a relaxation of the standards of proof normally required by the scientific community
- acts as a veiled form of trade protectionism, by circumventing the rules established by trade agreements and enforced by the WTO
- acts as a form of over-regulation that will lead to a loss of potential benefits and e.g. cause food shortages.

However, to accept the full force of all these criticisms would be to deny that there is any virtue in a precautionary approach, and few people would consider that that was an acceptable position. So what are the issues over which there can be reasonable agreement, and disagreement, in relation to the PP? A valuable analysis is provided by the report of an enquiry conducted by the Royal Society of Canada, and the following five issues are drawn largely from its analysis.[24]

13.5.1 Recognizing scientific uncertainty and fallibility

An important feature of the scientific method (1.5.1) is that it stresses the healthy sense of scepticism which should be brought to bear on all knowledge claims. This implies that there is often a possibility of error in estimating risks, and that the higher the risks the greater the required level of precaution. It is useful to distinguish here between the **laboratory scientist**, who can, and is required to, take time and effort to reduce uncertainty in affirming or rejecting a scientific hypothesis, and the **regulatory scientist**, who in making a decision rarely has the time or resources to reduce uncertainty. In the latter case, a sound interpretation of the PP would involve identifying the riskiest outcomes and taking measures to avoid them.

A commonly cited implication of this approach is the need to differentiate clearly between the **absence of evidence** and **evidence of absence** of a risk. The claim that 'there is no evidence that harm might result' can mean everything from 'no evidence of harm has been looked for' (i.e. a statement of ignorance) to 'thorough investigations have failed to discover any evidence of harm', which might qualify as 'evidence of absence'. Often, the above two interpretations have, erroneously, been conflated.

13.5.2 Presumption in favour of health and environmental values

In any meaningful interpretation of the PP, uncertainty is managed in a way that favours values such as health and environmental sustainability over other values, i.e. it is considered better to forgo important benefits by wrongly predicting harm than to experience serious harms by failing to foresee them. That is to say, according to the PP it is better to commit a **Type I error** (falsely predicting an adverse effect) than to commit a **Type II error** (falsely predicting no adverse effect when there is one). The standards of laboratory science generally require just the opposite – that it is a more grievous error to commit a Type I than a Type II error. The rules of evidence in courts of law show an equivalent bias to that of laboratory science, in that it is considered better that ten guilty people should be acquitted than that one innocent person should be wrongly convicted. Thus the PP violates the principle that generally governs both scientific research and criminal law.

This, of course, is the basis of the criticism (13.5) that the PP tends to restrict the development of new technologies by over-regulation. However, the PP need only establish a *presumption* in favour of safety over benefits, and weaker versions of the PP would apply proportionality rules, which took into account the costs of exercising caution. The weakest interpretations of the PP would indeed reduce the exercise to a simple risk–benefit analysis – which could ultimately undermine the point of the PP.

However, it is important to appreciate that there is a *trade-off* between Type I and Type II errors, because giving precedence to safety will entail costs in terms of surveillance, remedial management procedures, and possible lost economic opportunities, while giving precedence to innovation may carry losses due to damage inflicted on people, animals, and/or the environment.[25] Trade-offs are also encountered in assessing the relative risks of similar alternative practices, e.g. when deciding whether to employ a pesticide which is highly toxic but degrades rapidly and completely, or one that is only mildly carcinogenic but leaves residues in food.[26]

13.5.3 Proactive approaches to risk

A presumption of the PP is that it is reasonable to take proactive measures to avoid harm rather than to proceed on the assumption that any harms that occur can be dealt with afterwards (reactively). It follows from the PP then that certain measures are necessary in implementing a new technology, for example:

- research should be conducted to identify potential risks
- when there are uncertainties about risks, introduction of technologies should be delayed until reasonable confidence is acquired in reducing the risks to acceptable levels
- technologies should be designed to minimize risks.

13.5.4 Burden of proof

In cases of alleged criminal acts the prosecution bears the burden of proof, and must establish guilt 'beyond all reasonable doubt'. Proponents of biotechnology arguing for a weak version of the PP often suggest that the legal regulation of risk should adopt the same principle – that the technology should be presumed safe unless it can be proved otherwise. If the proof of risk were to be science-based in the strongest sense, it would follow the norms of laboratory science, usually defined in terms of a 95% confidence rule (i.e. with a probability of error of less than 5%). But this would imply that, to paraphrase the legal terminology (13.5.2), 'it is better that ten hazardous technologies should be employed to the detriment of human and environmental health than that one safe technology should be restricted.'

Because that conclusion is so obviously unacceptable, most versions of the PP shift at least some of the burden of proof (to establish that the technology is safe) onto the technology providers, or relax the standards of evidence required for the suspicion of unacceptable risks. Critics of the strong version of the PP argue that this puts an unacceptable burden on technology providers to *prove* safety, which is logically impossible – because it cannot be proved that harm will *never* occur. An alternative, weaker, version of the PP suggests that the standard used should appeal to the 'balance of evidence', which would only require proponents to prove that there is no *prima facie* case for saying that technology is unsafe. But even this position has been overlooked: the early scientific advice to the UK government on the risks of consuming BSE-infected beef (before the link between the cattle disease and the human form was established) is an example of a case where a *prima facie* health risk was discounted (Box 13.2).[27,28]

It is important to stress that precaution does not mean 'paralysis', but it does entail shifting the burden of narrowing the uncertainty range, and removing the theoretical unknowns, onto the shoulders of the technology providers. Nor is it inevitable that the burden of proof is born exclusively by either proponents or opponents of an innovation. A balance of burdens might be a fairer and more prudent outcome.[29]

13.5.5 Standards of acceptable risk

As noted in 13.2.1 and 13.2.3, the acceptability of risk is based on a number of value judgements. These include factors such as voluntariness, fair distribution of the risks and benefits, familiarity with the technology, and the trustworthiness of the risk manager (13.3). But even confining attention to physical measures, there are many instances where science is unable to specify whether a product or process is 'safe', because the criteria for safety (or 'acceptable risk') are themselves contestable.

13.6 Trust

In the absence of appropriate knowledge, or of the ability to understand the information available, lay people would seem to have to trust experts to make judgements for them in distinguishing between danger and safety. But this is a potential problem, because trust has to be earned and can easily be lost. In the UK and other DC, recent years have seen a growth of distrust in scientists and governments, largely because they have been seen as acting irresponsibly (or perhaps dishonestly) over issues like the safety of eating beef (later shown to be responsible for causing cases of vCJD when it was infected with BSE[30]), and the marketing of unlabelled GM foods (10.1.2). This has resulted in many people feeling so let down that they have an instinct to withdraw trust from all experts - with the result that many are instead placing their trust in alternatives to modern technologies. But in reality such strategies are just as open to (self) deception as others. For as philosopher Onora O'Neill puts it: *'Some partisans of the nicer, greenish remedies, and fashionable therapies extol herbal medicines, aromatherapy, strenuous forms of massage and manipulation, spiritual exercises, home births, and exotic diets, while offering no more than anecdotal evidence for their efficacy or even for their safety.'*[31]

So their reaction is simply not feasible or even coherent, because we *cannot avoid* trusting other people most of the time. To cope at all in modern society we are almost compelled to rely on many 'experts' (such as doctors, motor mechanics, airline pilots, food safety controllers, civil engineers, electricians, *et al.*). Risk may now be a feature of modern industrialized society (13.3.1), but it would be foolish to jump out of the frying pan into the fire.

Traditionally (perhaps up to 300 years ago), most individuals were governed by the nobility, priests and war lords, an arrangement that required them to have plenty of honour and faith, but little by way of trust. Now, when the majority value their autonomy and human rights within liberal democratic societies, trust has come to play a major role.[32] How, then, can bioscientists and biotechnologists, and those who wish to support them, build the trust of a sceptical public?

13.6.1 Trust and trustworthiness

An important distinction to be aware of at the outset is the difference between trust and **trustworthiness**. Since the widespread decline in public trust in the scientific establishment, efforts have been made to increase the trustworthiness of scientists and politicians by two types of strategy. First, there has been a move to establish rigorous regulations, policies and practices to ensure that certain ethical standards are met. Often these are incorporated into professional codes of practice. But in order to demonstrate that these are not just 'fine words' and that the rules are complied with, procedures have to be introduced for monitoring, inspection and the imposition of penalties for non-compliance. Prominent examples are the Home Office inspectorate, which is responsible for overseeing scientists' compliance with laws governing animal experimentation (8.8), and food standards officers, who perform a similar role for the food industry.

There is also a growing tendency to introduce systems of 'auditing' that are aimed at making public bodies more accountable, exemplified by the league tables introduced for judging the performance hospitals and police forces. Often these are associated with targets, designed to reward good performers and chasten weak performers. However, although this new 'audit culture' may have been introduced to improve trustworthiness, the fear of failure sometimes encourages an unhealthy emphasis on winning 'brownie points'; it has been argued that for such reasons: *'the very idea of restoring trust by increased audit is doomed'.*[33]

The second strategy for increasing trustworthiness is built on 'openness' or **transparency**. Transparency is certainly now a feature of most UK government committees, who publish the agendas and minutes of their meetings on the internet, and in several cases conduct most (e.g. the AEBC) or several (e.g. the FSA) meetings in public. For those with the time and know-how, gaining insight into the political decision-making process, e.g. over the licensing of GM crops or new regulations on reproductive technology, is a straightforward matter; often hard copies of committee reports are sent free on request.

Some procedures that have been introduced aim to combine both the audit and transparency criteria. For example, an attempt to build public trustworthiness in decision-makers is illustrated by the **Nolan Principles** (Box 13.3), which set standards for those serving in public life, e.g. on government advisory committees (14.3).[34] Of course, several of the principles might be regarded merely as aspirations, so there is little guarantee that all the standards are strictly observed.

BOX 13.3 THE NOLAN COMMITTEE'S SEVEN PRINCIPLES OF PUBLIC LIFE

Selflessness Holders of public office should take decisions solely in terms of the public interest. They should not do so in order to gain financial or other material benefits for themselves, their family, or their friends.

Integrity Holders of public office should not place themselves under any financial or other obligation to outside individuals or organizations that might influence them in the performance of their official duties.

Objectivity In carrying out public business, including making public appointments, awarding contracts, or recommending individuals for rewards and benefits, holders of public office should make choices on merit.

Accountability Holders of public office are accountable for their decisions and actions to the public and must submit themselves to whatever scrutiny is appropriate to their office.

Openness Holders of public office should be as open as possible about all the decisions and actions that they take. They should give reasons for their decisions and restrict information only when the wider public interest clearly demands.

Honesty Holders of public office have a duty to declare any private interests relating to their public duties and to take steps to resolve any conflicts arising in a way that protects the public interest.

Leadership Holders of public office should promote and support these principles by leadership and example.

In fact, despite the implementation of such measures, trust often remains low, and/or it is misplaced. The relationship between trust and trustworthiness can be represented by a simple 'trust grid':

	Trustworthy	Untrustworthy
Trust	SOUND	Gullible
Distrust	Cynical	SOUND

It is clearly sound advice to trust the trustworthy and distrust the untrustworthy. But many people, gullibly, trust the untrustworthy, and/or, cynically, distrust the trustworthy. In such circumstances, *being* trustworthy provides no guarantee that people's trust is apportioned in a sound manner. Moreover, even though elaborate systems of accountability and transparency are put in place, if people have a distrust of *these systems* then trust itself may still be hard to establish.[35]

In relation to biotechnology, an EU survey of public opinion indicated that although about 70% of Europeans had confidence in doctors, university scientists, consumer organizations and patients' organizations, less than 55% had the same opinion of scientists working in industry, newspapers and magazines, environmental groups, shops, farmers and the European Commission. Moreover, less than 50% had confidence in their own government and in industry.[36] As noted in 13.3, trust is often highly conditional on perceptions of whether the people seeking to gain trust have a vested interest in promoting a product or process, or downplaying its risks.

A factor that is sometimes overlooked is the psychological motivation of the expert. People regarded as 'experts' are expected to have authoritative opinions on their subjects, which will inspire the confidence of others. Experts rarely admit that they do not know the answer to problems posed in their field of expertise, for to do so would imply that they are incompetent. But the result is that their pronouncements, sometimes tentative and sometimes uninformed, tend to get vested with an authority which may not be justified. A historical example illustrates the point,[37] but the pronouncements of the Southwood Committee (Box 13.2), which did not include an expert on prion diseases, emphasizes the more recent problems of defining what the appropriate 'expertise' is.

13.6.2 Re-establishing trust

One response to the evident low level of public trust in the scientific establishment is that the answer is simply 'more of the same'. For example, engaging the public more frequently and more openly in discussions, debates, and citizens' juries about biotechnology will, it is argued, gradually re-build trust (a question examined more fully in 14.6).

But another approach suggests that a more radical change might be required in the ways the public learn about science and technology - i.e. primarily through the news media. (e.g. see Figure 13.1). At present, the media are subject to few of the controls

Daily Mail

THURSDAY, SEPTEMBER 28, 2000 Newspaper of the year 35p

So why did Tony get so hot under the collar? PAGES 32-33

WARNING OVER GM ANIMALS IMMUNE TO PAIN

GENETIC modification could lead to 'zombie' farm animals programmed to feel no pain or stress, Government advisers warned yesterday.

The disturbing image of ranks of 'animal vegetables' was painted by members of the new Agriculture and Environment Biotechnology Commission as

By **Sean Poulter**
Consumer Affairs Correspondent

part of an attempt to raise public awareness of the next GM battlefield – the Frankenstein Farmyards.

Despite the backlash over GM crops, most people are still ignorant of the rapid developments and ethical implications of genetic engineering of animals, they said.

Regulatory action is needed soon because current experiments will soon be turned into commercial reality. Laboratories are already working to produce chickens which have more breast meat and sheep which grow more wool.

Cows are being altered to ensure their milk is less fatty while pigs are being engineered to grow at incredibly fast rates.

One firm in the U.S. has said it will be ready to start supplying genetically-engineered salmon to fish farms around the world within a year. 'Super-animals'

Turn to Page 6, Col. 3

Terror of the honeymooners in ferry disaster

Honeymooners Marianne and Stephen Richards clung to each other for life after the sinking Greek ferry Express Samina tossed them into the stormy Aegean. Up to 100 died.

Full story – Pages 8 & 9

INSIDE: Keith Waterhouse 14, Nigel Dempster 41, Femail 43-56, TV 57-60, Letters 62-63, Coffee Break 71-73, City 74-76

Figure 13.1 An example of the hype in tabloid journalism. The article referred to a government committee's discussion document which noted the theoretical possibility of producing insentient animals by genetic modification (GM), but stressed that no such work was in process

which ensure that other professionals, such as doctors, scientists and civil servants behave trustworthily. If the latter fail to do so, they face stiff penalties - and politicians can be voted out of office. But newspaper journalists face few equivalent sanctions, and in an effort to increase sales irresponsible reporting of 'scare stories', particularly in certain tabloid newspapers, can have an unwarranted impact in impairing trust.

Clearly, control of the press is ethically problematical. A measure of a free society is its willingness to allow the expression of ideas which the government, and even the majority of the electorate, disagree with. Censorship is rightly regarded with misgivings, and most would argue that it can only be exercised justifiably in times of military conflict or to protect the innocent and vulnerable from abuse, e.g. from sadistic pornography.[38] Even so, according to Onora O'Neill: '*any newspaper or programme that presents some of its content as reporting has no unrestricted right to communicate untrustworthily, for example by substituting factoids and half-truths, rumours, and rubbish for reporting, knowing that at least some (and probably many) readers cannot hope to disentangle them from reliable reporting.*'[39] Arguing that unless standards of reporting are improved, we cannot expect any increase in public trust in medicine, science, and biotechnology, she makes the following suggestions for improvements:

- owners, editors, and reporters should be required to declare relevant interests and conflicts of interest, and relationships with lobbyists, political parties, companies, and campaigning organizations
- media should be required to publish credentials of reporters writing on technical topics, and to declare full information about payments made to obtain information
- where misinformation is published, corrections of equal prominence should be published
- penalties could be imposed for quoting stories shown to be libellous or invented.

The critical question concerns how the tension between autonomy and trust can be managed. Valuing the autonomy, especially of those who own and work for the media, more highly than trust may mean that scientists continue to be portrayed as untrustworthy. And, as a generalization, that would certainly be unjust.

THE MAIN POINTS

- Although modern technology confers many benefits, it is sometimes accompanied by risks of harm to people, animals, and the environment, which it is an ethical responsibility for scientists to seek to avoid or ameliorate.
- The novelty of modern biotechnologies may be accompanied by risks and uncertainties which are difficult both to assess accurately and to manage.
- Moreover, probabilistic risk assessments do not adequately account for many people's wider concerns, such as: voluntariness, trustworthiness, fair distribution of risks, challenges to the *natural order*, and adequate levels of compensation for harms suffered.

- Environmental concerns led to formulation of the Precautionary Principle (now applied more widely), but there are marked differences in the ways different people interpret the principle.
- The widespread scepticism of the general public over claims made for biotechnology suggests an urgent need for the scientific community to commit time and resources to rebuilding trust, and for additional controls on irresponsible reporting by some sections of the media.

EXERCISES

These can form the basis of essays or group discussions:

1. Critically discuss whether current food safety risk assessments in the UK are sufficiently, or too, rigorous (13.2.2–13.2.3).
2. Given their experience with GM crops, has the attitude of the USA Government to the application of the Precautionary Principle been justified? (13.4.1)
3. Will rigorous application of the Precautionary Principle stifle all technological innovation? (13.4)
4. Put yourself in the shoes of a member of the Southwood Committee advising the Government on the risks from consuming BSE-infected beef (before the link with vCJD was known). What would you have recommended to your fellow committee members, and why? (Box 13.2)
5. What measures can bioscientists take to re-build the trust of the general public? (13.6)

FURTHER READING

- *Citizen Science* by Alan Irwin (1995). London, Routledge. Examines the relationship between science, the public and environmental threats.
- *Calculated Risks: the toxicity and human health risks of chemicals in our environment* by Joseph V Rodericks (1992). Cambridge, Cambridge University Press. A readable text that dissects the procedures involved in toxicity testing.
- *Handling Uncertainty in Scientific Advice* (2004). Postnote 220. London, Parliamentary Office of Science and Technology. A concise account of developments in risk and uncertainty.

USEFUL WEBSITES

- **http://www.rsc.ca** *Royal Society of Canada* (2001) Elements of Precaution: recommendations for the regulation of food biotechnology in Canada.
- **http://europa.eu.int/comm/environment/** *European Commission* (2000) Communication on the Precautionary Principle.

- **http://www.bseinquiry.gov.uk/report/** *The BSE Inquiry: The Report* (2000): the official Government report on the BSE outbreak. Contains much that is of general relevance to the issue of biorisks.

NOTES

1. Sprent P (1988) Taking Risks. London, Penguin, p. 21
2. British Medical Association (1990) Living with Risk. London, Penguin, p. 28
3. Ravetz J R (1990) The Merger of Knowledge and Power. London, Mansell, p. 34
4. Parliamentary Office of Science and Technology (2004) Handling uncertainty in scientific advice. Postnote 220. http://www.parliament.uk/post
5. Rodericks J V (1992) Calculated Risks: the toxicity and human health risks of chemicals in our environment. Cambridge, Cambridge University Press, pp. 180–201
6. Nestle M (2002) Food Politics. Berkeley, University of California Press, p. 250
7. May R M (2000) Review of risk procedures used by the government's advisory committees dealing with food safety (Chair). London, DTI
8. Millstone E (1996) Food safety: the ethical dimensions. In: Food Ethics. (Ed.) Mepham B. London, Routledge, pp. 84–100
9. Food Standards Agency (2004) HACCP in meat plants. http://www.foodstandards.gov.uk
10. Economic and Social Research Council (1999) The Politics of GM Food: risk, science and public trust. ESRC Global Environmental Change Programme, University of Sussex, p. 8
11. Krebs J and Kakelnik A (1997) Cited in: Economic and Social Research Council (1999) The Politics of GM Food: risk, science and public trust. ESRC Global Environmental Change Programme, University of Sussex, p. 8
12. Irwin A (1995) Citizen Science. London, Routledge, p. 30
13. Slovic P (1998) Perception of risk. In: Risk and Modern Society. (Eds) Lofstedt R E and Frewer L J. London, Earthscan, pp. 31–43
14. Wildavsky A and Dake K (1998) Theories of risk perception: who fears what and why? In: Risk and Modern Society. (Eds) Lofstedt R E and Frewer L J. London, Earthscan, pp. 101–113
15. Kasperson R E, Renn O, Slovic P, Brown H S, Emel J, Goble R, Kasperson J X, and Ratik S (1998) The social amplification of risk: a conceptual framework. In: Risk and Modern Society. (Eds) Lofstedt R E and Frewer L J. London, Earthscan, pp. 149–162
16. Beck U (1992) Risk Society: towards a new modernity. London, Sage
17. Stirling A and Mayer S (1999) Rethinking Risk. University of Sussex, Science Policy Research Unit
18. Parker J (2001) Precautionary principle. In: The Concise Encyclopedia of the Ethics of New Technologies. (Ed.) Chadwick R. San Diego, Academic Press, pp. 341–349
19. United Nations Environment Programme (1992) Rio declaration on environment and development. http://www.unep.org
20. Food and Drugs Administration (2002) http://www.gao.gov/new.item/
21. European Commission (2000) Communication from the Commission on the Precautionary Principle: COM (2000) 1. Brussels, E C
22. Carr S (2002) Ethical and value-based aspects of the European Commission's precautionary principle. J Agr Environ Ethic *15*, 31–38

23. Pew Initiative on Food and Biotechnology (2003) US v. EU: an examination of the trade issues surrounding genetically modified food. http://pewagbiotech.org/resources
24. Royal Society of Canada (2001) Elements of Precaution: recommendation for the regulation of food biotechnology in Canada. Ottawa, Royal Society of Canada
25. Van den Belt H and Gremmen B (2002) Between precautionary principle and 'sound science': distributing the burdens of proof. J Agr Environ Ethic *15*, 103–122
26. Thompson P B (1995) Risk and responsibilities in modern agriculture. In: Issues in Agricultural Bioethics. (Eds) Mepham T B, Tucker G A, and Wiseman J. Nottingham, University Press, pp. 31–45
27. Phillips N A, Bridgeman J, and Ferguson-Smith M (2000) The BSE Inquiry: into BSE and variant CJD in the United Kingdom. Stationery Office. http://www.bse.org.uk
28. Report of the Working Party on Bovine Spongiform Encephalopathy (Southwood Committee) (1989). London, Department of Health and Ministry of Agriculture, Fisheries and Food
29. Van den Belt H and Gremmen B (2002) Between precautionary principle and 'sound science': distributing the burdens of proof. J Agr Environ Ethic *15*, 103–122
30. Phillips N A, Bridgeman J, and Ferguson-Smith M (2000) The BSE Inquiry: into BSE and variant CJD in the United Kingdom. Stationery Office. www.bse.org.uk
31. O'Neill O (2002) Autonomy and Trust in Bioethics. Cambridge, Cambridge University Press, p. 121
32. Lofstedt R E and Frewer L J (1998) Introduction. In: Risk and Modern Society. (Eds) Lofstedt R E and Frewer L J. London, Earthscan, p. 13
33. O'Neill O (2002) Autonomy and Trust in Bioethics. Cambridge, Cambridge University Press, p. 133
34. Nolan Committee (1995) www.archive.official-documents.co.uk/document/parliament/nolan
35. O'Neill O (2002) Autonomy and Trust in Bioethics. Cambridge, Cambridge University Press
36. Eurobarometer 58.0 (2003) Europeans and biotechnology 2002. http://www.lifesciencesnetwork.com/Repository/Eurobarometer2002
37. Mepham T B (1993) 'Humanizing milk': the formulation of artificial feeds for infants (1850–1910). Med Hist *37*, 225–249
38. Belsey A and Chadwick R (1992) (Eds) Ethical Issues in Journalism and the Media. London, Routledge
39. O'Neill O (2002) Autonomy and Trust in Bioethics. Cambridge, Cambridge University Press, p. 187

14

Politics and the biosciences

OBJECTIVES

When you have read and discussed this chapter you should:

- be aware of the ways in which the UK Government supports and administers research in the biosciences, in both research institutes and universities
- be aware of how science policy is influenced by the Government's various committees, commissions and international agreements
- appreciate the ways in which the public can be involved in influencing science policy and the parts played by the news media, pressure groups and ethical consumerism
- be familiar with two examples of *deliberative inclusionary process* concerning the growing of GM crops, one in the UK and one in India
- be able to formulate a view on how best to promote democratic decision-making in the biosciences

14.1 Introduction

It is a truism that science and technology are crucial elements of life in modern society. But this is a relatively recent state of affairs. It was only from the mid-nineteenth century that scientific education began to appear on the university curriculum; and Justus von Liebig, a pioneer of biochemistry in Germany, was probably the first person to introduce practical classes for undergraduates. But even by the early twentieth century there was no serious government interest in science. As so often happens, crisis forced change ('necessity is the mother of invention'), and it was the World War of 1914–18 that persuaded the UK government that greater investment in science and technology was urgently needed. The Department of Scientific and Industrial Research was established in 1916, but it was some years before the Medical Research Council (1920) and the Agricultural Research Council (1931) were to follow. So it is only for the last seventy years or so that UK governments, in the national interest, have taken financial responsibility for supporting research in the biosciences.

14.1.1 Public science and private science

Nothing succeeds like success. And the phenomenal growth of scientific knowledge over the last half century, that has transformed both our view of ourselves and our ability to alter the environment around us, has stimulated the quest for even greater scientific understanding and technological mastery. However, this has exposed a problem. Scientific research is expensive, and as more and more issues become open to scientific enquiry it would not be difficult to devote so much government money to research that little would be left for anything else.

By the 1960s it was becoming apparent that the, then exponential,[1] increase in Government-supported scientific activity had to be curbed, and soon Government policy encouraged private industry to pick up the tab for the *near-market* research from which it would reap financial benefits. The result is that research in the biosciences is now extensively supported both by governments and private industry, the former chiefly in the national interest, the latter for commercial reasons. These two types of funding interact, cooperatively and competitively, and the consequences have some important political implications.

The perspective adopted in this chapter is that politics and social ethics are closely related; that they are both aspects of what Aristotle thought of as *practical philosophy*. His *Nicomachean Ethics* (Box 2.3) was written *'not in order to know what virtue is, but in order to become good'*,[2] and an important means of facilitating this lies in ordering the affairs of a State to allow all its citizens to flourish.

Politics is about **power** - who exerts it and how it is exercised. In relation to the biosciences and biotechnology, politics also needs to take account of three other 'p's, viz. **property**, **priority** and **propriety**. That is to say, political decisions surround the questions of ownership, importance and public acceptability. In a truly just society (as described by Rawls: 3.3), valuable scientific discoveries and the technologies flowing from them, which were approved by society as a whole, would be available to all who needed them. At the other extreme, one could imagine that in a highly unjust society technologies that most people considered to be deeply unethical might be exploited by a small section of society, solely for personal gain. As Western democracies aim to manage science and technology in accordance with the principles of a just society, the purpose of this chapter is to examine the policies designed to meet that aim, and how successful they are in doing so.

14.2 UK Government policy in the biosciences

It is estimated that overall in the UK, 60% of the total funding on R&D is from private sources. But it is the research that is supported by public money which forms the basis of the discussion in this section. Public sector research is funded from the tax revenue, which is disbursed by the Treasury. The total amount the Government spent in civil (i.e. non-military) R&D in 2000 was about £5 billion; and the Government is currently increasing the Science Budget at an average annual rate of 10% in real terms. There are three main funding streams, which in 2000 supplied the approximate amounts shown [in square brackets]:

- the Department of Trade and Industry (DTI) [£650 million]
- the Department for Education and Skills (DfES) [£3600 million]
- other Government departments, such as the Department of Health (DoH), the Department for International Development (DFID) and the Department for the Environment, Food and Rural Affairs (DEFRA) [£700 million].[3]

Both the DTI and the DfES have much wider responsibilities, so that administration of research funding is the job of sub-departments, namely, the Office of Science and Technology (OST) and the Higher Education Funding Councils (HEFC), respectively (Figure 14.1). Government-funded research is largely carried out in two types of establishment - the **universities** and **research council institutes**, but in each case funding is also received from other sources (such as other Government departments, independent charities, commercial companies and the European Commission); and in some cases university departments and research council institutes collaborate in research projects and share facilities. (In fact, although the claim was made above that only public-funded research will be examined here, the close collaboration between public and private funders makes it impossible to completely disentangle them.)

The *mission* of OST is to *'lead for the Government in supporting excellent science, engineering, and technology and their uses to benefit society and the economy'*. Ministerial responsibility for OST lies with the Minister for Science (a parliamentary under-secretary in the DTI), while the head of OST is the Chief Scientific Advisor to the Government, who is responsible for developing and coordinating Government policy on science and technology, both nationally and internationally. (A useful guide to the Government's administration of science policy is provided by a British Council website.[4])

There are seven research councils, of which the three most directly concerned with the biosciences are the Biotechnology and Biological Sciences Research Council (BBSRC), the Natural Environment Research Council (NERC) and the Medical Research Council (MRC). To illustrate their general features, the following account focuses on the BBSRC.

14.2.1 The Biotechnology and Biological Sciences Research Council

The BBSRC is the principal funder of basic and strategic biological research.[5] Its mission is to:

- promote and support high-quality basic and strategic research relating to the understanding and exploitation of biological systems
- advance knowledge and technology, and provide trained scientists and engineers to meet the needs of users and beneficiaries (i.e. agriculture, bioprocessing, chemical, food, healthcare, pharmaceutical, and other biotechnology-related industries), thereby contributing to the economic competitiveness of the UK and the quality of life
- engage the public in dialogue, communicate research results and provide advice.

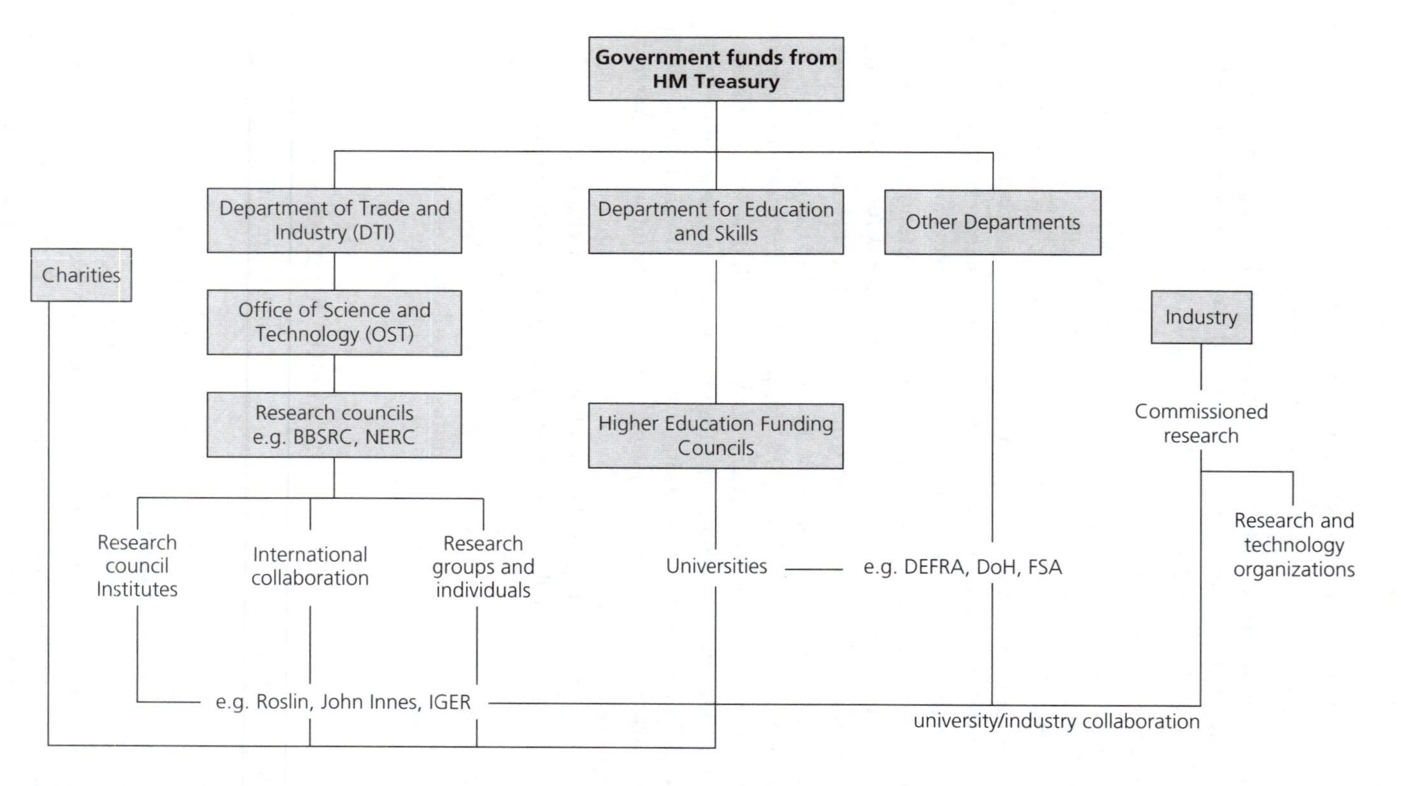

Figure 14.1 A simplified organogram of research funding in the UK. BBSRC, Biotechnology and Biological Sciences Research Council; NERC, Natural Environment Research Council; DEFRA, Department for the Environment, Food and Rural Affairs; DoH, Department of Health; FSA, Food Standards Agency; IGER, Institute of Grassland and Environmental Research. [Adapted from: British Council (2004) A guide to the organisation of UK science, engineering, and technology. http://www.britishcouncil.org/gost/orgindex.htm]

The BBSRC's total income for 2002/3 was £257 million. Funding is distributed through seven **research committees**, covering the following scientific areas:

- agri-food
- animal sciences
- biochemistry and cell biology
- biomolecular sciences
- engineering and biological systems
- genes and developmental biology
- plant and microbial sciences.

Research in the biosciences is carried out at eight BBSRC institutes which are located at sites throughout the UK (Table 14.1). For example, the Babraham Institute (which conducts research supporting biomedical, biotechnological and pharmaceutical sectors) is near Cambridge; the Roslin Institute (livestock genetics, breeding, welfare and biotechnology, and famous for the birth of the first cloned sheep, Dolly) is near Edinburgh; the John Innes Centre (plant and microbial research relevant to food, health, sustainable agriculture, and industrial innovation) is in Norwich, and the Institute for Grassland and Environmental Research (IGER) (sustainable grassland management and the wider managed environment) is at Aberystwyth. The total research institute funding from BBSRC in 2001/2 exceeded £73 million, and together with funding from other sources it exceeded £150 million. Each institute has a programme of research, overseen by its director, which conforms to BBSRC policy agreed by the Strategy Board and the appropriate specialist committee. Typically, these committees have about fifteen members, most of whom hold academic positions, but some work in industry.

An allocation of BBSRC funds is also made to those scientists and research teams in universities who are successful in the annual competition for research grants: in 2002/3 the success rate for grant applications was 30%. Such grants typically pay for salaries of

Table 14.1 Sources of income for the Biotechnology and Biological Sciences Council (BBSRC) research institutes (£ millions)

Research Institutes	BBSRC	Total
Institute for Animal Health	10.8	29.1
Babraham Institute	11.3	17.0
Roslin Institute	5.4	13.3
Institute of Arable Crop Research	13.2	27.5
Institute of Food Research	9.4	14.5
Institute of Grassland and Environmental Research	5.1	16.5
John Innes Centre	14.5	23.4
Silsoe Research Institute	3.9	10.5
TOTALS	73.7	151.8

research staff and technicians, postgraduate scholarships, some equipment and consumable materials. In 2002/3 research grants to universities amounted to £134 million. Salaries of lecturing staff and the costs of maintaining a well-equipped laboratory are provided by the HEFC, so that research in universities is financed by the so-called **dual-support system** (partly research council and partly DfES funding).

14.2.2 Other government bodies sponsoring research in the biosciences

The NERC's mission is:

- to promote and support research on terrestrial, marine, and freshwater biology and Earth, atmospheric, hydrological, oceanographic and polar sciences
- advance knowledge and technology to meet the needs of users and beneficiaries (including the agricultural, construction, fishing, forestry, hydrocarbons, minerals, process, remote sensing, water, and other industries), thereby contributing to the economic competitiveness of the UK, the effectiveness of public services and policy, and the quality of life.[6]

The MRC's mission is to:

- encourage and support high-quality research, with the aim of improving human health
- produce skilled researchers to advance and disseminate knowledge and technology to improve the quality of life and the economic competitiveness in the UK
- promote dialogue with the public about medical research.[7]

The importance of scientific understanding in the biosciences is so pervasive that many government departments also fund research activities. For example, DEFRA and DFID have responsibilities relating to agri-food research (relevant to the UK and LDC, respectively), and the FSA (14.3.3) for food safety; all three commission relevant research in the research institutes and universities – to the tune of approximately £40 million, £40 million, and £20 million p.a., respectively.[8]

14.3 Advice to government on science policy

Overall government science policy is subject to votes taken in the House of Commons on motions proposed by Government ministers, of whom the Minister for Science plays an influential role. The holder of that office at the time of writing was David Sainsbury, a member of the House of Lords, with particular interest in *'innovation policy, bioscience and chemicals (except GM foods)'* – the latter exception being due to a perceived conflict of interest in view of his family connection with J Sainsbury plc, the food retailers. The Chief Scientist at the time of writing was David King, a professor of chemistry.

Like the population at large, few members of the Houses of Parliament (Commons and Lords) have an advanced scientific education, which may be considered a disadvantage when political decisions have to be taken on complex scientific issues. In order to make up for these deficiencies, several parliamentary bodies conduct investigations and provide useful insights for parliamentarians on topical scientific issues. Features of the principal bodies are summarized below.

14.3.1 Parliamentary committees

The **Parliamentary Office of Science and Technology** (POST) is an office of the two Houses of Parliament (Commons and Lords), which aims to provide independent and balanced analyses of public policy issues related to science and technology. POST's remit is very wide, including: the environment, health, defence, transport, information technology and science policy. Its *Postnote* briefing papers are models of brevity and clarity.

The **House of Commons Science and Technology Committee** is appointed by MPs to examine the expenditure, administration and policy of OST and its associated public bodies. It is one of the departmental select committees.

The **Council for Science and Technology** (CST) is the government's top-level advisory body on science and technology policy issues, with a remit to advise the Prime Minister and the First Ministers of Scotland and Wales on strategic issues that cut across the responsibilities of individual departments. Due to be reconstituted in 2004, CST organizes its work around five broad themes (research, science and society, education, science and government, and technology innovation) and takes a medium- to longer-term approach.

The **Parliamentary and Scientific Committee** is a primary focus for scientific and technological issues, providing a long-term liaison between parliamentarians and scientific bodies, science-based industry, and academia. The main focus is on issues where science and politics meet, informing members of both Houses of Parliament of the relevance of scientific and technological developments to matters of public interest and to the development of policy.

The **House of Lords Science and Technology Committee** is an investigative committee, which represents a major forum of independent expertise, often drawing on some members' wide experience. The committee has a broad remit *'to consider science and technology'*. It works principally through inquiries undertaken by sub-committees, which are constituted afresh for each inquiry. Each inquiry produces a report, published together with the supporting evidence, which sets out the committee's findings and makes recommendations to the Government and others.

14.3.2 Advisory committees

In addition to the above sources of advice the Government can call on the expert opinions of a number of specialist committees (Table 14.2) together with that from the FSA (14.3.3), the HFEA (14.3.4), and a number of agencies with responsibility for aspects of the environment, such as English Nature and the Countryside Agency.

Table 14.2 Government scientific advisory committees (the list is not comprehensive)

Name of committee	Acronym
Advisory Committee on Animal Feedingstuffs	ACAF
Advisory Committee on Genetic Modification	ACGM
Advisory Committee on Microbiological Safety of Food	ACMSF
Advisory Committee on Novel Foods and Processes	ACNFP
Advisory Committee on Pesticides	ACP
Advisory Committee on Releases to the Environment	ACRE
Animal Procedures Committee	APC
Animal Welfare Advisory Committee	AWAC
Committee on Carcinogenicity of Chemicals in Food, Consumer Products and the Environment	COC
Committee on Mutagenicity of Chemicals in Food, Consumer Products and the Environment	COM
Committee on Toxicity of Chemicals in Food, Consumer Products and the Environment	COT
Farm Animal Welfare Council	FAWC
Gene Therapy Advisory Committee	GTAC
Genetics and Insurance Committee	GAIC
Scientific Advisory Committee on Nutrition	SACN
UK Xenotransplantation Interim Regulatory Authority	UKXIRA
Veterinary Products Committee	VPC

14.3.3 The Food Standards Agency

The FSA is an independent food safety watchdog set up by an Act of Parliament in 2000 to protect public health and consumer interests in relation to food. Although it is a Government agency, the FSA works at 'arm's length' from Government because it does not report to a specific minister and is free to publish any advice it issues. The FSA is accountable to Parliament through Health Ministers, and to the devolved administrations (Scotland, Wales, and Northern Ireland) for its activities within their areas. The FSA's key aims between 2001 and 2006 are to:

- reduce foodborne illness by 20% by improving food safety right through the food chain
- help people to eat more healthily
- promote honest and informative labelling to help consumers
- promote best practice in the food industry
- improve enforcement of food law
- earn people's trust.

Some committees listed in Table 14.2 report to the FSA (e.g. ACNFP and ACMSF).

14.3.4 The Human Fertilisation and Embryology Authority

The HFEA is a non-departmental government body that regulates and inspects all UK clinics providing IVF, donor insemination, or the storage of eggs, sperm or embryos. The Human Fertilisation and Embryology Act also requires the HFEA to regulate the creation, storage, and use of embryos in research throughout the UK. The Act was amended in 2001 to allow the use of embryos for stem cell research, and the HFEA now has the responsibility for regulating all embryonic stem cell research in the UK. Under the Act, any research must relate to one or more of the following purposes:

- to promote advances in the treatment of infertility
- to increase knowledge about the causes of congenital diseases
- to increase knowledge about the causes of miscarriage
- to enhance knowledge in the development of more effective contraception
- detection of genetic or chromosomal abnormalities before implantation
- to increase knowledge about the development of embryos
- to increase knowledge about serious disease or to enable any such knowledge to be applied in developing treatment for serious disease.

14.3.5 Government commissions

Following a comprehensive review of the regulatory and advisory framework for biotechnology, the Government established two biotechnology commissions. Especially in relation to public concerns over BSE and GM crops, the review concluded that the system for regulating individual products and processes needed to: (*a*) be more transparent, in order to gain public and professional confidence; (*b*) be more streamlined, in order to avoid gaps, overlaps and fragmentation; (*c*) ensure capacity to deal with rapid developments, and to take broad social and ethical issues fully into account. So, in 2000 the following were established:

- the **Human Genetics Commission** (HGC) – to advance these issues in the field of human genetics
- the **Agriculture and Environment Biotechnology Commission** (AEBC) – to have equivalent responsibility for all other areas of biotechnology (mostly relating to plants and animals).[a]

Other Government commissions address more defined issues and/or have a more limited life. An example of the former type is the **Sustainable Development Commission** (SDC) (12.5) and an example of the second type the **Policy Commission on the Future of Food and Farming** (12.4.2).

[a] A committee reviewing the AEBC's work in 2004 suggested that in future a new framework for strategic advice to the Government needed to be that of sustainable agriculture, including the extent to which GM might contribute to that end. The Government accepted the committee's advice to wind up the AEBC from April 2005 and would then announce the bodies to which its responsibilities would be transferred.

14.4 Non-government advisory bodies

Of course, advice is not the preserve of governments and their appointed experts: and it is important to the maintenance of democratic societies that a diversity of non-governmental organizations is also able to contribute new ideas for the formulation of science policies. In the same way that biodiversity provides resilience in the natural environment, so cultural diversity and its accompanying proliferation of new ideas, ensures intellectual resilience in addressing new social challenges. Many such organizations exist, so that the following is only a small selection of the more prominent or distinctive.

Scientific societies include: the Royal Society[9] (the world's first scientific society, founded in 1660); the Royal Institution of Great Britain[10] (established in 1799); and the British Association for the Advancement of Science[11] (founded 1831).

Professional bodies include the Institute of Biology[12] (a professional body for UK biologists, founded in 1950); all the separate disciplines comprising the biosciences also have their own professional societies, e.g. the Physiological Society, the Biochemical Society, and the Institute of Food Science and Technology.

Non-government organizations (NGO): the term is ambiguous, because many organizations that would not normally be classed as such do, strictly, conform to the definition. NGO are generally regarded as *pressure groups*, which seek to achieve political change by active campaigning, educational programmes, and, sometimes, high-profile public protests. Among the more prominent are Greenpeace and Friends of the Earth, who campaign on environmental issues; the Royal Society for the Prevention of Cruelty to Animals (RSPCA) and Compassion in World Farming (CIWF), who are concerned with animal welfare; and the Soil Association, which promotes organic farming. Some NGO are very proactive, and willing to break the law for their cause.

Bioethics councils: many of the organizations listed above address ethical issues, explicitly or implicitly; and bioethics in the guise of 'ethical, legal, and social issues' (or 'aspects') (ELSI or ELSA) has become a common theme in *science and society programmes* (14.6.1). But some bodies have been established specifically to address bioethical issues, e.g. Nuffield Council on Bioethics[13] and the Food Ethics Council.[14]

14.5 International bodies

The above discussion of the political arrangements in the UK, which generally have their equivalents in other countries, and especially in other DC, should not detract from the importance of international bodies (see Table 14.3). These exert influence at several levels, e.g. by impacting on UK domestic policy, trading relations with other countries, policies on international development in LDC, and global environmental sustainability. In some cases the UK Government is bound by international law (such as that

Table 14.3 Prominent international organizations

Organization	Acronym	Interest
World Trade Organization[15]	WTO	Liberalizing global trading
World Health Organization[16]	WHO	Promoting global health
UN Food and Agriculture Organization[17]	FAO	Sustainable farming and nutrition
UN Environment Program[18]	UNEP	Sustainable development policies
Codex Alimentarius Commission[19]	CAC	Harmonizing food safety standards

underpinning the WTO), while in others the Government has agreed to non-binding targets, such as those aimed at increasing aid to LDC. Most of these bodies have sub-committees whose remit is to address bioethical issues, and to engage with the world's publics.

14.6 Engaging the public in science policy decisions

The above account of the ways in which science policies are made, and the expert advice which governments can call on arriving at policy decisions, may seem to be both extensive and comprehensive. Certainly, for members of the public who are well versed in science and aware of new developments in the biosciences, there are many opportunities to express their opinions to decision-makers - e.g. by attending public meetings, writing to newspapers, or participating in web-based consultations. But such people are a very small proportion of the population, and it is hardly justifiable to claim that most people have only themselves to blame if they are unaware of recent developments. Even scientists are largely ignorant of developments in fields outside their own, so that it is understandable, and excusable, for most members of the public to lack the knowledge on which to make sound judgements.

But such is the pace of developments in science and technology, and so pervasive are the impacts for society - as evident in the many ethical questions discussed in this book - that there is little doubt that the scientific community must carry major responsibility for its actions and proposals. For when ethically problematical issues need to be addressed, there is a strong case for claiming that the *burden of proof* of justifying radical changes (be they proposals to produce GM crops or cloned babies) lies with those making the original discoveries and/or seeking to apply them - the bioscientists and biotechnologists. However, this presents a very significant challenge, because public trust in scientists, and in the politicians who advocate several novel biotechnologies, is at a low ebb (13.6). So, the only viable future strategy seems to lie in scientists engaging with the general public to a much greater degree than in the past, so that agreed science policies can be formulated, in advance, across society. The alternative, that of scientists and technologists launching biotechnologies without public consultation or approval, and hoping that market forces will determine their public acceptability, would seem to carry risks of many kinds – to consumers, the environment, animals, and future generations.

14.6.1 Public understanding of science

Earlier initiatives at promoting the public understanding of science (somewhat unfortunately abbreviated as PUS) were coordinated by a joint committee (CoPUS) of the Royal Society, Royal Institution and British Association for the Advancement of Science, established in 1986. The dominant attitude among scientists at the time was that the reason most of the public were apprehensive about developments in science and technology was that that they were abysmally ignorant of science (the so-called **deficit model**) – a situation best corrected by scientists making the effort to explain things more clearly, to allay unfounded fears.

However, in the years following BSE and the introduction of unlabelled GM crops, this 'top-down approach', which relies heavily on public trust in scientists, has lost much credibility. First, the public have begun to realize that often scientists are *not necessarily* trustworthy: they sometimes make assertions (e.g. about the safety of eating beef, or the insignificant impacts on the environment of growing GM crops), which later prove unfounded. Moreover, the deference which people formerly showed towards experts, of all kinds, has become undermined. Doctors, teachers, lawyers and politicians, as well as scientists – all have lost the almost universal respect they formerly received. Many people now believe that scientists are just as likely to be biased in their opinions as anyone else. This is for three reasons:

- because they often have a vested interest in a particular view which is markedly influenced by the source of their research funding (15.5.3)
- because they have been trained to work within a particular, often narrow, scientific paradigm (1.5.2)
- because of the *unknown unknowns* – i.e. we don't know what we don't know (13.4).

Consequently, scientific opinion now often fits what sociologist of science Barry Barnes calls the **market model**: you can choose the expert you want to promote your particular preference (e.g. pro-biotechnology or anti-biotechnology) and lay people can judge which to believe, as does a jury in a criminal court listening to the expert witnesses for the prosecution and the defence, respectively.[20] (The danger is, of course, that skilful presentation has no necessary connection with scientific truth or ethical validity.)

This, then, challenges the, earlier, implicit assumption that only the public needed to be educated, and that the scientists were not themselves in need of education in public values. For, as recognized in a House of Lords Science and Technology Committee report, the cultural change in public attitudes to science means that organizations such as scientific societies and research councils *'must respond to a new mood for dialogue'*.[21] This advice has been widely accepted, at least officially, and many scientific organizations and Government departments with responsibilities for science matters have now introduced **science in society programmes**. For example, the Royal Society's programme aims to achieve:

- openness
- two-way dialogue between all sectors of society
- responsiveness to public concerns and values.

But most scientists lack skills or experience in dialogue with the public (including use of the news and broadcasting media), so that there is a need for training in communication skills, and for making scientists aware of *'the social context of their research and its applications'*.[22] In particular, communication of risk and uncertainty (13.2.4) has a high priority. A concise way of putting this is to say scientists need to engage less in PUS and more in SUP (scientists' understanding of the public) and SUM (scientists' understanding of the media). These objectives are, of course, at the very core of bioethics.

In order to address such problems, the Science Media Centre was established in 2004 to *'promote the voices, stories, and views of the scientific community to the news media when science is in the headlines'*. The Centre is housed in the Royal Institution (14.4), but it is independent from it. Even so, as *'unashamedly pro-science'*, its goal is to *'secure science's license* [sic] *to practice* [sic] *not to restrict it'*.[23] (It apparently favours North American spelling.)

14.6.2 The role of the media

There is little doubt that the news media (principally, newspapers, magazines, radio, television and the internet) play an important role in shaping people's opinions on scientific issues (13.6.2). There have been few detailed studies of their influence but an exception is the analysis of, what was called at the time, **The Great GM Food Debate** in the UK in 1999, by academics John Durant and Nicola Lindsey.[24] Although this particular 'media storm' was triggered by events like the alleged harmful effects of GM potatoes fed to rats and the realization that unlabelled GM crops were being imported into the UK (10.1.2), it was fuelled by the decision of several newspapers to mount a campaign against GM foods. The debate, which was notable for its intensity and duration, led Durant and Lindsey to the following conclusions:

- even a single, unauthenticated, scientific claim can *'given in the right circumstances'* have a huge impact on public debate and opinion
- dealing with expert disagreement in socially sensitive areas of scientific research is extremely difficult
- when scientific or science-related issues become high-profile news, events can move very quickly indeed – and not always in directions that scientists expect
- when a science story is big enough, it will be handled by non-science reporters and others, up to and including newspaper editors – which can significantly change the way it is reported.

The exceptional public interest in the issue was due to the growing divergence, after 1996, between Government and industrial policy on GM food on the one hand, and public opinion, on the other. And, according to Durant and Lindsey: *'The real lesson of the great GM food debate... is that in a democracy any significant interest – science included – ignores public opinion at its peril.'*

Acknowledging that conclusion, academic and journalist Ian Hargreaves nevertheless points to the limitations of the study, in that it was confined to the printed media. For him, news is becoming more competitive (with incursions from the internet), less political (as the party political divide narrows), more consumer-oriented (as consumers take more interest in the values associated with purchase decisions), more rumour-prone (because the speed of reporting prevents checking of sources), and more global (with no geographical boundaries). This *'fast-changing, highly contingent, universe in which the old, apparently reliable top-down process of scientific information* [was] *cautiously reviewed at leisure and respectfully transmitted by a media which shares the values of the scientists to a trusting public, could not survive, even if it still existed, which it doesn't.'*[25]

So the media, in all their forms, play a powerful role in shaping and reflecting public opinions, and hence in the political process; and the ethical implications clearly demand serious attention.[26] In an attempt to provide reliable briefings for journalists who lack sound training in science, the Science Media Centre, in addition to serving the science community, also produces a 'science in a nutshell' series of leaflets designed to be *'useful for busy news-desks'* – although the admitted 'pro-science' bias (14.6.1) needs to be appreciated.

14.6.3 Public opinion surveys

A common means of eliciting public opinions on developments in the biosciences and biotechnology is to conduct polls: a form of **quantitative research**. Market research organizations, such as MORI (Market and Opinion Research International) and NOP (National Opinion Poll), are often commissioned to assess opinions at the time of heightened public concern, but routine studies are also conducted, notably by the European Commission. Published as **Eurobarometer** surveys,[27] the latter provide regular, detailed information on an EU-wide basis.

Such data have the advantage of being capable of statistical analysis, and of gauging major trends. As such, they inform politicians and members of the public alike of how their own perspective compares with others. But they are limited by the inability of respondents to clarify the questions asked or, often, to qualify their replies, and by the inability to assess the extent to which responses are well considered or honest.

14.7 Deliberative and inclusionary processes

More effective assessments of people's opinions are necessarily more expensive to conduct, in terms of both time and money, but they are likely to yield much more valuable information. They generally entail the active participation of members of the public and/or assigning the opinions expressed some political weight. A participatory arrangement may be defined as *'a method or activity that involves the public in a process of decision-making in society that affects or goes beyond mere participation in the voting procedures of representative democracy'.*[28] They can be said to have two justifications: (*a*) as a feature of the democratic process; (*b*) more pragmatically, as effective ways of solving anticipated or existing social problems.

The best-known forms are: **focus groups** (FG), **consensus conferences** (CC), **referendums** (RF), **citizens' juries, futures workshops** and **public consultations.** These vary in:

- the objectives (e.g. qualitative sociological research for FG; achieving consensus for CC; and a political decision for RF)
- the number of people involved (e.g. five to twelve for FG; twelve to fifteen for CC; all affected citizens for RF)
- the methodology (e.g. interviews by social scientists for FG; interviewing experts and writing a report for CC; voting for RF).

Despite these differences, all challenge the 'top-down' approach associated with PUS.

Perhaps the most truly democratic participatory processes are those classed as **deliberative and inclusionary processes** (DIPS). *Deliberation* entails careful consideration of the 'pros and cons' of any issue, while *inclusion* entails involving all affected interests in the decision-making process, and especially those whose views are often otherwise ignored. Two notable, but contrasting, exercises involving DIPS are discussed below. Both concerned the future of agriculture, and both had a significant focus on GM crops.

14.7.1 The GM nation? Debate in the UK

In 2001 the AEBC (14.3.5) recommended that in addition to the various strategic reviews of the possible commercialization of GM crops (10.7), *'There should be a wider public debate... to consider what role GM crops might have in UK agriculture in the future.'*[29] The Government agreed to this proposal and duly appointed a steering group to oversee the process. The aim was to *'promote an innovative, effective, and deliberative programme of debate on GM issues, framed by the public, against the possible commercial production of GM crops in the UK and the options for possibly proceeding with this'*, and to that end the Government provided £650,000 in support of the debate and its analysis and publication of the report.

The debate was conducted in June and July 2003, during which time an estimated 675 meetings were held throughout the UK, attended in total by about 20,000 people. For the typical local events (e.g. held in village halls) organizers could choose one or more items from the prepared 'toolkit', consisting of documentary *stimulus material*, an interactive CD-ROM and a video (made by a leading independent film-maker). Feedback from this **Open Debate**, in the form of completed questionnaire forms, was received from more than 30,000 people.

In addition to the Open Debate, there was a much smaller second strand to the GM debate, referred to as the **Narrow But Deep** (NBD) element, involving ten group discussion exercises and a total of seventy-seven people. Because it was realized that people who attend public meetings might not be representative of public opinion at large, the NBD involved inviting socio-economically balanced groups of people to attend two sessions, separated by two weeks and organized by professional facilitators. At the first session, participants were introduced to a range of GM issues and ways of exploring them, and asked to complete a questionnaire. At the second, after they had conducted their own research into the issues, they completed the questionnaire again, thus allowing assessment of the impact of increased understanding on their opinions.

It was concluded that the NBD participants were not a 'silent majority', with different values and attitudes from those who volunteered to join the Open Debate. In fact, the two groups shared many attitudes - recognizing that the GM debate encompasses many overlapping scientific, economic, social and ethical issues. *'They also have in common suspicion about the agenda and motives of the key decision-makers in GM - the government and the multi-national businesses.'*[30] The seven key messages the steering group considered the debate conveyed are shown in Box 14.1.

Feedback from the Open Debate showed *'a general pattern of caution, suspicion or outright hostility towards GM and GM crops or foods'*, with 54% never wanting to see GM crops grown in the UK. There were emphatic majorities in support of questions referring to risks of GM, or which made some negative comment about it. Only one question, that referring to the claim that some GM non-foods could have useful medical benefits, did not produce an outright 'anti-GM' majority.[31] Notwithstanding such apparently clear public reservations, the Government decided to approve the growing of GM maize in March 2003 (10.5.5).

Despite the unique nature of the GM Nation? debate, a detailed evaluation of its effectiveness, carried out by independent social scientists, identified numerous deficiencies. While it was assessed as 'good' in terms of *independence*, the assessments of *transparency, resources* and *representativeness* were all ranked 'poor', while *task definition* and *structured dialogue* were variable (ranging from 'good' to 'poor').[32] The *influence* exerted by the debate was listed as 'not yet known'. Moreover, the NGO GeneWatch argued that there was a lack of clarity on how results would be used by the Government; and it criticized the poor timing of the debate, which concluded before the results of the FSE (10.5.4) and other Government enquiries were published.[33]

BOX 14.1 THE GM NATION? DEBATE: KEY MESSAGES

- People are generally uneasy about GM.
- The more people engage in GM issues, the harder their attitudes and more intense their concerns.
- There is little support for early commercialization.
- There is widespread distrust of government and multi-national companies.
- There is a broad desire to know more and for further research to be done.
- Developing countries have special interests.
- The debate was welcomed and valued.

14.7.2 Prajateerpu: a citizens' jury in India

A claim frequently made for GM technology is that it has clear potential benefits for LDC in producing higher-yielding crops, which are more resistant to both biotic and abiotic stresses. This is the basis of Nuffield's claim that DC have a moral obligation to make GM technology available to those in LDC who want them (11.7).

An attempt to assess grassroots opinions on GM and alternative forms of agriculture was made in a citizens' jury exercise lasting five days held in the Indian state of Andhra Pradesh in June–July 2001. It was organized by experienced staff from Indian and UK institutes, and overseen by a distinguished panel who *'assessed the degree of fairness, trustworthiness and credibility'* of the process - and commented very favourably on the way it was conducted.

Nineteen jurors were appointed, in a way designed to represent as closely as possible the nature of the farming community in Andhra Pradesh - a state in which 70% of the population is engaged in agriculture (and of those 80% are small and marginal farmers and landless labourers), but who together own only 35% of the cultivated land. Women play a greater role than men in agricultural work, looking after almost 80% of the routine livestock management. Consequently, the jury appointed consisted of five men, thirteen women farmers, and one other woman (representing consumers): all the farmers grew crops and thirteen also kept livestock. Although mostly drawn from a very underprivileged section of society (*'living near or below the poverty line'*), the jurors were selected on the basis that they *'were likely to be articulate in discussions'*.[34]

The jurors were presented, initially through videos, and subsequently in workshops with specialists from academia and industry, with information on three scenarios for agriculture in Andhra Pradesh over the following twenty years:

- **Vision 2020**: backed by a loan from the World Bank (4.10.4), this proposes rapid modernization, entailing mechanization, and introduction of new (including GM) technologies, which will reduce the people working on the land from 70 to 40% by 2020
- **Export-based cash-crop model of organic production**: environmentally friendly farming, linked to international markets, and increasingly driven by supermarkets in DC wanting a cheap supply of organic produce
- **Localized food system**: a low external input agriculture based on increased self-reliance for rural communities, and with long-distance trade only in goods surplus to production or not produced locally.

In addressing these issues trained facilitators guided the jurors in a range of techniques of engagement, such as focus groups, plenary sessions, and reviewing evidence. Throughout, the proceedings were monitored for fairness by the *oversight panel*, recorded on film and reported on by members of the press. In producing their final verdict on the three scenarios, *'a remarkable degree of consensus emerged among the participants about a wide range of issues'* (Box 14.2).[35] The panel overseeing the proceedings were very impressed with the methodology used in the exercise and considered its use could well be extended to other issues.

The jury's verdict merits serious consideration by those who see technocratic solutions to social problems, such as poverty and malnutrition in LDC, as unproblematical. Despite their low level of formal educational achievement, the jurors were confident in

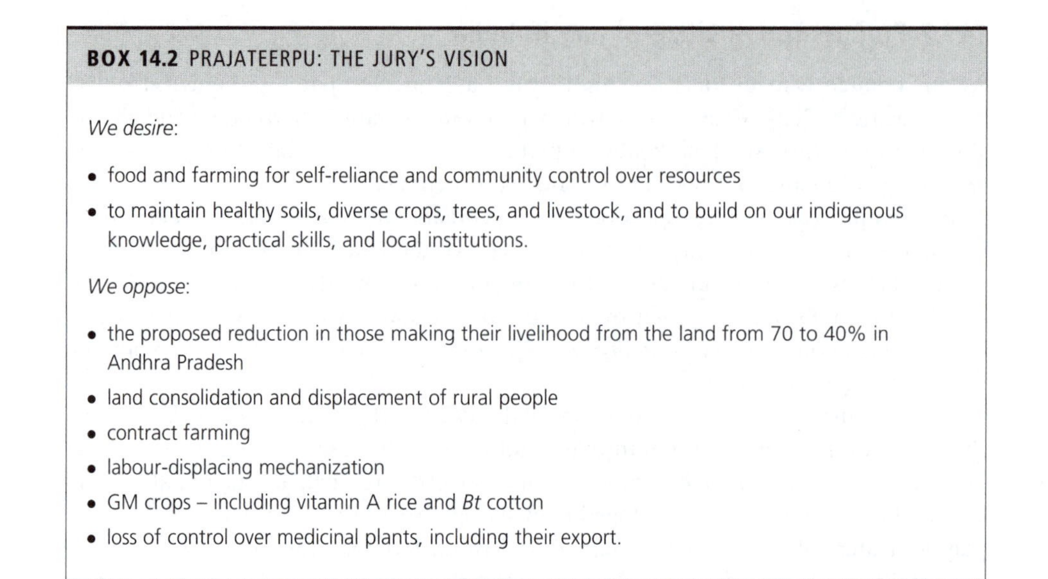

BOX 14.2 PRAJATEERPU: THE JURY'S VISION

We desire:

- food and farming for self-reliance and community control over resources
- to maintain healthy soils, diverse crops, trees, and livestock, and to build on our indigenous knowledge, practical skills, and local institutions.

We oppose:

- the proposed reduction in those making their livelihood from the land from 70 to 40% in Andhra Pradesh
- land consolidation and displacement of rural people
- contract farming
- labour-displacing mechanization
- GM crops – including vitamin A rice and *Bt* cotton
- loss of control over medicinal plants, including their export.

their ability to assess the practical implications of technological proposals. Like their equivalents in the UK GM debate (14.7.1), their experience of life had engendered a scepticism of 'bold conjectures' which might be considered in the best traditions of Popperianism (1.5.1). At the very least, it emphasizes that biotechnologies are of limited use if people are unconvinced of their value.

14.8 Ethical consumerism

Those who want to influence the course of biotechnological development in democratic societies often have a potent means of doing so, by means of their purchase decisions. This is most graphically illustrated by the widespread rejection of GM food by consumers in most European countries in the late 1990s, which resulted in the decision of supermarket chains to exclude GMO from their own-brand products, and in labelling laws to protect consumer autonomy.

In some cases this leads to *reduced* choice, e.g. when GM tomato purée was withdrawn from supermarket shelves even though it was popular with some consumers. However, the market is not always accommodating, even to the consumer choice of the majority. For example, in the USA milk produced with the aid of the GM hormone BST (Table 3.2) is unlabelled, despite indications that more than 80% of the public wanted this.[36]

Ethical consumerism has a long history. For example, in the form of boycotts it was used in the last century by Mahatma Gandhi in the political struggle for Indian independence, and by many in DC to challenge the regime of apartheid in South Africa.

With reference to environmental concerns, ethical consumerism came to prominence with the publication of the Green Consumer Guide in 1988 and the launch of the *Ethical Consumer* magazine in 1989.[37] These introduced the idea that consumers could make a difference in their everyday activities, through:

- **positive purchasing**, e.g. favouring products such as fair trade, organic food or cruelty-free products
- **negative purchasing**, e.g. avoiding products disapproved of, such as eggs from battery-caged chickens or pâté de foie gras (production of which entails force-feeding geese)
- **company-based purchasing** in which products of businesses as a whole may be boycotted, e.g. the widespread boycott of Nestlé brands by many student and religious groups as a protest against the company's perceived unethical practices in marketing baby foods, especially in LDC
- **discrimination in funding sources by charities**, e.g. in 2004 a leading cancer charity rejected a donation of £1 million from Nestlé for the above reasons.[38]

Some supermarkets place an emphasis on ethical trading, and *ethical banks* assure their investors that their money will be invested in sustainable, humanitarian projects and not be used to support activities such as the arms industry, pornography or tobacco sales. More direct forms of consumer action entail advertising campaigns (e.g. of animal welfare or environmental organizations), protest meetings and fund-raising events to finance promotional activities. Some who are sufficiently convinced of the rightness of their political stance, and prepared to take the consequences, resort to illegal activities, as when Greenpeace supporters uprooted GM crops in the *farm scale evaluations* conducted to test effects on biodiversity.

14.8.1 The McLibel Trial

A notable instance of public protest over perceived unethical practices in food marketing culminated in the so-called *McLibel Trial*. In 1986, a London Greenpeace group targeted McDonald's fast food restaurants by distributing a leaflet which accused the company of promoting Third World poverty, selling unhealthy food, exploiting workers and children, abusing animals, and destroying the Amazon rainforest. McDonald's decided to sue the activists who distributed the leaflet, Helen Steel and Dave Morris, for libel - marking the beginning of a court case which lasted more than three years (1994–97).

The judge ruled that the defendants had failed to prove most of their allegations and they were fined £60,000. But in 1999 three Court of Appeal justices overruled parts of the original McLibel verdict, and reduced the damages owed by Steel and Morris. In 2005 the EU Court of Human Rights gave its verdict that denying the defendants legal aid had *'violated their freedom of expression'*, and it seems likely that this will lead to a future change in the law in such cases.[39] McDonald's have since withdrawn their claim for costs.[40]

14.9 The ethical decision-making process in technology assessment

Although the *practice* of science itself can be ethically problematical (chapters 8 and 15) the ethical issues which most usually cause public concern are those arising from technological applications of scientific knowledge. We have seen that ethical consumerism is a means by which individuals can both act in accordance with personal ethical principles and seek to influence the course of public events. As a strategy it is thus reactive rather than proactive, because it is only applicable 'after the fact' of commercialization. So if public DIPS (14.7) are to influence the course of biotechnological development, organizational and conceptual tools need to be established by which the outcomes of these consultations are fed into the democratic decision-making process at an early stage. One way, at least in principle, of influencing technological outcomes in this 'upstream' fashion is by use of **foresight programmes**.

Foresight is designed to provide challenging visions of the future, so ensuring effective current strategies. It does this by providing a core of skills in science-based futures projects and high-level access to leaders in government, business and science. For example, in the UK the *Brain Science, Addiction and Drugs Project* is a programme designed to provide a vision of how scientific and technological advancement may impact on our understanding of addiction and drug use over the next twenty years. Areas of science included are:

- *genetics*: looking at the hereditary nature of addiction. Over the next fifty years the genes involved in the various aspects of addiction are likely to be identified.
- *brain science*: will it be possible to develop molecules to block or destroy the relevant receptors and what might this mean? What type of pharmacological or other interventions may become possible?
- *behavioural change*: what is our best understanding of and framework for considering addictive behaviour and successful interventions?[41]

14.9.1 Political decision-making

At present, it is not altogether clear how the results of foresight panels will be used. Consequently, building on concepts discussed elsewhere in the book, Figure 14.2 proposes a scheme by which ethical assessment in foresight exercises might impact on the process by which scientific knowledge is translated into technological application.[42] Ideally, the ethical framework at the core of this process (of which the ethical matrix discussed in this book is an example) needs to satisfy two types of requirement. It should provide:

- a rationale for the **procedures** used in ethical assessment, e.g. defining the qualifications of the persons considered competent to perform the assessment; the manner in which they are made aware of the scientific and ethical issues; and the ways in which their opinions are recorded and 'weighed'

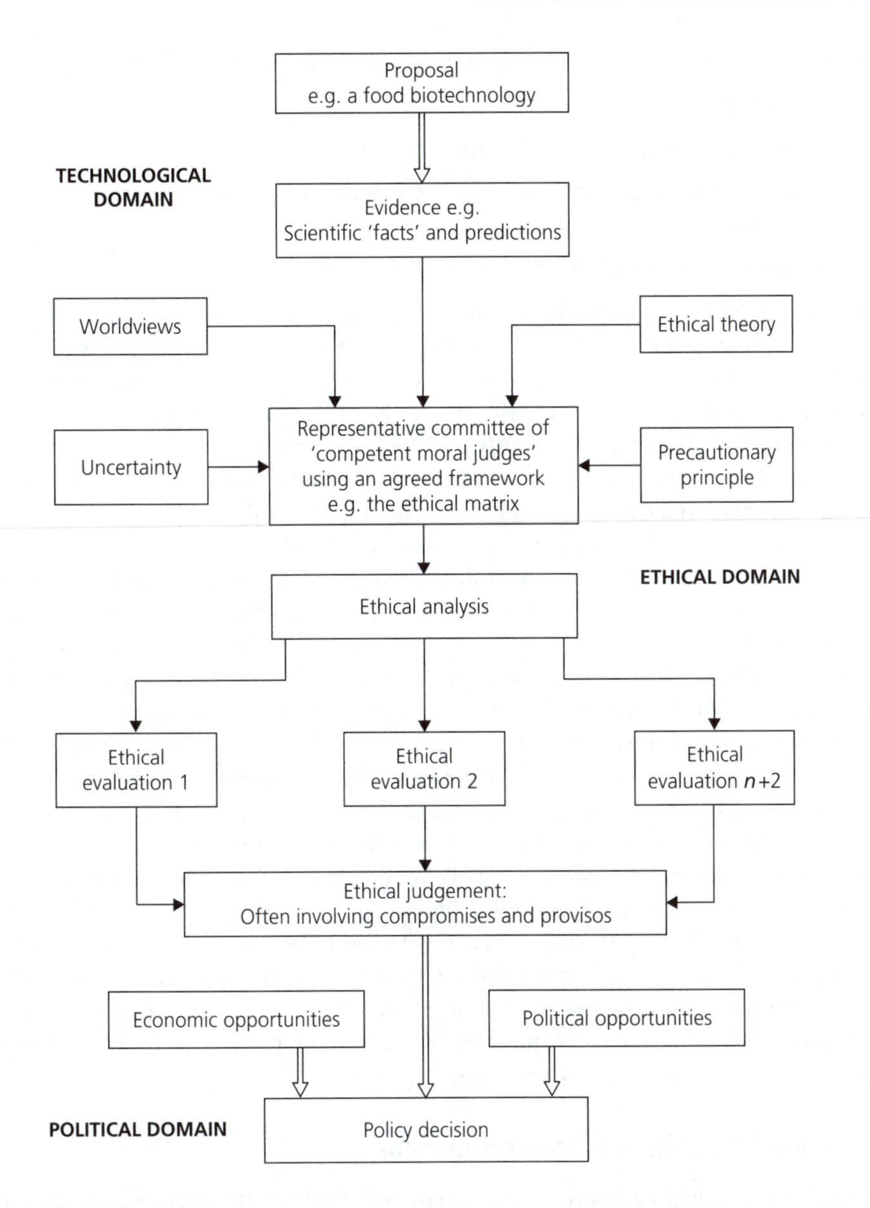

Figure 14.2 A proposed scheme for involving ethical evaluations in the policy-making process relating to biotechnologies. (From Mepham, 2000)

- a **substantive** basis, identifying the ethical principles which are to structure the ethical analyses, e.g. in the case of the ethical matrix (3.5) the *prima facie* principles derived from the common morality.

Figure 14.2 emphasizes that the outcome of deliberations in committees is unlikely to produce a consensus over the ethical analysis of an issue. As an ideal, Rawls proposed, in an early publication,[43] that such decisions should be made by committees of *competent moral judges*, who had the following characteristics:

- normal intelligence
- reasonable knowledge of world affairs
- a capacity to 'reason', i.e. see both sides of a question, making allowance for personal bias
- an imaginative appreciation of other people's predicaments.

Certain constraints were attached to the judgement procedure, namely: (*a*) the judge must be immune from, and have no vested interest in, the judgement; (*b*) which must only be arrived at after a careful enquiry into the facts of the case; (*c*) be delivered with appropriate certitude; and (*d*) be stable. It may be, however, that this is a very tall order. Is it possible to find people sufficiently aware of any technological issue, but who have not adopted an ideological stance on it? Perhaps the best that is practically achievable is the appointment of a *balanced* committee, who take seriously the recommendations of Nolan Committee on Standards in Public Life (Box 13.3).

In the ethical matrix, inclusion of both utilitarian and deontological principles aims to avoid any bias towards either worldview. But this does not necessarily imply that the middle ground is the best position to occupy. In fact, it is often untenable, e.g. when a decision has to be made *either* to permit *or* ban a practice. For example, in the case of crop biotechnology in the UK, according to philosopher Onora O'Neill: *'there may be nothing intrinsically wrong with an agricultural policy and food system that accepts GM food crops or with one that rejects them, but (given the undisciplined habits of plants) a mixed policy might not be feasible.'*[44]

Moreover, it has been claimed that in assessing scientific and technological issues ethically, utilitarianism is a questionable basis for decision-making. This is because at the frontiers of knowledge, we are usually largely 'in the dark'. The frontiers of knowledge are, of course, also the 'frontiers of ignorance'.[45] So, given that a range of analyses will probably result, the ethical *judgement* will need to be taken by an arbiter, who will most probably be a government minister in matters which have legal implications. Figure 14.2 also indicates that ethical judgements are often subject to political and economic constraints. Policies cannot be implemented if they cannot win a majority in a parliamentary vote, or if they prove too expensive.

14.9.2 **Evidenced-based decision-making**

In reaching political decisions, governments often claim these are based on sound evidence, rather than on mere subjective opinion. The case is illustrated by the UK Government's decision over the licensing of GM crops (10.5.5). While agreeing to the AEBC's recommendation for a public debate on GM crops (14.7.1), the Government also set up a scientific committee (the SRP, see 10.4.1) and an economic 'cost-benefit analysis' (10.7), to complement results from the FSE (10.5.4). But according to policy analyst Ruth Levitt, this appeal to 'evidence' is based on two, questionable, assumptions: namely, that:

- it is possible to identify all the relevant evidence
- this (scientific/economic) evidence is more valid than (mere) expressions of opinion by non-experts in society.[46]

A more positive characterization of public opinion is that it represents evidence of a different sort – i.e. evidence of society's 'values and interests' rather than just prejudices.

By counting only certain types of evidence as admissible, Levitt argues that the Government felt justified in overruling the widespread public scepticism revealed by the GM Nation? debate (14.7.1) and agreeing to the licensing of one GM crop. She also argues that, even in its own terms, the uncertainties of the scientific and economic evidence on which the Government's decision was based would not seem to justify the decision made. These seem to be cogent arguments, which invite a response from those who support the Government's decision to approve the growing of GM maize in the UK.

THE MAIN POINTS

- The UK Government spends hundreds of millions of pounds annually on research in the biosciences (and private industry even more), which is conducted in universities and research institutes with the aim of benefiting 'society and the economy'.
- Science policy-making is a complex process which is the responsibility of the Office of Science and Technology and the research councils (e.g. BBSRC), with specialized inputs from bodies such as the FSA.
- Numerous parliamentary and advisory committees provide expert advice, and longer-term strategic advice is provided, among others, by the Human Genetics Commission and Agriculture and Environment Biotechnology Commission.
- High-profile concerns (e.g. BSE and GM foods) have led to moves to increase public involvement in science and technology policy, exemplified by the GM Nation? debate in 2003 – which, however, involved only a small proportion of the electorate.
- The ethical challenge, in a world of globalized trade and multicultural values, is to discover means by which, in democratic societies, the public interest can influence the course of bioscience and biotechnology for the common good.

EXERCISES

These can form the basis of essays or group discussions:

1. Commissions such as the HGC and the AEBC were established to advise the Government on ethical issues. Discuss the impact you consider they have had and whether their influence could, or should, have been greater (14.3.5).
2. By comparison with the former *deficit model*, what are the advantages and disadvantages of Barnes' concept of the *market model* by which the public now assesses scientists' accounts of their work and its value? (14.6.1)
3. It is widely acknowledged that sections of the tabloid press often engage in 'scaremongering' tactics over novel technologies (13.6.2). Is this a price that must be paid for having a free press in democratic countries? (14.6.2)

4. Compare and contrast the GM Nation? debate in the UK with the Prajateerpu in Andhra, India (14.7).

5. In making a decision on the growing of GM crops in the UK, to what extent should the Government take account of public opinion as 'evidence of public values', alongside scientific data on food safety, biodiversity, etc.? (14.9.2)

FURTHER READING

The reports of two consultation exercises, one in the UK and one in India, provide useful comparisons, of both participants and methodologies:

- *GM Nation? The Findings of the Public Debate* (2003). London, Department of Trade and Industry
- *Prajateerpu: a citizens' jury/scenario workshop on in food and farming futures for Andhra Pradesh, India* by Michel P. Pimbert and Tom Wakeford (2002). London, IEED and Sussex, IDS.

  A third report provides an analysis of the coverage of concerns over GM food in the UK printed media in 1999.

- *The 'Great GM Food Debate'* by John Durant and Nicola Lindsey (2000). Parliamentary Office of Science and Technology, London, POST.

USEFUL WEBSITES

- **http://www./britishcouncil.org/science/gost/index.htm** *British Council*: guide to the organization of UK science, engineering, and technology. An invaluable portal to Government Committees.
- **http://www.parliament.uk/parliamenrtary_office/post.cfm** *Parliamentary Office of Science and Technology*: regularly produces concise, readable reports.
- **http://ethicalconsumer.org** *The Ethical Consumer*: the online version of the bimonthly *Ethical Consumer* magazine. Reviews social and environmental impacts of products and the ethical records of companies making them.

NOTES

1. De Solla Price DJ (1963) Little science, Big Science. New York, Columbia University Press
2. Cambridge Dictionary of Philosophy (1999) (2nd edition). (Ed.) Audi R. Cambridge, Cambridge University Press, p. 50
3. Office of Science and Technology (2003) SET statistics. http://www.ost.gov.uk/setstats/index

4. British Council (2004) Guide to the organisation of UK science, engineering and technology. http//www.britishcouncil.org.gost/
5. Biotechnology and Biological Sciences Research Council (2004) http://www.bbsrc.ac.uk
6. Natural Environment Research Council (2004) http://www.nerc.ac.uk
7. Medical Research Council (2004) http://www.mrc.ac.uk
8. Department of Trade and Industry (2003) www.ost.gov.uk/research/forwardlook03/
9. The Royal Society (2004) http://www.royalsoc.ac.uk
10. The Royal Institution of Great Britain (2004) http://www/rigb.org
11. British Association for the Advancement of Science (2004) http://www.the-ba.net
12. The Institute of Biology (2004) http://www.iob.org
13. Nuffield Council on Bioethics (2004) http://www.nuffieldbioethics.org
14. Food Ethics Council (2004) www.foodethicscouncil.org
15. World Trade Organization (2004) http://www.wto.org
16. World Health Organization (2004) http://www.who.int/en
17. UN Food and Agriculture Organization (2004) http://www.fao.org
18. UN Environment Program (2004) http://www.unep.org
19. Codex Alimentarius Commission (2004) http://www.codexalimentarius.net
20. Barnes B (2002) The public evaluation of science and technology. In Bioethics for Scientists (Eds) Bryant J, Baggott la Velle, and Searle J. Chichester, Wiley, pp. 19–36
21. House of Lords Science and Technology Committee Third Report (2000). Science and Society. http://www.publications.parliament.uk./pa
22. House of Lords Science and Technology Committee Third Report (2000). Science and Society. http://www.publications.parliament.uk./pa
23. Science Media Centre (2004) http://www.ScienceMediaCentre.org
24. Durant J and Lindsey N (2000) The Great GM Food Debate (report to the House of Lords Select Committee on Science and Technology). London, Parliamentary Office of Science and Technology
25. Hargreaves I and Ferguson G (2000) Who's Misunderstanding Whom? London, Economic and Social Research Council, p. 46
26. Belsey A and Chadwick R (Eds) (1992) Ethical Issues in Journalism and the Media. London, Routledge
27. Eurobarometer (2002) http://europa.eu.imt/comm/public_opinion
28. Ethical Bio-TA Tools (QLG6-Ct-2002-02594) Interim Report (2004) The Hague, LEI, p. 31
29. Agriculture and Environment Biotechnology Commission (2001) Crops on Trial. London, DTI, p. 25
30. GM Nation? Findings of the Public Debate (2003) London, DTI, p. 42
31. Ibid., pp. 32–33
32. Understanding Risk Team (2004) A Deliberative Future? An independent evaluation of the GM Nation public debate. Norwich, Centre for Environmental Risk
33. Mayer S (2003) GM Nation? Engaging People in a Real Debate? Buxton, GeneWatch UK
34. Pimbert M P and Wakeford T (2002) Prajateerpu: a citizens' jury/scenario workshop on food and farming futures for Andhra Pradesh, India. London, IEED and Sussex, IDS, p. 8
35. Ibid., p. 73

36. Smith B J and Warland R H (1992) Consumer responses to milk from BST-supplemented cows. In Bovine Somatotropin and Emerging issues. (Ed.) Hallberg M C. Boulder, Westview Press, pp. 242–264
37. Ethical Consumer (2004) http://ethicalconsumer.org
38. Frith M (2004) Breast cancer charity rejects £1m. http://news.independent.co.uk/uk/health
39. BBC News (2005) McLibel pair win legal and case. (15.2.05) http://newsvote.bbc.co.uk
40. Schlosser E (2001) Fast Food Nation. London, Penguin, pp. 247–249
41. Foresight programme. http://www.forsight.gov.uk/servlet/
42. Mepham T B (2000) The role of food ethics in food policy. P Nutr Soc, *59*, 609–618
43. Rawls J (1951) Outline of a decision procedure for ethics. Philos Rev *60*, 177–197
44. O'Neill O (2002) Autonomy and Trust in Bioethics. Cambridge, Cambridge University Press, p. 168
45. Holdsworth D (1995) Ethical decision making in science and technology. In: Introducing Applied Ethics. (Ed.) Almond B. Oxford, Blackwell, pp. 130–147
46. Levitt R (2003) GM Crops and Foods. Evidence, Policy and Practice in the UK: a case study. London, ESRC Centre for evidence based policy and practice, Queen Mary College, London

15

Bioethics in the laboratory

OBJECTIVES

When you have read and discussed this chapter you should:

- be aware of the norms of science described by Merton and the social context of scientific research
- appreciate how the norms are specified as ethical standards for bioscientists and the particular duties associated with studying living organisms
- understand the need for bioscientists to observe scientific integrity in terms of truthfulness, diligence, objectivity, circumspection and collegiality
- understand the need for bioscientists to show respect for their subjects of study, both human and non-human
- understand bioscientists' duties in terms of professional responsibilities, educational responsibilities and social responsibilities

15.1 Introduction

This chapter focuses on ethical issues that confront those engaged in the practice of science – as researchers, lecturers or students. To a large degree, what is discussed in the rest of the book affects bioscientists in their role as citizens as much as practitioners of science. The difference here is that the ethical concerns stem from the actual practice of scientific research and education.

Science may be said to have two sorts of goal, which can be labelled **epistemic** and **practical**. Epistemic goals are concerned with the acquisition of accurate knowledge – the quest for objective truth – whereas practical goals are pursued to acquire knowledge relevant to particular human desires and to achieving those desires. To a large degree (but not entirely) these two goals correspond to the distinction made between **pure** (or **basic**) and **applied** science, i.e. knowledge for its own sake and *useful* knowledge respectively. But in practice the distinction is rarely clear cut. What an endocrinologist discovers when researching the physiological factors stimulating release of insulin from the rabbit pancreas could well be relevant to medical therapy, making it impossible in advance to label this research pure or applied. The distinction might usually have more to do with the motivation or the sources of research funding. Even so, distinguishing between epistemic and practical aims can be useful in that it facilitates discussion of different ethical impacts.

15.1.1 The nature of the problem

In pursuing epistemic goals, scientists have responsibilities to discover, describe and explain the truth to the best of their abilities. Because the very hallmark of science is its reputation for objectivity and reliability, anyone seeking to attain the status of 'scientist' has clear moral obligations to maintain these standards. To that degree, science may be considered a *profession*,[1] even though early scientists, like Harvey (who discovered the circulation of the blood) and Lavoisier ('the founder of modern chemistry'), were talented amateurs whose objectivity was a matter of personal integrity.

But precisely because science is a human activity, mostly carried out by ordinary people rather than paragons of virtue, it is vulnerable to the consequences of human moral weakness, e.g. in the forms of fabrication, falsification and plagiarism of experimental data. It might be assumed that examples of such unethical behaviour would be rare, both because they would be soon be discovered if committed and because the shame attending exposure of such dishonesty would be a significant deterrent. Even so, a number of notorious cases of such behaviour are to be found in the annals of science. However, what might be more problematical are the (probably much more numerous) minor deceits, and other forms of corruption, that escape easy detection.

Ethical issues relating to the practical aims of science concern matters such as:

- the social acceptability of the aims
- the impacts of the research methodology on the welfare, rights, and fair treatment of the subjects of the research (whether humans or non-humans)
- the wider consequences of the research and its application (e.g. as technology) for society, including future generations, and the biosphere.

Several such issues have been discussed in chapter 1 and chapters 4–12. Although practical issues overlap with those of an epistemic nature, the main focus of this chapter is on the latter.

15.2 The Mertonian norms

According to scientist John Ziman, *'The most tangible aspect of science is that it is a social institution.'*[2] Although, in principle, individual scientists have quite a lot of freedom in what they do and how they do it, their thoughts and actions only have scientific meaning in the larger social scheme. When people work together in a community, they inevitably conform to a set of *professional rules*, which determine how they behave. Anyone who falls foul of the rules is liable to be excluded from the community.

An illuminating analysis of the rules governing scientific research was provided by the sociologist of science Robert Merton, who suggested that *'the prescriptions, proscriptions, preferences and permissions'* that scientists feel bound to follow could be encapsulated in a set of general *norms*.[3] These have subsequently been labelled: *communalism, universalism, disinterestedness, originality* and *scepticism*.[4] They are not generally presented as ethical principles but rather as traditions, which are transmitted from established scientists to new members of the profession, principally by custom and example. But, in a sense, they constitute the *ethos* of science (a word with the same Greek root as 'ethics'), which is built into the scientific conscience of individual scientists. The significance of each of the five norms may be briefly summarized as follows.

Communalism. This refers to the requirement that the results of academic research should be considered as *public* knowledge. Data should not be kept secret, but at the first reasonable opportunity disseminated through the usual channels of scientific communication: at conferences, at seminars, and especially in papers submitted to scientific journals. Printed communications contribute to the scientific archive, the stock of communally built knowledge on which further progress in science depends.

Universalism. This refers to the democratic nature of scientific activity, which does not tolerate discrimination on the basis of gender, race, religion, or social or academic standing. Only the validity of the scientific observations made, and the rationality of the conclusions drawn, are deemed to count.

Disinterestedness. This word (that causes confusion when misused in place of 'uninterestedness') refers to a condition in which the scientist has no vested interest in the results of their scientific investigations, e.g. because they, or their family or friends, might gain or lose financially by a certain outcome; or because they 'have an axe to grind'.

Originality. This is a requirement for the research to contribute something new, e.g. by presenting new data, proposing a new theory or offering a new explanation for existing observations. At best, it appeals to the ideal of fostering creativity; at the minimum, it urges avoidance of plagiarism.

Scepticism. This represents the necessary counterbalance to the drive for originality, epitomizing the Popperian edict to attempt to refute any novel claim (1.5.1). Most formally, scepticism is built into the refereeing process for scientific publications submitted to academic journals. This so-called **peer-review** is carried out confidentially, neither the author nor the referee knowing the other's identity, and is intended to ensure that all claims as to the validity of data, rationality of argument, and plausibility of any deductions made can withstand the critical scrutiny of independent experts in the same field. But scepticism often extends beyond the formal reviewing process when published data or theories are exposed to the comments of fellow scientists with competing interpretations.

15.2.1 Academic and industrial science

The initial letters of the five norms outlined above spell CUDOS (the Greek word, usually spelt *kudos*, means 'fame as a result of achievement'), which is regarded as the reward received by scientists who practise such norms effectively. The norms encompass several of the ethical principles discussed earlier, such as those based in utilitarian and deontological theories (2.5–2.6), but they combine in a particularly powerful way when applied to the practice of science, which is, after all, about the pursuit of truth. So, it would seem to follow that good science depends as much on high ethical standards as it does on high standards of intellectual, technical and mathematical competence.

Of course, the reality is that because these norms are *ideals*, in practice, they are often only partially observed - or even sometimes flouted. But to the extent that they are not observed, science as an enterprise is likely to suffer. It is also important to recognize that the norms can be expected to be followed more readily in some situations than in others. For example, the conditions of employment of scientists working for a commercial company (say, on drug development) are such as to impede observance of many of the norms. Thus, communalism is prevented because it is important for companies to maintain commercial confidentiality if they wish to prevent their rivals stealing a march on them; as epitomized by patents and other forms of intellectual property rights (Box 4.2). Universalism is challenged by the hierarchical structure of businesses. Disinterestedness and (rigorous) scepticism are almost impossible to maintain when commercial success (and the scientist's job security) are at stake.

Originality is not necessarily compromised by the commercial objectives, and might even be encouraged by it, as efforts are directed to seek new solutions to practical problems. On the other hand, only those solutions will be supported which show the promise of being commercially viable, which may be a significant constraint in respecting this norm. Cynically, it might be concluded that the norms are not only ideals but also idealistic.

In fact, so different are the norms of academic and industrial science that John Ziman has suggested that for the latter the acronym PLACE (defining the goal and reward of leading scientists in industry) is more appropriate than the Mertonian CUDOS.[5] The letters stand for **proprietary**, **local**, **authoritarian**, **commissioned** and **expert**, the qualities characterizing industrial research, which are largely antithetical to the ideals of academic science.

Now, it is just possible that these two types of scientific activity (academic and industrial) could co-exist if a rigid barrier were drawn between them: they could, as it were, run in parallel and only touch at formal points when the differing ground rules were recognized and respected. But, whether or not this situation ever existed, a meaningful distinction between the two types of research may now, in the early twenty-first century, be almost impossible. For example, academic scientists receive funding from commercial companies (sometimes under confidentiality contracts), industrial scientists hold honorary academic positions in universities, and government research institutes collaborate with multinational companies. The notional 'barrier' is thus highly porous, and the interpenetration of the two ideologies extensive. It is instructive to consider, with reference to specific examples familiar to the reader, the extent to which PLACE has eroded CUDOS.

The reasons for this blurring of the distinction between academic and industrial science are readily understood. It was seen in 14.1.1 that by the 1980s government funding for science in many DC began to decline, and commercial companies were encouraged to share the financial burden. It was reasoned that in the modern world a thriving national economy is dependent on an active scientific and technological research base, so that it is in society's interest to attract successful companies and foster their collaboration with academia. The mission statement of OST (14.2), which has overall responsibility for the Government's science policy, is to support *'excellent science, engineering, and technology and their uses to benefit society and the economy'*,[6] indicating that the epistemic and practical goals of science are conflated without any suggestion of a conflict of interests. In subsequent sections we shall explore whether this arrangement has any serious ethical implications.

The Mertonian norms, and the extensively modified versions of them applicable in industrial science, represent generalized (if idealized) sociological accounts of the ethos of scientific research. But, as guiding principles for the practising scientist they lack specificity. So what do they mean for bioscientists working in the laboratory, the field or the office? Box 15.1 proposes ten standards of ethical conduct. Five of them relate to the integrity of the working scientist in conducting research, two to the treatment of their subjects (human and non-human), and three to wider ethical duties that derive from the professional status of bioscientists. The subsequent discussion explores the meaning of the standards and illustrates ways in which they may be observed, and infringed.

BOX 15.1 STANDARDS OF ETHICAL CONDUCT IN THE BIOSCIENCES

Scientific integrity

- *Truthfulness*: rejecting all fabrication, alteration or misrepresentation of data
- *Diligence*: working with care, minimizing errors and self-deception
- *Objectivity*: avoiding bias and conflicts of interest
- *Circumspection*: arriving at conclusions only after careful deliberation
- *Collegiality*: sharing results, methods and ideas with other scientists

Respect for subjects of study

- *Respect for human subjects*: fully respecting subjects' rights, dignity and welfare
- *Respect for non-human subjects*: showing appropriate respect to non-human subjects in accordance with their welfare, rights and dignity

Professional duties

- *Professional responsibilities*: playing a responsible role in the scientific community and in its relations with wider society
- *Educational responsibilities*: showing willingness to inform others (in formal education and as the general public) of the significance of one's work
- *Social responsibilities*: using scientific knowledge and technique to benefit society in just and socially acceptable ways

15.3 Scientific integrity

Regardless of the subject of study, certain personal qualities are essential to the conduct of authentic research in the biosciences. Each is discussed below.

15.3.1 Truthfulness

This might be considered the most important of all standards, because failure to respect it undermines a major goal of science - to establish objective truth. The whole enterprise of science is based on the assumed trust which can be placed in the honesty of all scientists in reporting their results. Fabricating (inventing) data, altering data, or presenting data in ways that are intentionally dishonest - all fall foul of this standard. **Plagiarism** is the act of representing another person's work or ideas as if they were one's own. All these are ethical infringements of a deontological nature (2.6), since they originate in an *intention* to deceive. But, as is true of other such principles, flouting the rule of truthfulness also has serious *consequences*, because if dishonesty became universalized science would soon lack any credibility. Regrettably, however, cases of dishonesty are by no means unknown in the biosciences.

In their, now classic, exposé of scientific fraud and deceit, *Betrayers of the Truth*, William Broad and Nicholas Wade detailed numerous cases of dishonest behaviour by scientists. Some of the more notorious are:

- Sir Cyril Burt, the distinguished British psychologist, whose research on the role of heredity in intelligence, published in 1969, was fabricated
- Thereza Imanesh-Kari, a co-author of Nobel Prize-winner David Baltimore, whose data in a paper published in the journal *Cell* in 1986 are believed to have been fabricated (the so-called 'Baltimore Affair')
- Mark Spector, a postgraduate student at Cornell University, working in the laboratory of the distinguished biochemist Ephraim Racker, whose theory of cancer causation was based on fraudulent data.

In each case, and many others that could be cited, exposure of the fraud was delayed by the reluctance of others to admit the problem was real. But only rarely (as far as is known) are people able to get away for long with blatant dishonesty, as in the case of Elias Alsabti, who published sixty papers in medical journals in the 1970s - all of them plagiarized from others' work.[7]

Why did they do it? Perhaps the reason, but certainly not the excuse, lies in the pressures to which scientists feel exposed. The main objective of science is *explanation*: there is little kudos in simply cataloguing natural events. So presenting a persuasive, rational theory which appears to explain diverse observations effectively is what drives most scientists. Significantly, it is also what earns them recognition (even fame), improved opportunities for more research funding, and better prospects for job security or promotion. Unfortunately, experimental data do not always conform neatly to a scientist's pet theory - possibly because the theory is false or possibly because the design and/or execution of the experiment prove inadequate. But in the race to be first to get a publication out supporting the theory, some scientists are tempted to cut corners - by being selective in citing data, omitting the 'odd' results and embellishing the ones that conform to the theory. Getting away with it once, they may be tempted to do so again.

It seems that dishonesty is not confined to junior scientists. We have noted that Mendel was not above suspicion (1.4.1), and it appears that even such luminaries as Darwin and Newton were not blameless.[8] But no one should imagine that this gives licence to dishonesty. Quite apart from the demands of personal integrity, the standards of accuracy required in modern-day science are much higher than those often deemed adequate in the past - and the possibilities for detecting abuses much greater. Among the complex of motives and factors that determine the personal behaviour of scientists, appeal to Aristotelian virtue theory (2.7) may be considered crucial. Each scientist has to ask him- or herself the questions: 'Could I live with the fact that I might receive credit for work that was built on deceit?' and 'Do I want to be responsible for corrupting the scientific endeavour, the value of which depends on the absolute integrity of scientists?'

Despite such strictures, it has to be acknowledged that judgement often plays a critical role in ethical behaviour. For example, in writing a research paper decisions almost always have to be made as to what to include and what to omit: editors are unlikely to accept for publication excessively long papers. This means that some selectivity must be practised, and doing so accurately and even-handedly is a matter of ethical judgement.[9]

15.3.2 Diligence

The importance of diligence in achieving accuracy in both practice and reasoning is obvious: slipshod behaviour is unlikely to produce reliable data. However, we need to recognise the factors that might challenge this standard. Chief among these is the 'pressure to publish', stemming from scientists' anxiety to buttress their list of scientific publications, in order to improve their chances of personal advancement. In such circumstances due diligence may become a victim of the desire to simply churn out data.

A related misdemeanour is the phenomenon of the *least publishable unit*, a term used to describe the minimal amount of scientific work which can be published as a separate scientific paper. The consequence is that *'Instead of publishing one comprehensive paper that ties the work together, a researcher will publish four or five shorter ones. Sheer volume, he reasons, will help his career.'*[10] Alternatively, the same work may be published in different journals, under slightly different titles, to the same effect. Some scientists even joke openly of doing a 'cut and paste job'.

15.3.3 Objectivity

We have seen that in the past biology has often been exploited for political ends. A prominent example is the rejection of Mendelism in the Soviet Union in the 1940s, on the grounds that it did not conform to the Marxist belief that permanent change could be achieved by environmental influences (1.4.3).

Nowadays, the most critical conditions are usually those in which there is a **conflict of interests** due to commercial factors.[11] Say, for example, that a scientist is employed by an imaginary biotechnology company that is developing an injectable product designed to stimulate lean growth rates in pigs (let's call it *Porcilene*). Meat from this source might be produced more efficiently in economic terms, and be a healthier product for consumers, because of its lower fat content. The scientist has an obvious interest in the commercial success of *Porcilene*, but clearly also wishes to report experimental data accurately. However, if research results suggest that *Porcilene* has adverse effects on animal welfare, reporting this fact conflicts with the desire for commercial success (and even possibly the scientist's job security). In such circumstances, the scientist's integrity demands accuracy in reporting results - resisting any temptation to be influenced by personal interests. The Mertonian norm of *disinterestedness* remains critical in scientific research.

But this standard encompasses more than just methodological accuracy. It is important to adopt a frame of mind that is aware of any personal bias that might influence one's results, and this also applies to collecting data, analysing them and proposing a theory to explain them.

15.3.4 **Circumspection**

Truthfulness, diligence and objectivity may be said to contribute to, and complement, the standard of circumspection. This is the cautious, well-considered approach, which supports only those conclusions that are well grounded in evidence. In Mertonian terms, this standard is based in the sceptical attitude that characterizes the true scientist, for whom hype and optimistic speculation are corrosive influences that have no place. However, it is another standard that is increasingly challenged by some significant pressures on modern-day scientists.

In the first place, in the highly competitive 'market' for research grants, self-publicity has virtually become normal practice for scientists. Whereas in the past scientists were almost invisible, as they worked away quietly and patiently in their laboratories, it now seems obligatory for scientists to advertise themselves to attract attention (and funding), have their own websites and publicize the perceived benefits of their research. The assumption is that those who don't trumpet their achievements haven't made any. Of course, the downside of all this self-publicizing is that it devalues the currency. The audience will, rightly, become increasingly sceptical, and assume that many claims may be taken with a pinch of salt.

But the second effect is that in the scramble for funding, there is a tendency to exaggerate the significance of the proposed research: no longer is it deemed sufficient to produce scientifically interesting and reliable results. Instead, it is thought necessary to make claims for the social or economic benefits - in line, for example, with research council aims (14.2.1). And the more your competitors promote their work, the more you will find it necessary to promote yours. Neither of these two pressures is conducive to circumspection.

15.3.5 **Collegiality**

Prominent among Merton's norms are those depending on the health of the scientific community - *communalism* and *universalism*. Both, of course, have intrinsic merits - there is usually more pleasure to be gained from sharing your ideas with people of a similar interest, and challenging and being challenged by colleagues who share a common aim of understanding a scientific problem. And when formal barriers are broken down, by admitting into the 'club' all who can contribute fertile ideas regardless of their status in the academic hierarchy, the sense of collegiality can be its own reward.

But quite apart from the pleasure to be derived from being part of this lively community, the mutual sharing and respectful criticism of ideas is an essential element in the *quality control* process. Testing of ideas is an indispensable element of scientific progress. Whether you appeal to the dialectical process advocated by Socrates or to Popper's prescription of 'attempted refutation' (1.5.1), exposure of new ideas to critical examination by others is the soundest method of establishing their value. Suppression of criticism (as for example in the Soviet Union of the 1940s) is a certain recipe for self-deception.

In such terms, industrial science labours under a severe disadvantage. Because secrecy is one of the most important considerations for companies aiming to develop new products, exposure of ideas to those with no vested interest is a missing element in the assessment process. If academic scientists are now increasingly encouraged into acts of self-promotion, this has, of course, long been standard practice for commercial companies. Financial profits are the bottom line for commercial companies, and however much industrial scientists may share their academic counterparts' interest in science for its 'own sake', when the chips are down, collegiality (and objectivity and circumspection) are ranked lower than commercial imperatives. Even more problematical is the blurring of the distinction between academic and industrial research (15.2.1).

15.4 Respect for the subjects of study

Over and above the above ethical concerns associated with the practice of scientific investigation, the study of biology has been claimed to make particular ethical demands as a result of its subject matter. According to bioethicist Tristram Englehardt, our sense of respect for life reflects several different values, such as:

- the affirmation of life's intrinsic value
- the value of the diversity of life
- the dignity of life, which should not be violated
- respect due to a sympathy with living organisms, disposing us not to cause unnecessary suffering.[12]

These are additional to the respect for other *persons*, which is the essence of Kant's philosophy (2.6). But, as Englehardt points out, there are a variety of ways in which to value and respect life, so '*this problem places persons in a cardinal position and makes all general approaches to respecting life a compromise between competing moral visions*'.[13]

15.4.1 Respect for human subjects

Although this book is not directly concerned with medical issues, it is important to consider the *Nuremberg Code*, which was drawn up in the wake of the experiments performed by Nazi scientists during the 1939–45 World War, since it marked the beginnings of bioethics as a discipline. The experiments were performed on prisoners of

BOX 15.2 THE NUREMBERG CODE

- voluntary consent of human subjects is essential
- experiments should be anticipated to yield fruitful results for the good of society
- experiments should be based on prior results of animal experiments
- experiments should be designed to avoid all unnecessary physical and mental suffering and injury
- no experiments should be performed when there is an *a priori* reason to expect death or disabling injury
- the risks taken should never exceed the humanitarian importance of the problem addressed
- proper precautions should be taken to prevent even remote possibilities of injury, disability or death
- experiments should be conducted by fully qualified scientists exercising the highest degrees of skill and care
- the subject should be free to terminate the experiment when desired
- the scientist in charge should be prepared to terminate the experiment if necessary to avoid injury, disability or death

war and civilians, including Jews and 'asocial persons', as part of a planned government programme. Some of the barbarities inflicted are described in 1.6.1. The Code sought to establish principles that should govern any future human experimentation, based on the premise that vital knowledge was unprocurable by other means and that '*the principles should satisfy moral, ethical and legal concepts.*'[14] The main principles are summarized in Box 15.2.

More recently, the World Medical Association's *Declaration of Helsinki* (1964, but updated several times since then) has also sought to define ethical standards for research on human subjects, which are relevant in non-medical research and class experiments, as well as in medical research.[15]

But the Nazi experiments were unfortunately not an isolated case of institutional ethical malpractice. For example, at the Tuskagee Institute, Alabama, USA, a study was conducted, over the years 1932 to 1972, on 399 black men suffering from late-stage syphilis. Sponsored by the US Department of Health, the research project involved comparing the health status of the infected, but untreated, patients with that of 200 men of similar age who were not infected. The study lasted for forty years, even though an effective treatment for syphilis became available from the mid-1940s. When it was finally terminated in 1972, one-hundered men had died of syphilis, forty wives had contracted the disease, and nineteen children had been born with congenital syphilis. However, the US Government did not admit wrongdoing until 1997, when at an official ceremony, attended by five of the surviving eight men, President Clinton formally apologized.[16]

15.4.2 Milgram's experiments

Harm is not, however, just a matter of physical pain: it can also include the harm of deception. One of the more notable cases of subjects being wilfully deceived occurred in the psychological experiments performed in the USA in the 1970s by Stanley Milgram

BOX 15.3 THE POWER OF SCIENTISTS' AUTHORITY

In 1974, **Stanley Milgram** described psychological experiments he had conducted at Yale University over many years *'to find when and how people would defy authority in the face of a clear moral imperative'*. The experiments involved three people: a ***teacher***, a ***learner*** and the experimenter himself. Teachers were recruited by advertising in a local newspaper and paid a small fee for their services, and although the roles of teacher and learner were supposedly determined by lottery, in fact it was rigged so that the naïve subject of the psychological experiment was always given the job of 'teacher'.

The purported experiment concerned the effects of punishment on learning, and involved giving electric shocks to the 'learner', who was located in the next room but within earshot of the other two. The teacher was asked to give electric shocks of increasing voltage (in 15 V steps), by pressing an array of thirty switches, marked from slight (15 V), through *strong, intense*, and finally *danger: severe shock* (450 V), when the learner failed to remember some memory tasks. The learner, strapped to a chair, (who was an accomplice of the experimenter and actually received no electric shocks at all) made noises corresponding to the pain he was supposedly getting from the electric shocks. These ranged from a mild grunt for lower voltages to agonised screams, and cries to be set free, for voltages over 330 V. However, the point of the 'real experiment' was to see how far the teacher would go when the learner was apparently suffering acute and life-threatening shocks.

When administering the 'shocks' any teacher who questioned the experimenter as to whether he should proceed, in the light of the learner's apparent distress, was told, in a cool, emotionless way: *'you have no other choice – you must go on'*. Many psychologists questioned beforehand about the likely results thought that less than 20% would continue to the half-way point. In fact, 80% continued to this point, while 60% actually threw the final switch at 450 V.

A large part of the obedience to expert authority was associated with the fact that the experimenter was perceived as a *scientist*. When Milgram arranged for the experimenter to pretend that he was a lay person, the completion rate for the experiment dropped to 20%.

(Box 15.3).[17] The experiments illustrate the type of dilemma which often faces scientists involved in biological research. Undoubtedly, Milgram's results were a startling revelation of the authority wielded by scientists – although it is important not to generalize too freely, since the studies were performed at a specific location and specific time which may not be typical of people in other situations. But were they ethically acceptable? In a utilitarian analysis, the benefits of this knowledge might be deemed to outweigh the psychological harm inflicted on the subjects. By contrast, a deontological stance, giving due respect to the subjects' right to know the nature of the experiment and to avoid deceit, would condemn such experiments – as indeed would application of the Nuremberg Code.

15.4.3 **Respect for non-human subjects**

The considerations bearing on the use of sentient animals in research were discussed in chapter 8. While for many people the welfare of the animals is a critical factor, there are also increasing concerns over threats to their intrinsic value. Animal experimentation in the UK is governed by the Animals (Scientific Procedures) Act (8.4), which is reputedly one of the most stringent instruments of legal control in the world. However, complacency would be inappropriate, and concerns over the increasing use of GM animals and the pressures for greater use of primates in medical research pose urgent ethical questions (8.7.2).

15.5 Professional duties

The biosciences, like other branches of science, are professions. To train, qualify and practise as a bioscientist entails the formal requirements of higher education, official certification and, usually, membership of a specialist professional body. But these privileges come with certain responsibilities.

15.5.1 Professional responsibilities

As members of a profession, bioscientists acquire responsibilities to maintain its viability, preserve its professional standards and ensure that its importance is recognised by the wider society. Senior scientists often play leading roles in academic societies, sit on expert subject panels or serve on government advisory committees. But many responsibilities are also taken on by scientists lower in the academic hierarchy.

An important role for many scientists is refereeing papers submitted for publication in academic journals. In principle, the process is generally completely anonymous, in that the identity of neither the paper's author nor the referee is revealed to the other. But that said, if a field is sufficiently specialized there may be several clues as to the author's identity, especially if, as is usual, the paper itself refers to earlier publications.

Refereeing papers has both benefits and costs: it sometimes provides valuable insights into recent scientific developments, but if performed conscientiously it takes time (for no financial rewards) which might be better spent on one's own work. Sometimes a referee may find that the editor has misjudged his expertise, presenting him with a paper beyond his competence to assess adequately. For example, it is important that papers explain the methods used in sufficient detail for others to repeat the study and hence check on the reliability of the results. But the specialised nature of much research means that the referee may find that methodological descriptions are unfamiliar, and hence unclear to him. The temptation must often be to *assume* the author's practical competence, because to admit to the editor that one is not up to the job of refereeing the paper entails of loss of face. The referee's sense of pride can thus allow substandard papers to join the 'body of scientific knowledge'.

The confidentiality which underpins the system also presents various temptations, which are sometimes exploited by unscrupulous referees to their own advantage. For example, the referee is privy to information and ideas known only to, usually, one other referee and, in theory, to the journal's editor (although the latter is probably unlikely to have the time or inclination to read all submitted papers). This situation is clearly open to abuse, e.g. if the referee were to plagiarise the ideas, or to delay sending the referee's report until his own related work had been submitted to another journal. The same factors might come into play in cases where an application for a project grant is refereed. When selection committees discuss grant applications, scientists who have a vested interest in a particular proposal are required to withdraw from the meeting room. But this does not prevent deals being struck in advance of the meeting, to the mutual advantage of two or more parties.

There are numerous other ways in which covert activities can be resorted to in order to gain unfair advantage.[18] One such is *honorary authorship*, when people who had no substantial involvement in, or responsibility for, a piece of research are credited with co-authorship as a way of winning some favour or repaying a debt. It would be reassuring to think that such self-serving arrangements were exceptional aberrations, but this is something which, by its nature, is almost impossible to assess accurately. What is certain is that the role of professional scientists makes rigorous ethical demands on them, simply because the required standards are especially high, and yet there are numerous opportunities for their abuse.

But sometimes scientists are faced with the problem of what to do when they discover others' misdemeanours. In such circumstances responsible scientists are placed under an ethical obligation to report instances of misconduct to the appropriate authorities, adopting the role of **whistleblower**. The practice is personally risky, since vested interests may jeopardise the security and well-being of whistleblowers themselves - and this is doubtless why fraud often goes unreported for so long (15.3.1). Philosopher David Resnik suggests the following guidelines for prospective whistleblowers:

- have morally good motives; being sure only to report illegal, unethical, or immoral activities and not to try to advance one's own career
- have well-documented evidence, rather than relying on personal observation or hearsay
- report to the relevant authorities, only appealing to other bodies in the last resort
- carefully deliberate on the proposed action, avoiding rushed judgements.[19]

15.5.2 Educational responsibilities

The health of an academic discipline clearly depends on its established members attracting competent recruits and inspiring them to engage in research to advance the knowledge base. One of the principal ways in which knowledge is transmitted is through publications, as academic research papers and review articles aimed at the scientific community, and through textbooks for students. Publication might almost be regarded as an ethical requirement, justifying society's investment in the research. As we have noted, it is a requirement amply fulfilled by most scientists, not least as a means of securing status and future research funding.

But the more usual understanding of education relates to instruction and training provided for students at both undergraduate and postgraduate levels. One contentious area is the use of animals in practical classes, either for dissection exercises or to provide material for physiological studies or demonstrations. The traditional view is that a proper education in the biosciences often entails hands-on experience, and that those unwilling to perform such work ought not to sign up for such courses. On the other hand, it might be the case that use of animals is now less important because new technology, in the form of videos, computer programs and realistic models, obviate the need. A new, conservationist, outlook might suggest that in purely utilitarian terms (2.5) the educational benefits of using live (or dead) animals rarely outweigh the costs. It is thus important for lecturers to assess the significance of animal studies to the course objectives, and to make clear to prospective students what will be expected of them. In any case **conscientious objection** needs to be taken seriously.

Although academics bear titles that relate to their teaching role (as lecturers, readers or professors), in reality it is usually their distinction as researchers that wins them professional advancement. Consequently, some senior staff pay little attention to teaching, which they delegate to junior staff. It is more usual for senior academics to take an interest in supervising postgraduate students, who contribute to what usually counts for more, the research output of the department. And, of course, the bigger the research team of students, the more ambitious research projects can be. The possible downsides of this are that there can be a conflict of interests – between research students receiving a broad, high-quality, personalized education on the one hand, and, at the other extreme, merely supplying an 'extra pair of hands' in a research project mainly serving the interests of industrial sponsors. For example, one wonders what the significant contribution of each co-author could have been in a paper published in 2003, which had 157 co-authors![20]

15.5.3 Social responsibilities

Virtually all bioscientists are trained, at least largely, at the taxpayers' expense. Whether or not they, or their family, paid entirely for their educational tuition, they are beneficiaries of the State educational and scientific establishments that underpin modern democratic societies. In such terms, a strong case can be made for their having moral responsibilities to use their knowledge and skills for the benefit of society, i.e. in the *public good.*

However, this objective encounters problems when the distinction between the public and private spheres becomes blurred. In his, now classic, analysis *Biotechnology: the university–industrial complex*, Martin Kenney described how, in the USA in the 1980s, the collaboration between university professors and industry led to the emergence of the **academic-entrepreneur.**[21] Approached by private companies anxious to recruit their expertise and experience, the academics welcomed the opportunity to secure research funding far in excess of anything they could expect from government sources. Collaboration with industry often brought new laboratories, state of the art equipment, more technical assistance, additional research studentships, and increased income for both the academics and their universities (the latter through patent fees).

Even by the 1980s, some research grants from multinational companies were very large: for example, the Monsanto Company provided US$23.5 million to both Harvard Medical School and Washington University, while Hoechst funded the Massachusetts General Hospital to the tune of US$70 million.[22] Writing almost twenty years ago,

Kenney speculated, *'It must have been very difficult for professors who were receiving $30,000 to $60,000 a year to resist the lure of capital gains of the order of millions of dollars.'*[23]

However, the threats to the Mertonian norms are all too obvious. Academic scientists used to working in the open, rigorously sceptical environment, in which their colleagues were prepared to share and challenge their findings, were suddenly ruled by a different ethos, in which secrecy, corporate loyalty, and financial profitability were paramount (see 15.2). In subsequent years, links with private industry have spread throughout most of the biosciences, and giving rise to what Ziman calls **post-academic science**, an uneasy compromise between academic and industrial norms.[24] The resulting challenges to scientists' integrity (15.3) are substantial.

Representing biotechnology to the public. One test of scientists' integrity is the way in which they chose to present scientific ideas to the general public. It is not unusual for those wishing to promote new forms of technology in democratic societies to encounter resistance. The classical illustration is the Luddite Uprising of 1811–12, in which a group of Nottinghamshire craftsmen (led by a self-styled 'King Ludd') destroyed new textile machinery, whose introduction was losing them their jobs. The term *luddite* is now applied more generally, and disparagingly, to those who oppose scientific and technological innovation. When the claimed benefits of a biotechnology are more speculative and/or seem likely to challenge established ways of living and thinking, proponents of the technologies often resort to rhetorical language to convince the doubters.

Sociologist Sheldon Krimsky claims that the resulting verbal contest between proponents and opponents for control of the symbolic meaning of biotechnology leads to a process of *myth-making*.[25] He stresses that the term *myth* is not introduced to denigrate either side of the dispute, but rather, *'judiciously, to signify a cultural story that embodies hope, expectations, moral attitudes, fears, or positive visions of modernity. Myths are constructions out of reality that transcend the real into a virtual world of expectations.'* We can list a few such **techno-myths** (and the corresponding **anti-myths**) to illustrate the point. Thus, it is claimed that: 'Biotechnology will:

- provide natural (or unnatural) products
- offer greater (or less) control over nature
- contribute to more (or less) biodiversity
- be friendly (or unfriendly) to the environment
- feed (or not feed) the World's hungry
- provide (or not provide) cures for major diseases.'

The visible outcome of such differing attitudes is evident in the advertising campaigns of companies such as Monsanto, on the one hand, and of environmental pressure groups, such as Greenpeace, on the other. Without discussing here the merits or demerits of the different arguments advanced, it is clear that important factors are at work, which are external to the practice of science. Clearly, the value-systems of the different players exert a powerful influence, one way or another, over public attitudes. To some degree, the techno-myths promoted might be the result of unconscious factors, the proponents and opponents presenting their respective arguments in the way that seems to them most accurate.

BOX 15.4 THE BERG LETTER

In 1971, Nobel Prize-winner **Paul Berg** (discoverer of tRNA (transfer RNA)) suggested to a research student that she insert some SV40 DNA fragments into *E. coli*, the common gut bacterium. SV40 is an oncovirus – one that transforms normal cells to cancer cells. But when discussing the proposed experiment, informally, at a conference, she encountered alarm. The possibility of a carcinogenic bacterium spreading throughout the world's population seemed an unjustifiable risk, a view shared by Berg himself when it was brought to his attention. Chastened by this experience, Berg and colleagues (all prominent molecular biologists) wrote a letter to the leading weekly science journals *Nature* and *Science*, suggesting that there should be a voluntary moratorium on all similar experiments until the risks had been assessed. Known as the *Berg letter*, it also called for a major international conference to discuss the biosafety issues surrounding use of GMO.

The conference was duly held at Asilomar, California in 1974. It established an international consensus on the GMO ban and led to the drawing up of a set of rules for the use of GM bacteria (the National Institutes of Health Guidelines). With increased experience, the guidelines have since undergone substantial revision. But the Berg letter is often hailed as a shining example of scientists acting with social responsibility.

However, this conclusion is sometimes less credible. Consider, for example, the case of pesticide advertising to farmers in agricultural magazines. In terms of eradicating insect pests, these have proved highly effective over the last sixty years. But, latterly, critical questions have been raised about effects on environmental sustainability and the social risks associated with their use. Initially, science was harnessed as a means of dominating nature, which was thought of as an enemy to be conquered. Now, however, green attitudes are widely acknowledged, with nature viewed as a resource which needs to be protected. The effect on pesticide advertising has been dramatic. In the 1940s–1960s, *science* was the dominant theme in advertisements, in terms of product names like Simazine, Isotox, and Lindane. By the 1970s, *control* became the dominant theme, represented by products with names like Prowl, Marksman, Bullet and Warrior. But since the early 1990s, agrochemical companies have reinvented pesticides as *nature attuned*, using names for their products such as Harness, Resource, Harmony and Accord.[26] In that the modes of operation are essentially unchanged, the representations might well be viewed as examples of 'spin'.

Acting in socially responsible ways Because bioscientists are custodians of specialised knowledge, which is largely incomprehensible to most non-scientists, they carry particular responsibilities for protecting human and environmental safety. Such considerations are especially important, for example, in relation to GMO, which could in theory, colonise the biosphere - conceivably to detrimental effect. As many of the examples discussed above refer to 'cautionary tales', it is appropriate to refer to a case which reflects more favourably on scientists' behaviour (Box 15.4).[27]

15.5.4 Business ethics

Many bioscientists work for commercial companies, for whom standards of behaviour come under the heading *business ethics*. It is not uncommon for people to suggest that ethics and business are incompatible: that life in the world of business is inherently aggressive and ruthless, and that sharp practice is not just to be expected, but actually amounts to 'good business'. However, this is a view that has been firmly rejected by philosophers such as Robert Solomon,[28] who argues that capitalism has not succeeded with brute strength or because it has made people rich, but because it has produced responsible citizens and (albeit unevenly) prosperous communities. Moreover, the *'idea that business can operate free of any moral constraints is nonsense'.*[29]

Business ethics is open to different interpretations, depending on the ethical theory given prominence. For Solomon, espousing a form of Aristotelian virtue theory (2.6), important considerations are *'the corporation as a community, the search for excellence, the importance of integrity and sound judgement, and a more cooperative and humane vision of business'.* Nevertheless, the tension between such objectives and the 'bottom line' (profitability) is certain to present many dilemmas. Moral issues are often accentuated by the power exerted by multinational companies. For example, it has been alleged that multinational companies compete unfairly in LDC, because they can borrow money from local lenders at favourable rates (leaving little for local firms), and *'because they do not carry their fair share of the cost of social development, which imposes greater burdens on local industries'.*[30]

15.6 Playing God?

Of all the ethical challenges facing humanity the prospect of a future ability to create life 'from scratch' must be one of the most problematical. If the widespread concerns over the possibility (and, to date, unconfirmed reports) of human babies cloned by nuclear transfer are a guide (5.5.3), attempts to create life may unleash an unprecedented backlash against modern biology – with scientists accused of 'playing God'.[31] But before reacting to such an allegation, we need to consider what precisely it might mean to 'create life'. In 1937, the distinguished biochemist N W Pirie wrote an article entitled *'The meaningless of the terms "life" and "living"'*,[32] and when we try to pin it down life *is* extraordinarily difficult to define.

In 2002, the polio virus was produced by chemically stitching together strands of DNA ordered over the internet,[33] and in 2003 Craig Venter reported the creation of a bacterial virus. The latter entailed development of a technique for rapid recombination and transfer of many genes at once. According to bioethicists Arthur Caplan and David Magnus, *'Today scientists can "synthesise" a small genome to create a virus. Soon they will move to larger genomes, and eventually to bacteria and possibly even to new genomes for animals and plants. It is conceivable that human beings will some day contain artificially synthesised chromosomes in their cells'* (a prospect discussed in 6.6.3). They identify three risks in such approaches:

- adverse impacts on human safety and the environment
- potential uses in bioterrorism
- issues of ownership, and hence possible monopoly control.[34]

But perhaps this list is altogether too utilitarian. Does it not also need to consider deeper questions about the implications of such programmes for our understanding of ourselves, and of our relationship to the universe we inhabit?

Strictly speaking, such developments do not 'create life', if only because viruses are not technically alive – they have to commandeer other cells in order to replicate. But this may be splitting hairs, in that the potential to create *entirely new* life forms might be said to be tantamount to creating life. For example, work at the Scripps Institute in the USA is exploring the introduction into living organisms of novel amino acids. It is claimed: *'This work effectively removes a billion year constraint imposed by Nature on the chemist's ability to manipulate protein structure and function, and will lead to proteins and possibly organisms with new functions not restricted by the twenty common amino acid building blocks.'*[35]

It could be argued that the attempt to model and create novel organisms simply represents the culmination of the reductionist agenda of the biosciences that spanned the twentieth century. Many biologists now, almost instinctively, see life as the operation of the genes and the metabolic processes they control 'writ large', so that all 'improvement', whether by removing problems or enhancing desirable traits is most effectively pursued at the subcellular level. But some people might well feel diminished by such an attitude, because *'a reductionist understanding of life is not satisfying to those who believe that dimensions of the human experience cannot be explained by an exclusively physiological analysis'.*[36] Perhaps just as importantly, from an epistemological perspective, assuming that 'it all boils down to DNA' is liable to distort, and seriously limit, our understanding of life.

15.6.1 The meaning of life

So it is appropriate to end this book in much the same way as it began, with reference to philosophical matters – because the philosophy biologists adopt is certain to affect the way they approach their work and treat the subjects of their study. Whatever attitude is adopted in researching life's processes, it will be underpinned by certain assumptions. They may be explicit, and hence recognized for their advantages and disadvantages, or – which is less satisfactory – implicit, so that they guide the biologists' reasoning and practice without them realizing it.

The history of biology reveals that scientists' understanding of *life* has progressed through a series of stages, under the influence of the dominant ideologies of the time (see Box 1.1). Following the seventeenth-century Scientific Revolution, it seemed that *life* was best explained as a mechanical process, subject to exactly the same laws of physics as the material world at large. Thus, Descartes believed that animals (lacking souls) were *just* automata, reacting to external stimuli but devoid of any feeling (7.1.1). This **physicalism** (given further scientific support in the nineteenth century by advances in biochemistry and biophysics) sought to explain *life* by reducing it to the physicochemical processes observed in the biochemist's test tube or on the physiologist's kymograph.

But to many biologists **reductionism** was unable to explain many of the distinctive properties of living organisms – such as evolution, self-replication, self-regulation, growth and differentiation. The reaction to physicalism was **vitalism**, which is the belief that living organisms possess a special vital force that is not found in inert matter.[37]

Often the vital force was located in the cell 'protoplasm', whose physico-chemical properties were thought to be beyond any conceivable chemical analysis. In that vitalism challenged the naïvety of the physicalist view that *life* was purely a mechanical process, it performed an important role. But it became associated with many reactionary metaphysical beliefs, and despite the support of several leading biologists, it also was ultimately shown to be flawed.

From about the 1930s, the emergence of a new paradigm (1.5.2), **organicism**, provided a more coherent explanation for both physicochemical basis of *life* and its distinctive properties. In essence, organicism (or **holism**, which is almost synonymous) stresses the fact that the distinctive property of *life* is not so much its composition but its *organization*. For example, knowing what the human body is made of will tell you little about how it works, because what is crucially important is how it is put together. Moreover, and just as critically, new properties *emerge* from the interaction of the organism's component parts and processes; and for the most part, these emergent properties are intrinsically unpredictable. As historian of biology Ernst Mayr puts it: *'Organisms are fundamentally different from inanimate matter. They are hierarchically ordered systems with many emergent properties never found in inanimate matter; and, most importantly, their activities are governed by genetic programs containing historically acquired information, again something absent in inanimate nature.'*[38]

Biology is thus an *autonomous* science, not reducible to physics or chemistry. It follows that in studying *life* there is a need to resist simplistic explanations, which from their different perspectives, might satisfy the physicist or the chemist. Moreover, complete reductionism undermines all possibility of individual freedom of action (2.1.2), which makes nonsense of our attempts to both understand and change the world.

We also need to be aware of the fact that, in the words of the Nobel-Prize-winning immunologist Peter Medawar, science is the *'art of the soluble'*:[39] that is to say, scientists only seek to tackle problems they believe they can solve. But this leaves a large territory of (at least, currently) 'insoluble' problems – living with which demands a sound ethical strategy. In such circumstances, at the 'frontiers of ignorance', it may be that application of a utilitarian calculus as a guide to ethical reasoning has severe limitations (14.9.1). Pursuit of biological knowledge mindful of respect for others as individuals (2.6.1) might provide a more adequate protection of human dignity than would a philosophy that locates human nature in the genes.

THE MAIN POINTS

- A set of norms has been proposed to govern academic science, which have been represented as: communalism, universalism, disinterestedness, originality and scepticism; however, these are difficult to fully observe in pursuing industrial science.
- Standards applicable to the biosciences focus on: scientific integrity, respect for subjects (human and non-human) and professional duties.
- Scientific integrity entails truthfulness, diligence, objectivity, circumspection, and collegiality – but infringements have been committed, e.g. as fraud, fabrication, plagiarism, by self-serving individuals.

- Respect for human subjects is governed by codes ensuring consent and care, but, historically, utilitarian reasoning has often over-ruled human rights.
- As citizens with specialized knowledge and expertise, bioscientists have particular responsibilities to society, which place them under duties to their profession; to students and the general public in educational contexts; and to society as a whole in providing objective, disinterested advice in political and commercial contexts.

EXERCISES

These can form the basis of essays or group discussions:

1. Discuss the claim that the 'pursuit of science is ethically neutral'.
2. What ethical issues are raised by research collaboration between university and industrial scientists (15.2.1)?
3. Analyse Milgram's experiments (Box 15.3) from the perspectives of utilitarian, deontological and virtue ethics (chapter 2).
4. Can one expect scientists to observe higher ethical standards than members of society at large? (15.3)
5. If it were possible, should scientists create a new, freely reproducing organism? (15.6)

FURTHER READING

- *The Ethics of Science: an introduction* by David B Resnik (1998). London, Routledge. An account focusing on ethics for the practising scientist.
- *The Ethical Dimensions of the Biological and Health Sciences* (2nd edition), edited by Ruth Ellen Bulger, Elizabeth Heitman, and Stanley Joel Reiser (2002). Cambridge, Cambridge University Press. An anthology of articles on responsible science aimed at students and practising researchers in the biosciences.
- *Betrayers of the Truth: fraud and deceit in the halls of science* by William Broad and Nicholas Wade (1982). London, Century. Something of a 'classic' of notoriety.

USEFUL WEBSITES

- **http://www.unmc.edu/ethics/links.html** *On line sources for Research Ethics*: a portal for several sites on research ethics (mainly in the USA).
- **http://www.wma.net/e/policy/b3.htm** *World Medical Association*: Declaration of Helsinki – which governs research involving human subjects.

- **http://www.hse.gov.uk/research/ethics** *UK Health and Safety Executive*: this provides research ethics committee links.

NOTES

1. Resnik D B (1998) The Ethics of Science: an introduction. London, Routledge, pp. 34–52
2. Ziman J (2000) Real Science. Cambridge, Cambridge University Press
3. Merton R K (1973) The Sociology of Science. Chicago, Chicago University Press
4. Ziman J (2000) Real Science. Cambridge, Cambridge University Press, p. 44
5. Ziman J (2000) Real Science. Cambridge, Cambridge University Press, p. 78
6. Office of Science and Technology (2004) http://www.ost.gov.uk
7. Broad W and Wade N (1982) Betrayers of the Truth. London, Century
8. Ibid.
9. Whitbeck C (1998) Research ethics. In: Encyclopedia of Applied Ethics. (Ed.) Chadwick R. San Diego, Academic Press. Vol. 3, p. 840
10. Broad W and Wade N (1982) Betrayers of the Truth. London, Century, p. 53
11. Bulger R E (2002) The Scientist and industry. In The Ethical Dimensions of the Biological and Health Sciences (2nd edition). (Eds) Bulger R E, Heitman E, and Reiser J R. Cambridge, Cambridge University Press, pp. 281–289
12. Englehardt H T Jr (1997) In: The Ethics of Life. (Eds) Noble D and Vincent J-D Paris, UNESCO publishing, p. 26
13. Ibid., pp. 26–27
14. The Nuremberg Code (2002) [1949] In: The Ethical Dimensions of the Biological and Health Sciences (2nd edition). (Eds) Bulger R E, Heitman E, and Reiser J R. Cambridge, Cambridge University Press
15. World Medical Association (2002) Declaration of Helsinki. http://www.wma.net/e/policy/b3.htm
16. Youngson R M (1998) Scientific Blunders. London, Robinson, pp. 210–213
17. Barnes B (1985) About Science. Oxford, Blackwell, pp. 72–89
18. Resnik D B (1998) The Ethics of Science: an introduction. London, Routledge, pp. 179–205
19. Ibid., p. 26
20. Ota T *et al.* (2003) Complete sequencing and characterization of 21,243 full-length cDNAr. Nat Genet *1285*, 40–45
21. Kenney M (1986) Biotechnology: the university-industrial complex. New Haven, Yale University Press
22. Ibid., p. 56
23. Ibid., p. 97
24. Ziman J (2000) Real Science. Cambridge, Cambridge University Press, p. 67
25. Krimsky S (1995) In: Issues in Agricultural Bioethics. (Eds) Mepham T B, Tucker G A, and Wiseman J. Nottingham, Nottingham University Press, pp. 1–18
26. Kroma M M and Flora C B (2003) Agric Human Values *20*, 20–35
27. Goodfield J (1977) Playing God: genetic engineering and the manipulation of life. New York, Random House
28. Solomon R C (1992) Ethics and Excellence: cooperation and integrity in business. New York, Oxford University Press

29. Jackson J (1998) Business ethics: overview. In: Encyclopedia of Applied Ethics. (Ed.) Chadwick R. San Diego, Academic Press. Vol. 1, pp. 397–411
30. De George R T (1990) Business Ethics (3rd edition). New York, MacMillan, p. 403
31. Grey W (1998) Playing God. In: Encyclopedia of Applied Ethics. (Ed.) Chadwick R. San Diego, Academic Press. Vol. 3, pp. 525–528
32. Marquand J (1968) Life: its nature, origins and distribution. London, Oliver and Boyd, p. 89
33. Cello I J, Paul A V, and Wimmer E (2002) Chemical synthesis of polio virus cDNA: generation of infectious virus in the absence of a natural template. Science *297*, 1016–1018
34. Caplan A L and Magnus D (2003) New Life Forms: new threats, new possibilities? Hastings Centre Report 33(6), 7
35. The Schultz Lab (2004) Research at the interface of chemistry and biology. http://schultz.scripps.edu/research
36. Cho M K, Magnus D, Caplan A L, and McGee D (1999) Ethical considerations in synthesising a minimal genome. Science *286*, 2087–2090
37. Goodfield G J (1960) The Growth of Scientific Physiology. London, Hutchinson
38. Mayr E (1997) This is Biology. Cambridge, Mass., Bselknap Press
39. Medawar P B (1967) The Art of the Soluble. London, Methuen

Coda

In 1989, molecular biologist and Nobel laureate Max Perutz wrote of the *humanizing influence of science*, referring to the ways in which progress in medicine, agriculture, population control and energy supply had transformed human life from the 'nasty, brutish, and short' experience of our forefathers into one of material and cultural wealth today. Yet, recognizing that science often exacts a price, he questioned whether there might not be some inevitable downsides. In short, *'Is scientific research the noblest pursuit of the human mind, from which springs a never-ceasing stream of beneficial discoveries, or is it a sorcerer's broom that threatens us all with destruction?'*[1]

Undoubtedly, the tension between the two faces of science and technology has heightened in recent years, fuelled by high-profile issues like GM crops, therapeutic cloning, BSE and global climate change. The result is that scientists can no longer ignore the political consequences of their research. Like it or not, they are now placed under an obligation to consider, justify and explain to an often sceptical public the reasons why they feel it right to pursue their work. Bioethics is thus not a luxury but an indispensable feature of the biosciences in the twenty-first century.

But bioethics is not something that can be 'mugged up' at short notice: not a matter of choosing a few fine words to satisfy research council administrators that a grant application has been 'checked out for ethics'. What is really called for is a sea change in attitudes; a realization that the, often innocent, activities of 'men in white coats' have had profound affects on the social relations between people – and while many changes might be welcome, others are not. For Richard Sclove: *'An engaged citizenry must become critically involved with the choice, governance, and even design of technolog(ies)'* to ensure that only those *'that are compatible with preserving or strengthening their society's democratic nature'* are adopted.[2] As is increasing recognized in science and society programmes, bioscientists have to come out of the laboratory and engage with the general public. And in doing so, they would do well to consider Nicholas Maxwell's concept of the *philosophy of wisdom* as a corrective to the current emphasis on 'neutral' knowledge.[3]

This book has tried to introduce to students some of the bioethical issues they will face, and ways of thinking about them in a rational and constructive manner. In doing so, it has challenged the *scientism* that unfortunately persists in some quarters. Those who wish to use science and technology ethically need to be able to provide sound reasons for their views. Lest the challenging style suggests otherwise, it is perhaps worth saying that the author is decidedly *pro-science*. But he believes that the wise pursuit and application of the biosciences in democratic societies will depend on a new generation of scientists who are genuinely alert to ethical concerns.

NOTES

1. Perutz M F (1989) Is Science Necessary? Essays on science and scientists. London, Barrie and Jenkins, p. 3
2. Sclove R E (1992) The nuts and bolts of democracy: democratic theory and technological design. In: Democracy in a Technological Society. (Ed.) Winner L. Dordrecht, Kluwer, p. 139
3. Maxwell N (1984) From Knowledge to Wisdom: a revolution in the aims and methods of science. Oxford, Blackwell

INDEX

K

L

M

N

O

P

X

Z